Essentials of Molecular Biology

Essentials of Molecular Biology

Editor: Gloria Doran

R CALLISTO REFERENCE

www.callistoreference.com

Callisto Reference,
118-35 Queens Blvd., Suite 400,
Forest Hills, NY 11375, USA

Visit us on the World Wide Web at:
www.callistoreference.com

ISBN: 978-1-63239-967-0 (Hardback)

Cataloging-in-Publication Data

Essentials of molecular biology / edited by Gloria Doran.
    p. cm.
Includes bibliographical references and index.
ISBN 978-1-63239-967-0
1. Molecular biology. I. Doran, Gloria.
QH506 .E87 2018
572.8--dc23

Table of Contents

Permissions

List of Contributors

Index

Preface

Molecular biology studies biological activities that occur on a molecular level in a cell. Proteins, DNA and RNA are the primary molecules studied under this field. The interactions along with biosynthesis are closely evaluated in molecular biology. While understanding the long-term perspectives of the topics, the book makes an effort in highlighting their impact as a modern tool for the growth of the discipline. Those in search of information to further their knowledge will be greatly assisted by this book.

This book is a comprehensive compilation of works of different researchers from varied parts of the world. It includes valuable experiences of the researchers with the sole objective of providing the readers (learners) with a proper knowledge of the concerned field. This book will be beneficial in evoking inspiration and enhancing the knowledge of the interested readers.

In the end, I would like to extend my heartiest thanks to the authors who worked with great determination on their chapters. I also appreciate the publisher's support in the course of the book. I would also like to deeply acknowledge my family who stood by me as a source of inspiration during the project.

Editor

Characterization of Two Orthologs of REVERSION-TO-ETHYLENE SENSITIVITY1 in Apple

Aide Wang[1,2] & Kenong Xu[1]

[1] Department of Horticulture, Cornell University, New York State Agricultural Experiment Station, Geneva, New York, USA

[2] Present address: Department of Horticulture, Shenyang Agricultural University, Shenyang, Liaoning, China

Correspondence: Kenong Xu, Department of Horticulture, Cornell University, New York State Agricultural Experiment Station, Geneva, NY 14456, USA. E-mail: kx27@cornell.edu

Abstract

The *Arabidopsis* RTE1 (REVERSION-TO-ETHYLENE SENSITIVITY1) and its tomato ortholog GREEN RIPE (GR) are important in plant ethylene signaling, response and/or fruit ripening. To investigate the potential role of *RTE1*-like genes in apple (*M. ×domestica* Borkh.) fruit ripening, two orthologs of RTE1 were identified in the apple genome, designated *MdRTE1a* and *MdRTE1b*. Using yeast two-hybrid (Y2H) approach, a cDNA prey library was constructed from a pool of maturing/ripening fruit of 'Golden Delicious'. Screening the library with MdRTE1a identified two putative NAC (NAM/ATAF1, 2/CUC2) proteins, designated MdNAC1 and MdNAC2, that interacted with MdRTE1a in yeast cells. MdRTE1b, a closely related paralog of MdRTE1a, was also confirmed to be interactive with MdNAC1 and MdNAC2 using the Y2H system. The expression of *MdRTE1a* and *MdNAC1* was high in young fruit but low in more developed and ripening fruit. This pattern of low expression remained largely consistent and was not affected by treatment with 1-methylcyclopropene (1-MCP), an inhibitor of ethylene perception, or ethephon, an ethylene releasing compound. However, the expression of *MdRTE1b* and *MdNAC2* rapidly increased during fruit ripening and was highly inhibited by 1-MCP. These data suggested that MdRTE1a, MdNAC1 and their possible interactions in apple cells may play roles in growth, whereas MdRTE1b, MdNAC2 and their putative interactions are involved in ethylene signaling, response and fruit ripening. The possible interactions between an RTE1-like protein and a NAC protein in plant cells have not been reported thus far. Its implications in ethylene signaling and response are discussed.

Keywords: apple, fruit ripening, ethylene, yeast two-hybrid (Y2H), REVERSION-TO-ETHYLENE SENSITIVITY1 (RTE1), NAC (NAM/ATAF1, 2/CUC2)

Abbreviations

ACC: 1-aminocyclopropane-1-carboxylic acid

ACO: ACC oxidase

ACS: ACC synthase

CTR1: CONSTITUTIVE TRIPLE RESPONSE1

EIN2: ETHYLENE INSENSITIVE2

ERF1: ETHYLENE RESPONSE FACTOR1

ETR1: ETHYLENE RESPONSE1

1-MCP: 1-methylcyclopropene

NAC: NAM/ATAF1, 2/CUC2

qRT-PCR: Real-time quantitative reverse transcription polymerase chain reaction

RAN1: RESPONSIVE-TO-ANTAGONIST1

RTE1: REVERSION-TO-ETHYLENE SENSITIVITY1

Y2H: yeast two-hybrid

1. Introduction

The phytohormone ethylene plays critical roles in many biological processes in plants, including plant growth and development, plant response to biotic and abiotic stresses, and ripening of climacteric fruit such as tomato and apple. Biosynthesis of ethylene has been well defined in the Yang cycle that mainly involves two enzymes: 1-aminocyclopropane-1-carboxylic acid (ACC) synthase (ACS) and ACC oxidase (ACO) (Adams & Yang, 1979; Yang & Hoffman, 1984). There are two systems of ethylene production in plants: System 1 occurs during plant/fruit growth and development, and System 2 is exclusively in floral senescence and fruit ripening (Barry & Giovannoni, 2007). There are at least five *ACS* (*MdACS1-5*) and four *ACO* (*MdACO1-4*) genes in apple (Kondo, Meemak, Ban, Moriguchi, & Harada, 2009; Wiersma, Zhang, Lu, Quail, & Toivonen, 2007) and these genes appear to operate similarly in the two systems for ethylene production. *MdACS1* and *MdACO1* are considered System 2 genes as their expression is highly correlated with the ethylene production burst in ripening apple fruit.

Much insight into the ethylene signal transduction and response has been gained from studies in *Arabidopsis*. The linear cascade of ethylene signal transduction, which includes ETR1 (ETHYLENE RESPONSE1) and other ethylene receptors, RAN1 (RESPONSIVE-TO-ANTAGONIST1), CTR1 (CONSTITUTIVE TRIPLE RESPONSE1), EIN2 (ETHYLENE INSENSITIVE2), EIN3 and EILs (EIN3 Like) and ERF1 (ETHYLENE RESPONSE FACTOR1), has become more complex as recent evidence suggests that regulators in other pathways, as well as mechanisms of feedback and elements of protein turnover machinery are involved in ethylene signal transduction and response (Kendrick & Chang, 2008; Stepanova & Alonso, 2009). An important finding is the discovery of REVERSION-TO-ETHYLENE SENSITIVITY1 (RTE1), which is a membrane protein highly conserved in plants and specifically interacts with ETR1 to regulate ethylene signaling in *Arabidopsis* (Dong et al., 2010; Resnick, Rivarola, & Chang, 2008; Resnick, Wen, Shockey, & Chang, 2006; Rivarola, McClellan, Resnick, & Chang, 2009; Zhou, Liu, Xie, & Wen, 2007). Moreover, GREEN RIPE (GR), an ortholog of RTE1, is directly involved in both ethylene signaling and fruit ripening in tomato in a primarily fruit tissue specific manner (Barry & Giovannoni, 2006; Barry et al., 2005).

In apple fruit, major elements in the linear model of ethylene signaling have been identified. There are six apple ethylene receptor genes: *MdETR1*, *MdETR1b*, *MdETR2*, *MdETR5*, *MdERS1* and *MdERS2*. Studies of these ethylene receptors have focused on understanding their roles in apple shelf life by quantifying the levels of their mRNA and/or protein accumulation (Tatsuki & Endo, 2006; Tatsuki, Endo, & Ohkawa, 2007; Tatsuki, Hayama, & Nakamura, 2009; Wang et al., 2009; Wiersma et al., 2007). The transcript levels of *MdETR1* and *MdERS1* were also evaluated in a study of ethylene biosynthesis and perception in apple fruitlet abscission (Dal Cin, Danesin, Boschetti, Dorigoni, & Ramina, 2005).

Several members of the negative regulator CTR1 are likely encoded by the apple genome, as multiple *CTR1-like* genes exist in tomato (Adams-Phillips et al., 2004; Lin, Alexander, Hackett, & Grierson, 2008; Zegzouti et al., 1999). To date, only *MdCTR1* has been described in apple. The conserved 3' end of the *MdCTR1* coding region (AY670703) was cloned by Dal Cin et al. (2005). Its full length cDNA was characterized by several alternative splicing variants (DQ847147-DQ847149) (Wiersma et al., 2007). Expression of *MdCTR1*does not change widely in ripening fruit of several apple cultivars studied, including 'Golden Delicious' (Dal Cin et al., 2007; Dal Cin, Rizzini, Botton, & Tonutti, 2006; Wiersma et al., 2007), 'Sunrise' (Wiersma et al., 2007) and 'Delicious' (Li & Yuan, 2008).

Efforts dedicated to MdEIN2 have been limited and its full-length cDNA clone is not available. In contrast to the uniqueness of the *EIN2* gene in *Arabidopsis*, two *MdEIN2* genes, *MdEIN2A* (CV084546) and *MdEIN2B* (CO066480) were designated to distinguish the two distinct pools of apple EST (expressed sequence tag) sequences in GenBank that are all similar to the C-terminus of EIN2 (Wiersma et al., 2007). As expected, the transcript levels of *MdEIN2A* and *MdEIN2B* undergo little variation in fruit, either pre- or post- harvest (Wiersma et al., 2007).

The three *EIN3-like* genes, *MdEIL1* (GU732484), *MdEIL2* (GU732485) and *MdEIL3* (GU732486), were studied in apple (Tacken et al., 2010). Expression of *MdEIL1* and *MdEIL3* remained at a constant level during fruit ('Royal Gala') development. However, mRNA levels of *MdEIL2* increased progressively along with the fruit development and peaked at harvest with a relative expression ratio greater than 10-fold, suggesting that *MdEIL2* may contribute more to ethylene sensitivity during fruit ripening than other EIN3-like genes (Tacken et al., 2010).

MdERF1 and MdERF2 are the first two primary target transcription factors isolated and characterized in apple. Expression of the two *MdERFs* is not only specific to ripening fruit (predominantly for *MdERF1*, exclusively for *MdERF2*), but also positively regulated by ethylene, strongly suggesting that they are involved in fruit ripening

(Wang, Tan, Takahashi, Li, & Harada, 2007). A screening of 38 putative *MdERF* genes for their expression patterns during 'Royal Gala' fruit development and in response to exogenous ethylene treatment identified 17 of these genes were predominately expressed at 35 DAF (days after full-bloom), two primarily expressed at 132 DAF, and two (*AP2D29* and *AP2D19*) were most sensitive to 4 h ethylene treatment (Tacken et al., 2010).

However, studies of the apple orthologs of RTE1 or GR have not been documented. To investigate their potential role in apple fruit ripening and firmness retention, two apple orthologs, designated MdRTE1a and MdRTE1b, of RTE1 were identified in the draft sequence of the apple genome (Velasco et al., 2010) and MdRTE1a was used to screen a cDNA prey library constructed from 'Golden Delicious' (GD) based on yeast two-hybrid (Y2H) system in this study. Instead of identifying the expected ETR1-like ethylene receptors, two putative members of the large gene family of NAC (NAM/ATAF1, 2/CUC2) transcription factors, designated MdNAC1 and MdNAC2, were found to interact with MdRTE1a and MdRTE1b in the yeast cells. qRT-PCR analyses demonstrated that the expression of *MdRTE1b* and *MdNAC2* was highly regulated by ethylene while that of *MdRTE1a* and *MdNAC1* remained largely consistent at low levels in maturing/ripening fruit, and therefore unresponsive to ethylene signaling.

2. Materials and Methods

2.1 Y2H Library Construction

Fruit of 'Golden Delicious' (GD), the genome of which has been sequenced (Velasco et al., 2010), were chosen for the Y2H library construction. The fruit were collected from the experimental orchards in Geneva, New York at three time points: d -19 (18 September 2009, 5 fruit), d 0 (8 October 2009, commercial harvest, 5 fruit) and d 6 (14 October 2009, 25 fruit). The 25 fruit harvested at d 6 were kept at room temperature (24°C) for a series of five samplings (5 fruit per sample) at d 6, d 11, d 16, d 21 and d 26. For each sample, the five fruit were cut into slices, frozen in liquid nitrogen and stored at -80 °C prior to RNA isolation.

Total RNA was isolated from each of the seven GD fruit samples (d -19, d 0, d 6, d 11, d 16 and d 21 and d 26) as described previously (Lay-Yee, Dellapenna, & Ross, 1990). For the Y2H library construction, mRNA was isolated from the total RNA of 980 µg that were evenly pooled from the seven samples (140 µg per sample) using the Ambion (Austin, Texas, USA) Poly(A) Purist Kit. The library was constructed with Make Your Own Mate & Plate Library System (Clontech, Mountain View, CA, USA) following the manufacturer's instructions. Briefly, the first-strand cDNA was synthized from 1 µg mRNA with the Oligo-dT primers (CDSIII Primer, Table 1). The cDNA was amplified by Long Distance PCR (LD-PCR) with 5' PCR primer and 3' PCR primer (Table 1) for 20 cycles to generate double-strand cDNA, which was then purified by the CHROMA SPIN TE-400 column. After purification, the double-strand cDNA of 3.4 µg was ligated into SmaI-linearized pGADT7-Rec cloning vector (Clontech) and subsequently transformed into yeast strain Y187 to generate the cDNA library for Y2H assay using Yeastmaker Yeast Transformation System 2 (Clontech) according to the manufacturer's instruction.

2.2 Construction of the Bait Constructs and Transformation of Yeast Cell

To identify the apple orthologs of RTE1, its protein sequence (NP_180177.2) was blast searched against the *Malus* EST databases in GenBank using the TBLASTN program. An apple UniGene Mdo.6837 (UGID: 1406185), consisting of 16 *Malus* ESTs (expressed sequence tags), was found that was similar to RTE1 (71.1% identity over all 242 amino acids of RTE1). Inspecting the 16 ESTs indicated that they represent two closely related genes: One group of seven ESTs is for MDP0000225233 (apple gene/protein ID at GDR (Genome Database for Rosaceae), http://www.rosaceae.org/), a predicted gene on chromosome 15, designated *MdRTE1a*; the other group of 9 ESTs for gene MDP0000341171 on chromosome 8, designated *MdRTE1b*. *MdRTE1a* and *MdRTE1b* share 94.1% identity in deduced amino acid sequences (Figure 1A).

The full length coding sequence of *MdRTE1a* was PCR amplified with primers KT7_RTE1F (attached with an EcoRI site at 5' end) and KT7_RTE1R (attached with a BamHI site at 5' end) (Table 1) from the cDNA of the GD fruit to create the MdRTE1a bait construct. The PCR products were restricted with EcoRI and BamHI, and cloned into pGBKT7 vector (Clontech) that was already linearized with the same two enzymes. *MdRTE1b* was similarly amplified and cloned, but with primers KT7_RTE1F2 and KT7_RTE1R (Table 1). The primer bases discriminating *MdRTE1a* from *MdRTE1b* were found in KT7_RTE1F and KT7_RTE1F2 (Table 1). The resulting recombinant plasmids pGBKT7-RTE1a and pGBKT7-RTE1b were confirmed by sequencing with T7 primer (Table 1) and then transformed into yeast strain Y2HGold using Yeastmaker Yeast Transformation System 2 (Clontech) with minor modifications. Briefly, 5 ml 1× YPDA liquid medium was inoculated with a fresh yeast (Y2HGold) colony and incubated at 30°C with shaking at 250 rpm for no more than 14 h. The cells were collected by centrifugation at 700 ×g for 5 min, and then washed once with 5 ml sterile water and once with

750 μl 1.1× TE/LiAc (lithium acetate) solution (1.1 ml of 10× TE and 1.1 ml of 10× LiAc (1 M) in 10 ml H$_2$O). The washed cells were collected by centrifuging at 700 ×g for 5 min, and then resuspended in 50-100 μl 1.1× TE/LiAc solution for transformation. A mix of 50 μl of cells, 100 ng of construct DNA, 5 μl of Yeastmaker Carrier DNA at 10 μg/μl, and 500 μl of PEG/LiAc (containing 40% PEG 3350, 1× TE and 1× LiAc) was made and incubated at 30°C for 2-3 h with gentle shaking, and then 20 μl DMSO (dimethyl sulfoxide) was added to the mixture and incubated at 42°C for 15 min. The cells were gently mixed every 5 min during the incubation, and later were collected by centrifugation at 16,000 ×g for 15 s. The pellets were resuspended in 500 μl 1× YPDA and incubated at 30°C for 60-90 min with shaking at 160 rpm. The cells were harvested again by centrifuging at 16,000 ×g for 15 s and finally resuspended in 0.9% of NaCl solution (200-400 μl) and plated on the single dropout (SD/-Trp) plates for evaluation of transformation efficiency and positive bait colonies.

Table 1. List of primer names and their sequences

Primer name	Sequence (5'-3')
CDS III	ATTCTAGAGGCCGAGGCGGCCGACATG-d(T)$_{30}$VN*
5' PCR primer	TTCCACCCAAGCAGTGGTATCAACGCAGAGTGG
3' PCR primer	GTATCGATGCCCACCCTCTAGAGGCCGAGGCGGCCGACA
KT7_RTE1F[a]	GCC<u>GAATTC</u>ATGATGGAGCTAAAGGAAGCTTATG
KT7_RTE1F2[a]	GCC<u>GAATTC</u>ATGATGGAGCTGAAAGAAGCGTATG
KT7_RTE1R[b]	GCC<u>GGATCC</u>CTAGCAGCAAGGTCTTG
T7	TAATACGACTCACTATAGGG
ex_RTE1aF	GAAGCTTATGACATCGAGCGTA
ex_RTE1aR	GCTACGGAACCAAATGTGAAA
ex_RTE1bF	GAAGCGTATGACATTGAGCATT
ex_RTE1bR	ACGGCACCAAATGCAAAG
ex_NAC1F	TTCAGCTGTCGCAGAATGAT
ex_NAC1R	GCATCTTCCCCTAGTCAGCAG
ex_NAC2F	TGTTCAGCTGTCACAGGATGAC
ex_NAC2R	CATGCTTCTTCCCCTAATCACTAA
Actin F	GGCTGGATTTGCTGGTGATG
Actin R	TGCTCACTATGCCGTGCTCA
Y2H_amp F	CTATTCGATGATGAAGATACCCCACCAAACCC
Y2H_amp R	GTGAACTTGCGGGGTTTTTCAGTATCTACGATT

[a]The primer bases discriminating *MdRTE1a* from *MdRTE1b* are highlighted in gray;

[a,b]The bases underlined show the restriction site of EcoRI and BamHI, respectively;

*N=A, T, G or C; V=A, G or C.

2.3 Y2H Library Screening

The bait clone of construct pGBKT7-RTE1a was used to initially screen the Y2H library with the GAL4-based Matchmaker Gold Yeast Two-Hybrid System (Clontech) according to the manufacturer's instructions. Briefly, 4 ml of the bait strain overnight cultures (SD/-Trp liquid medium), 1 ml of library aliquot (1.2×10^7 cells) and 45 ml 2× YPDA medium were combined in a 2 l flask and incubated at 30°C for 20 h with shaking at 50 rpm, and then checked for the formation of zygotes under a Zeiss (Oberkochen, Germany) Axioskop phase contrast microscope (40×). When zygotes were present, the cells were collected by centrifugation at 1,000 ×g for 10 min, and resuspended in 10 ml of 0.5× YPDA medium. The mated cell cultures were spread on the SD/-Trp/-Leu/X/A or DDO/X/A plates (containing double dropout SD medium that lacks tryptophan and leucine, but is supplemented with X-α-Gal, the substrate of α-galactosidase encoded by the reporter gene *MEL1*, and Aureobasidin A, a new yeast antibiotic that kills all cells of non-hybrids and acts as a selection marker in the Matchmaker Gold Yeast Two-Hybrid System) and kept at 30°C. Positive colonies appeared within 5 d and were streaked onto plates of quadruple dropout SD medium lacking tryptophan, leucine, histidine and adenine while containing X-α-Gal and Aureobasidin A, i.e. the SD/-Trp/-Leu/-His/-Ade/X/A, or QDO/X/A plates, for high stringency screening. Genuine positive hybrid colonies appeared in 5 d on these QDO/X/A plates. The plasmids were rescued from the colonies survived on QDO/X/A plates using Easy Yeast Plasmid Isolation Kit (Clontech), which were then transformed into *E. coli* strain JM109 and plated on LB plate with 100 mg/ml ampicillin to select the prey plasmids. To reveal the identity of the prey plasmids, their inserts were sequenced with T7 primer (Table 1).

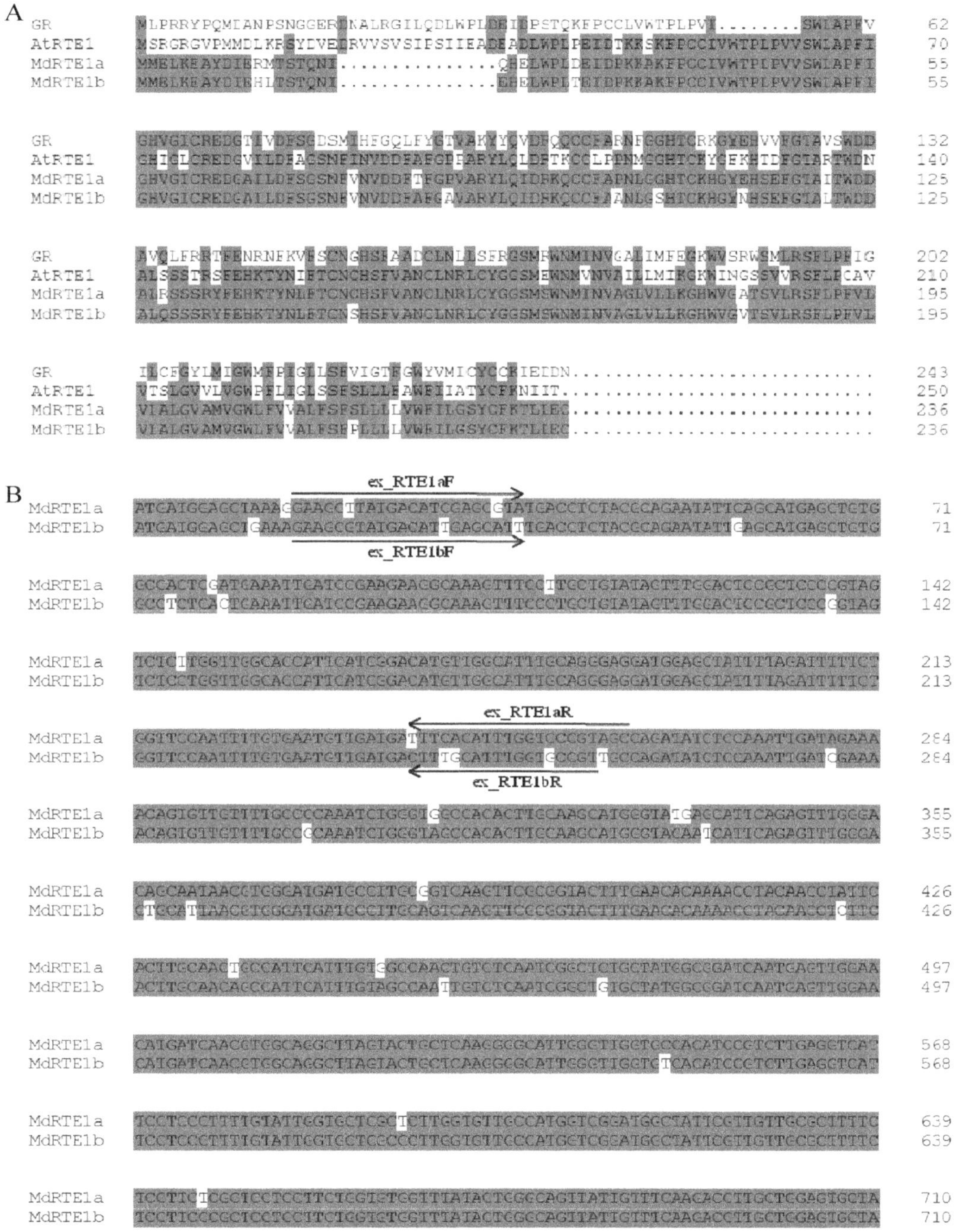

Figure 1. Sequence information of MdRTE1a and MdRTE1b

A: Alignment of predicted amino acid sequences of GR, RTE1, MdRTE1a, and MdRTE1b. The accession numbers of the aligned sequences are ABD34613.1 (GR, tomato), NP_180177.2 (AtRTE1, *Arabidopsis*), MDP0000225233(MdRTE1a, apple), and MDP0000341171 (MdRTE1b, apple). Amino acid residues that are conserved (present in more than three of the four accessions) were highlighted in gray. **B**: Alignment of coding region of *MdRTE1a* and *MdRTE1b*. The last base "G" in the stop codon was not shown at the end of the sequences. The primers used for gene expression assays were indicated with arrows.

Table 2. Date (2010) and number of 'Golden Delicious' fruit harvested for qRT-PCR assay and/or evaluation of fruit ethylene production and firmness

Variety	Date/ Treatment	Number of fruit sampled																	
	Days after Harvest														0	5	10	15	20
GD	DAF	7	15	22	28	35	42	56	70	84	98	112	126	133	140				
	Date	5/11	5/19	5/26	6/1	6/8	6/15	6/29	7/13	7/27	8/10	8/24	9/7	9/14	9/21	10/3	10/8	10/13	10/18
	Normal	>10	>10	>10	5	5	5	5	5	5	5	5	5	5	25/5[a]	5	5	5	5
	1-MCP														20/0[b]	5	5	5	5
	Ethephon														20/0[b]	5	5	5	5
Fuji	Date														10/13	10/18	10/23	10/28	11/4
	Normal														25/5[a]	5	5	5	5

[a]The 25 fruit were divided into five groups of five. One group was sampled immediately at harvest as control and each of the rest groups were sampled every five days as shown.

[b]The 20 fruit were treated and no fruit were sampled after immediate treatment. These fruit were divided into four groups of five and then sampled one group every five days as shown.

2.4 Co-transformation of Yeast Cells

To further confirm the interaction, plasmids of 100 ng of bait and 100 ng of positive prey clones were co-transformed into the Y2HGold yeast strain using Yeastmaker Yeast Transformation System 2 (Clontech) with minor modification as described in section 2.2 *Construction of the bait constructs and transformation of yeast cell*. The co-transformed cells were placed on the DDO/X and DDO/X/A plates.

2.5 Fruit Sampling and Treatment for Gene Expression Assay and/or Evaluation of Fruit Ethylene Production and Firmness

The fruit of GD were sampled over the entire growth, development and maturing stages in 2010. Starting from 7 days after full-bloom (7 DAF, 11 May 2010) through commercial harvest (140 DAF, 21 September 2010), a total of 14 samples were collected (Table 2). For the 7 to 22 DAF samples, the whole fruit were pooled with the petiole removed, and at least ten fruit were used for each sampling date. For the rest of the samples, only the cortex tissues were used and at least five fruit were pooled for each sample.

At commercial harvest, in addition to the five fruit sampled regularly, another 60 fruit of GD were harvested and divided into three batches of 20 fruit each for the following treatments, which were all conducted under room temperature (24°C). The first 20 fruit were treated with 1 µl l⁻¹ 1-MCP (Agrofresh, Philadelphia, PA, USA) for 24 h at the Cornell Postharvest Research Facility, and then placed in a regular room. The second 20 fruit, at the same time, were dipped into 500 mg /l ethephon solution (Sigma-Aldrich, St. Louis, MO, USA) for 10 s and then sealed in a container holding for 24 h at room temperature. The third batch was the untreated control. Fruit sampling was similar in these three treatments: one group of five fruit was sampled at 5, 10, 15 and 20 d after commercial harvest (Table 2).

In addition, 25 fruit of 'Fuji' were also sampled at commercial harvest (13 October 2010). Without being treated, the 'Fuji' fruit were sampled in the same manner as the untreated GD fruit, and used as a low ethylene production cultivar control (Table 2). The flowers and leaves of GD were sampled at 7 DAF (11 May 2010). All samples were immediately frozen in liquid nitrogen and stored at -80 °C for RNA isolation, except for the fruit harvested at commercial harvest that need to be treated and stored as described above and to be evaluated as follows.

```
MdNAC1  GCCACAACCAATGGGAGGCGAACATCTTGTACAACCAGAGCATACTCATGATGCCATCTTTGTCAAACAAAATGCCCGAGTCCAAGTTTGA   630
MdNAC2  ...............................................................................GGAGTTCAAGTTTGA   15

MdNAC1  ATCAGATCGTAACTATCCCTTCATCAAAAACGCGAGTCACATCATGTTCGGCAGCATACCGGCTCCACCTGCATTTGCTTCAGAGTTCCC   720
MdNAC2  GTCAGATGGTAACTACCCCTTCATCAAAAACGCAAGTCACATGATGTTCGGCAGCATACCTGCTCCACCTGCATTTGCTTCCGAGTTCCC   105

MdNAC1  TGCTAAACAGGCAATGCTTAGGTTAAATTCTGCTGCATCGTCTTCCAGTTCGCTTCATCTTACTCCAGGCATGATTAGAATTACACACAT   810
MdNAC2  TTGTAAAGAGGCAATCCCTAGGTTAAATTCTGCTCCAGCATCTTCCAGTTCGCTTCATCTCACTCGTCCATCGATTACAATTACACACAT   195

                                                                       ex_NAC1F
MdNAC1  AACTTCAAGTGACAACAGGATGGACTGGTCTTTTGGAAAGGATGGGTTGTCAACCTTGTGTTTCTCGTTCAGCTCTCGCGCAGAATCATGG   900
MdNAC2  AACTTCAAGTGACAACAGGATGGACTGGTCTTTTGGAAAGGATGGGCTTGTCAACCTTGTCTTTCTCGTTCAGCTCTCACAGGATCACGG   285
                                                                       ex_NAC2F

MdNAC1  AAACTCTGGCAATCTCGGTGCCGGTCGGTGGCTCACTCTCTCGGAAAGACCGGATGTGTGGCTCATGCGGCGATGGTTTTTGTTTATGTTTCT   990
MdNAC2  AAACTCTGGCAATCTCGGTGCCAGTCGGTGGCTCACTCTCTCGGAAAGACCGGATGTGTGGCTCATGCGGCGGCTGGTTTTTGTTTATGTTTTT   375

MdNAC1  CTGGGTTCTGTTTCTTTCTATGAGTTTCAAAATCGGAAGCTACATCTGTACCAGCTCAAGCTTCGACAGTAATCGTGGCTTTCTAACCCTA   1079
MdNAC2  CTGGGTCCTATTCTTTCCATCAGTTTAAAAATCGGGAGCTATATCTGTACCAGGTCATTCTTGTACACTAAAGGTGGTCTTCGAGCCTA   465

                            ex_NAC1R
MdNAC1  TCTCAGGATCTTGTAGTAAACTTCCCGCTGACTACGCGAAGATGCTACTTCTTGCAAACTCGATATTTAGTTAATTGTTATACTGAC   1169
MdNAC2  TCTCAGGATCTTGTAGTGAGCTTCCTTAGTGATTAGGGGAACAAGCATGTTCTTGCAGGCC..ATATTTAGTTAATTGGTACACTGAC   553
                            ex_NAC2R

MdNAC1  TGCAGAACTAATTATTTACACAAACCGTTT...TAACATAGCCTAAAGTGTGTGAT..ATTTTGAATAGCTATATACTGAATCTGAATGA   1254
MdNAC2  CCGTTCAACTAATTATTTATACAAACCGTTTGCATAACATTGCCTAACGTGCCATGTTTCTATTTTGAATAACTATATACTGAAT......GA   637

MdNAC1  ATGTAAAACTGAATTTTTGTTCC......................AAAAAAAAAAAAAAAAAAAAA....................   1297
MdNAC2  ATGTAAAACTGAATTTTTGTTTCAGTAAAATTATTTGATTGTAACGAAAAAAAAAAAAAAAAAAAAAAAAAAAAAAAA.............   714
```

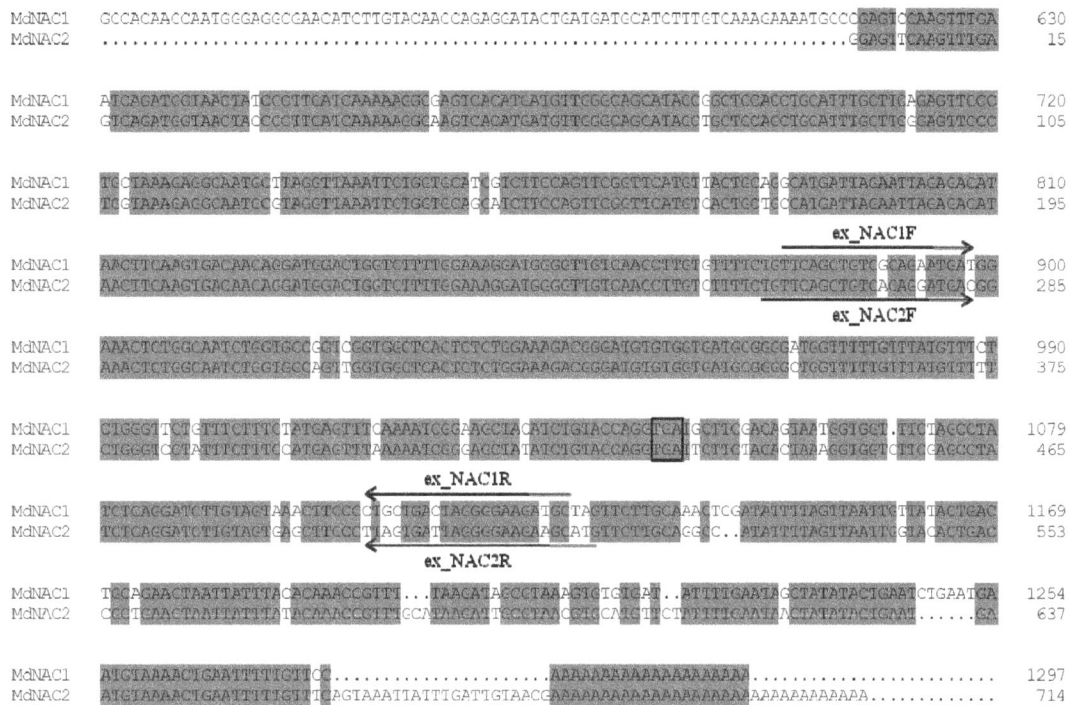

Figure 2. Comparison of the cDNA sequences at the 3'end of MdNAC1 (int9) and MdNAC2 (int10)

Identical nucleotides were shaded in gray. The stop codon was boxed. The primer sequences used for gene expression assays were indicated with arrows.

2.6 Evaluation of Fruit Ethylene Production and Firmness

For each sampling point of five fruit, fruit were weighed individually and enclosed in an air-tight container of 1.89 l (Model 7K24, Rubbermaid, Atlanta, GA. USA) equipped with septa for 1 h at 24 °C. For each fruit, one milliliter (1 ml) of the gas was sampled through the headspace of the container using a BD (Franklin Lakes, NJ, USA) syringe (BD Cat. #309602). Ethylene concentration was measured with a gas chromatograph HP 5890 series II (Hewlett-Packard, Palo Alto, CA, USA) equipped with a flame ionization detector.

Flesh firmness was measured with a penetrometer (FT 327, McCormick, Facchini, Alfonsine, Italy) fitted with an 11.1 mm-diameter probe. Four skin discs (approximately 2.5 cm in diameter) were removed from opposite sides of each fruit. For each of the four cut surfaces, the probe was pressed into a depth of 8 to 9 mm in a single smooth motion.

2.7 Real-time Quantitative Reverse Transcription Polymerase Chain Reaction (qRT-PCR)

RNA was isolated using the ZR Plant RNA MiniPrep kit (ZYMO Research, Irvine, CA, USA) according to the manufacturer's instruction. Three independent reverse transcription reactions, which began with 1 µg of total RNA, were performed for each of the RNA samples using the Superscript III RT (Invitrogen, Carlsbad, CA, USA) by following the manufacturer's manual. The resulting first stand cDNA was diluted by 10 times, and then used as templates for qRT-PCR assays with the four sets of gene specific primers (Table 2): Primers ex_RTE1aF and ex_RTE1aR are for *MdRTE1a*, ex_RTE1bF and ex_RTE1bR for *MdRTE1b*, ex_NAC1F and ex_NAC1R for *MdNAC1*, and ex_ NAC2F and ex_NAC2R for *MdNAC2* (Table 2). The primer bases discriminating *MdRTE1a* from *MdRTE1b*, and *MdNAC1* from *MdNAC2* were indicated (Figures 1B, 2).

qRT-PCR was performed with the Roche (Indianapolis, IN, USA) LightCycler 480 Real-Time PCR System. For each reaction, a final volume of 20 µl was used, which contained 5 µl of the cDNA dilutions, 0.5 µM of the forward and reverse primer, and 1× SYBR green master mix (Roche Cat. # 04707516001). The qRT-PCR program included an initial denaturation step of 10 min at 94°C, 45 cycles of amplification using 10 s at 94°C, 25 s at 55°C, and 25 s at 72°C, and a dissociation stage of 5 s at 95°C, 60 s at 60°C, and 15 s at 97°C. The apple actin gene (EB136338) was amplified by primers Actin F and Actin R (Table 1) and used as a reference to the target genes. Expression quantification and data analysis were performed by LightCycler 480 Software (Version

1.5) using the comparative cycle threshold method (Pfaffl, 2001). Comparison of means and test of statistic significance were performed with the JMP9.0 software (SAS Institute, Cary, NC, USA).

3. Results

3.1 Characterization of the Y2H Library of 'Golden Delicious'

The Y2H library was constructed with the size fractioned cDNA (>0.3 kb) from the maturing/ripening GD fruit. Titration of the library on SD/–Leu plates indicated a titer of 1.2×10^7 cfu/ml. A PCR survey of the insert size of 156 randomly selected clones using primers Y2H_amp F and Y2H_amp R (Table 1) found a mean size of 829 ± 313 bp, ranging from 300 through 1700 bp (data not shown), meeting the size expectations for the Y2H library. Sequencing analysis of the first 62 clones showed that 47 (75.8%) had at least one significant hit in GenBank while 15 (24.2%) had no significant similarities (with an expect threshold at 10^{-9}). The top five abundant genes surveyed in the library encode *M. domestica* metallothionein (MT)-like proteins including AMT2 (8.1%) and AMT1 (4.8%), MdACO1 (4.8%), auxin-repressed protein-like protein (4.8%) and abscisic stress ripening like protein (3.2%).

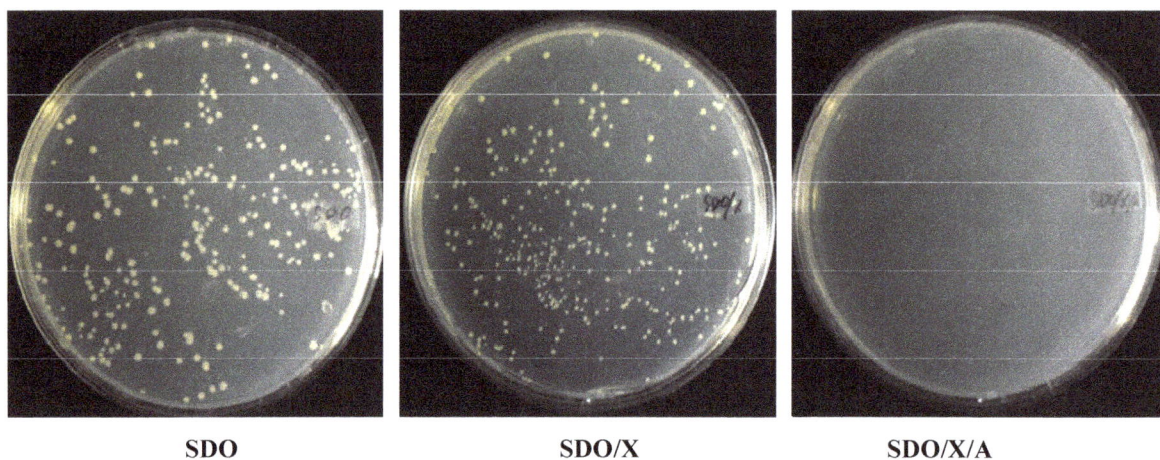

| SDO | SDO/X | SDO/X/A |

Figure 3. Auto-activation test of pGBKT7-RTE1a

Recombinant plasmid pGBKT7-RTE1a was transformed into yeast strain Y2HGold. Cells were spread on plates of single dropout medium (SD/-Trp, SDO), SDO plus X-α-Gal medium (SDO/X), and SDO/X plus Aureobasidin A medium (SDO/X/A). Colonies were able to grow on the SDO and SDO/X plates, but not on the SDO/X/A plate. In addition, there were no blue colonies observed on the SDO/X plate. These growth characteristics indicate that pGBKT7-RTE1a does not have auto-activation activity and are suitable for Y2H library screening.

3.2 Screening the Y2H Library for the Putative Interactors of MdRTE1a

The entire coding region (predicted) of the *MdRTE1a* cDNA was inserted into the vector pGBKT7 to construct the bait construct pGBKT7-RTE1a. Before screening the Y2H library, the coding sequence of *MdRTE1a* was sequence confirmed to be in frame in the construct and the bait was proven to be free of auto-activation activity and toxicity when transformed into yeast cells (Figure 3). A total of 12 million cells (estimated) from the Y2H library were screened with the bait pGBKT7-RTE1a; the mating efficiency was estimated as 6.6 %. Despite being grown on the low stringency plates (DDO/X/A), there were only 35 positive colonies obtained from the screening. Sequencing analysis of the plasmids rescued from the 35 clones indicated that the 35 sequences could be aligned into three unique cDNA contigs. To further test the interactions, the hybrid colonies of the representative clones 'int9' (standing for 24 clones), 'int10' (for 8 clones) and 'int22' (for 3 clones) were streaked onto the high stringency plates (QDO/X/A). The hybrid colonies of clones 'int9' and 'int10' survived well while that of clone 'int22' did not grow, suggesting clones 'int9' and 'int10' might genuinely interact with MdRTE1a.

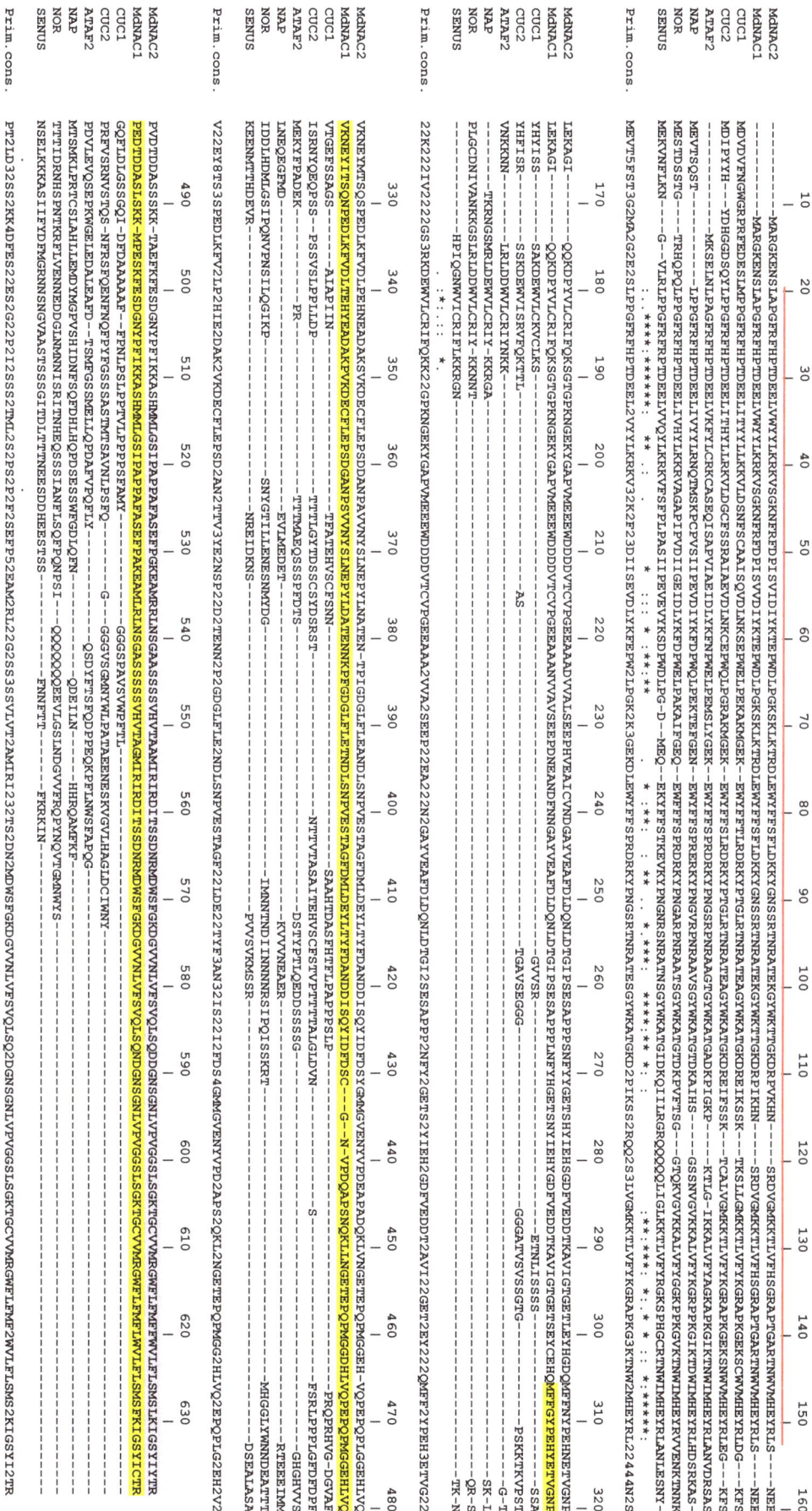

Figure 4. Alignment of the predicted amino acid sequences of MdNAC1 and MdNAC2 with other plant NAC proteins

The NAC amino acid sequences used in the alignment include NAP (CAA10955.1), ATAF2 (NP_680161.1), CUC1 (NP_188135.1) and CUC2 (NP_200206.1) from *Arabidopsis*, NAC-NOR (AAU43922.1) and SENU5 (CAA99760.1) from tomato, and MDP0000457996 (MdNAC1) and ADL36811 (MdNAC2) from apple. The NAC domain was overlined and the conserved amino acid residues were indicated with asterisk. Deduced amino acid sequences of 'int9' and 'int10' were highlighted in yellow in MDP0000457996 (int9) and purple in ADL36811 (int10), respectively.

Blast searches against GenBank and the apple genome at GDR (Genome Database for Rosaceae) found that 'int9' is nearly identical to the C-terminus of both MDP0000457996 and MDP0000695714, the two duplicated genes/proteins closely located on chromosome 16 (thus likely to be allelic genes from the two haploid genomes), whereas 'int10' almost perfectly matches the C-terminus of protein ADL36811 in GenBank, or MDP0000621646 on chromosome 2. All these proteins contain a conserved NAC (<u>N</u>AM/<u>A</u>TAF1, 2/<u>C</u>UC2) domain at their N-terminal region, characteristic of the NAC family of transcription factors (Figure 4). Thus clones 'int9' and 'int10', despite of their sequences do not contain the NAC domain (Figures 2, 4), represent at least two putative NAC transcription factors in apple, designated MdNAC1 (for MDP0000457996 and MDP0000695714) and MdNAC2 (for ADL36811 and MDP0000621646), respectively. MdNAC1 and MdNAC2 are highly related and share 93.2% of identity in their coding sequences. However, the 3'-UTR sequences of 'int9' and 'int10' are identical only by 88.5% (Figure 2).

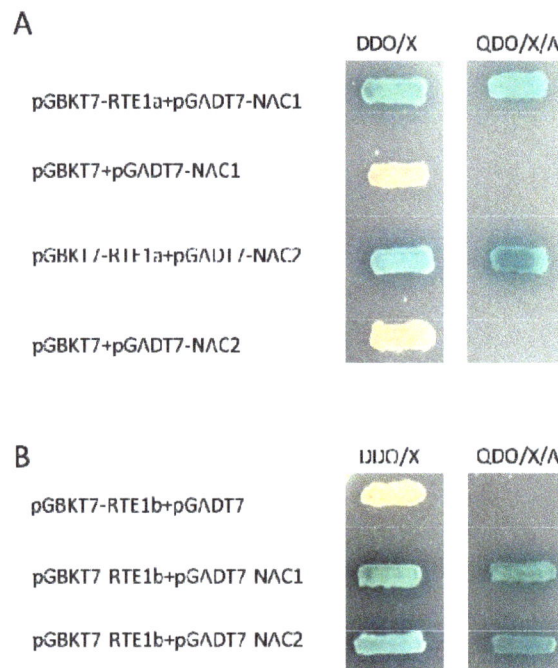

Figure 5. Confirmation of the interaction between MdRTE1 and MdNAC proteins in yeast cells

DDO/X: SD/-Trp/-Leu/X (X-α-Gal). QDO/X/A: SD/-Trp/-Leu/-His/-Ade/X/A (Aureobasidin A). **A**: Co-transformation of pGBKT7-RTE1a and pGADT7-NAC1, pGBKT7-RTE1a and pGADT7-NAC2, pGBKT7 (empty vector) and pGADT7-NAC1 (CK), and pGBKT7 and pGADT7-NAC2 (CK). **B**: Co-transformation of pGBKT7-RTE1b and pGADT7-NAC1, pGBKT7-RTE1b and pGADT7-NAC2, and pGBKT7-RTE1b and pGADT7 (empty vector) (CK).

3.3 Confirmation of Protein-protein Interactions in Yeast Cells

To further confirm the interaction between MdRTE1a and the two NACs, pGADT7-NAC1 (coding region from int9) and -NAC2 (coding region from int10) were respectively co-transformed with pGBKT7-RTE1a into yeast strain Y2HGold. The transformant cells were grown on the DDO/X and QDO/X/A plates. The blue colonies on both plates (Figure 5A) indicated the positive interaction between pGBKT7-RTE1a and pGADT7-NAC1 or -NAC2. As controls, pGADT7-NAC1 or -NAC2 were also co-transformed with pGBKT7 empty vector into the yeast cells. But their colonies did not turn blue on DDO/X plates and were not able to grow on QDO/X/A plates (Figure 5A), suggesting that pGADT7-NAC1 or -NAC2 alone is not able to activate the reporter genes. These results demonstrated that the interaction between MdRTE1a and MdNAC1 or MdNAC2 protein is positively confirmed in the yeast cells.

As described in the Materials and Methods section, MdRTE1a and MdRTE1b share 94.1% identity in deduced amino acid sequences (Figure 1A). To investigate if MdRTE1b interacts with the two NACs, we constructed a recombinant plasmid of pGBKT7-RTE1b and co-transformed with pGADT7-NAC1 and -NAC2, respectively.

The experiment showed that MdRTE1b also interacts with MdNAC1 and MdNAC2 in the yeast cells (Figure 5B), similar to MdRTE1a (Figure 5A).

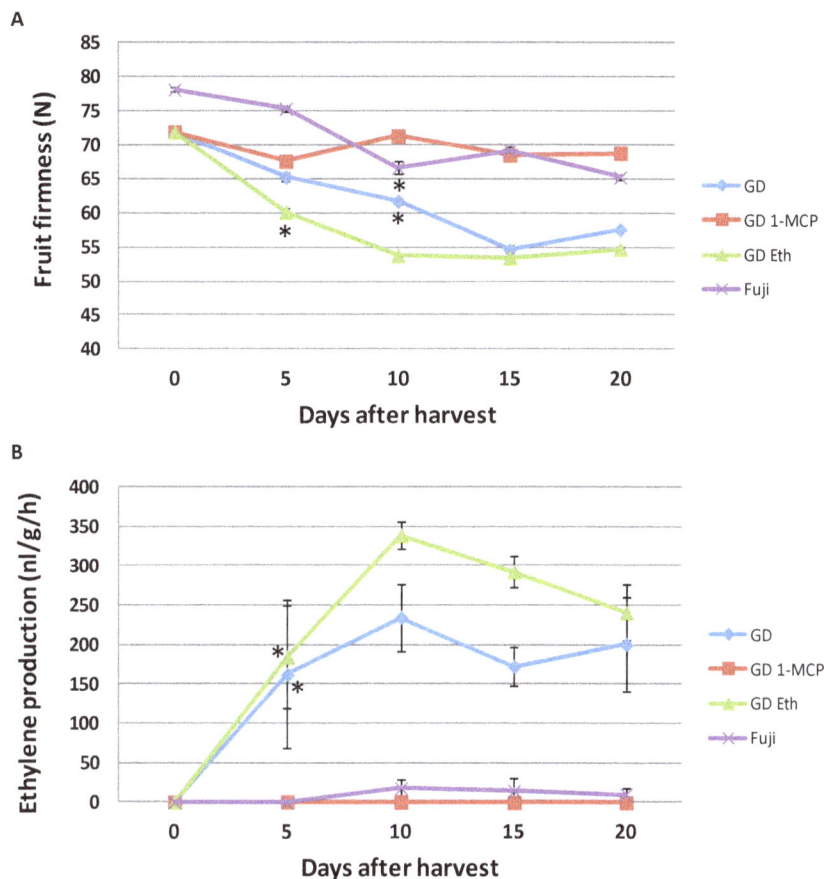

Figure 6. Changes in firmness and ethylene production of apple fruit during storage (24°C)

The bars represent SD measured in five fruit. *The first measurement showing significant change from d 0 in fruit firmness or ethylene production (P<0.01). **A**: Fruit firmness of 'Golden Delicious' and 'Fuji': GD (untreated), GD 1-MCP (treated with 1-MCP), GD Eth (treated with ethephon) and 'Fuji' (untreated). **B**: Ethylene production.

3.4 Fruit Firmness and Ethylene Production

During postharvest storage of 20 d at room temperature (Figure 6A), the firmness of the untreated GD fruit decreased from 71.9 N to 57.6 N, and this significant drop in firmness occurred in 10 days. The ethephon treated GD fruit lost more firmness (from 71.9 N to 54.7 N) and quicker as the significant reduction of firmness was observed in five days. However, the GD fruit treated with 1-MCP did not soften significantly. The untreated 'Fuji' fruit showed a similar softening trend as the GD fruit, but had much higher flesh firmness during the test.

The ethylene production (Figure 6B) of the untreated GD fruit increased significantly in 5 d when stored at room temperature, and a burst of ethylene occurred at 10 d. Ethephon treated GD fruit produced ethylene in a similar pattern to the untreated fruit, but at a higher level. 1-MCP treatment almost blocked ethylene production in GD fruit. Ethylene production in 'Fuji' was low and there was no significant increase during the storage.

3.5 Expression Profile of MdRTE1a and MdRTE1b

To understand the expression profile of *MdRTE1a* and *MdRTE1b* at various stages of fruit, GD fruit were sampled from 7 DAF (days after full-bloom) through harvest (140 DAF) and further through a storage of 20 d (Table 2, Figure 7). The expression of *MdRTE1a* was high at the early stages of fruit growth and development and then decreased through harvest although there was an increase at 35 DAF and another increase at 133 DAF (Figure 7A). During the storage, the expression of *MdRTE1a* was largely consistently at low levels in the untreated GD fruit, and this pattern of low and consistent expression was not affected in the 1-MCP or ethephon treated GD fruit (Figure 7B). In addition, *MdRTE1a* expression in the untreated GD fruit did not show much difference from that in 'Fuji' although 'Fuji' had higher firmness and lower ethylene production than GD (Figure 6). To assess the expression of *MdRTE1a* in different apple tissues, RNA from young leaves, flowers (at full bloom) and fruit (at harvest) of GD were used. *MdRTE1a* was expressed lowest in fruit, medium in flowers and highest in leaves (Figure 7C). These results suggested that *MdRTE1a* is likely a gene to function in growth rather than a gene that is related to fruit ripening and regulated by ethylene.

The expression of *MdRTE1b* was markedly different from that of *MdRTE1a*. It began with a level that was the highest (at 7 DAF) in the dataset (Figure 7D), and then followed by a sharp reduction (15-22 DAF). The lowest expression levels were observed from 28 DAF through 133 DAF, the vastly major period for fruit growth and development. However, the expression of *MdRTE1b* was significantly increased at harvest (Figure 7D). During the storage, *MdRTE1b* expression was elevated even higher in the untreated GD fruit (Figure 7E). In the ethephon treated GD fruit, *MdRTE1b* was similarly expressed as in the untreated controls, but with slightly lower levels at d 5 and d 10, and slightly higher levels at d 15 and d 20 (Figure 7E). This suggested that there might be a negative feedback regulation in place, i.e. excessive ethylene may result in lower or insignificant promotion of *MdRTE1b* expression. In the 1-MCP treated fruit, however, the expression of *MdRTE1b* was severely suppressed (Figure 7E). The expression of *MdRTE1b* in 'Fuji' fruit was lower than that in GD fruit except for those at d 0 and d 10 (Figure 7E). Among the three organs, fruit had higher expression of *MdRTE1b* than flowers and leaves did (Figure 7F). These data, therefore, indicated that *MdRTE1b* is most likely an apple ripening related gene and involved in ethylene signaling and response.

3.6 Expression Profile of MdNAC1 and MdNAC2

The expression pattern of *MdNAC1* was similar to that of *MdRTE1a*, i.e. high at the early stages of fruit growth and development and then declined through harvest (Figure 7G). During postharvest storage, *MdNAC1* expression in the untreated GD fruit was low and was not affected by 1-MCP or ethephon treatment although there was a slight deviation from this trend at d 20 (Figure 7H). Moreover, *MdNAC1* expression was similar in the fruit of both 'Fuji' and GD (untreated), and was higher in flowers and leaves than in fruit (Figure 7I). These results suggested that *MdNAC1* may play roles in apple growth, but unlikely in fruit ripening.

The expression of *MdNAC2* was similar to that of *MdRTE1b*, i.e. high at the initial stages of fruit growth and development and then become low and consistent through harvest (Figure 7J). In the untreated GD fruit, *MdNAC2* expression greatly increased at d 5 and maintained at high levels throughout the storage (Figure 7K). *MdNAC2* expression was highly inhibited in the 1-MCP treated GD fruit; but the expression levels were also much lower in the ethephon treated fruit than in the control (Figure 7K). This indicated that *MdNAC2* expression is regulated by ethylene and excessive ethylene may also have a negative feedback regulation on the expression of *MdNAC2*, similar to what has been suggested for *MdRTE1b* above. In addition, the expression of *MdNAC2* was much higher in the untreated GD fruit than in 'Fuji' during the storage (Figure 7K), which was positively correlated with the fact that GD fruit produced much more ethylene than 'Fuji' fruit did during the same period of storage (Figure 6B). These results suggested that *MdNAC2* is likely a fruit ripening related gene and is involved in the ethylene signaling and response although its expression was higher in leaves and flowers than in fruit (Figure 7L).

4. Discussion

4.1 Construction and Application of the Y2H Library of 'Golden Delicious'

The Y2H system is primarily used to screen a complete library of proteins for interactions with a specific protein of interest. Since its initial description (Ma & Ptashne, 1988), the Y2H system has been used to detect and analyze protein-protein interactions in many studies (Bruckner, Polge, Lentze, Auerbach, & Schlattner, 2009). In this report, we constructed a Y2H library from the maturing/ripening fruit of GD. Based on the limited sequence survey of 62 clones, five genes are the most abundant in the library, including *AMT2* (8.1%), *AMT1* (4.8%), *MdACO1* (4.8%), auxin-repressed protein-like protein (4.8%) and abscisic stress ripening like protein (3.2%). The high abundance of these genes appears to be correlated with the developmental and ripening stages of the fruit used for constructing the library. In apple, *AMT1* and *AMT2* have been reported to be highly expressed during the late fruit development and ripening stages especially for *AMT2* (Reid & Ross, 1997). High expression of metallothionein genes during these stages has been observed in pear (Itai, Tanabe, Tamura, & Tanaka, 2000) and pineapple (Moyle, Fairbairn, Ripi, Crowe, & Botella, 2005). In addition, two tomato metallothionein (MT)-like protein coding genes, *LeMT(A)* and *LeMT(B)*, are ethylene responsive (Whitelaw, LeHuquet, Thurman, & Tomsett, 1997). It is known that *MdACO1* is largely responsible for the burst of ethylene production in ripening fruit, consistent with its high abundance in the library. Screening the Y2H library enabled us to identify proteins, MdNAC1 and MdNAC2 interacting with MdRTE1a and MdRTE1b in the yeast cells, suggesting that the Y2H library does have applications for screening protein-protein interactions involved in ethylene signaling and response in maturing/ripening apple fruit.

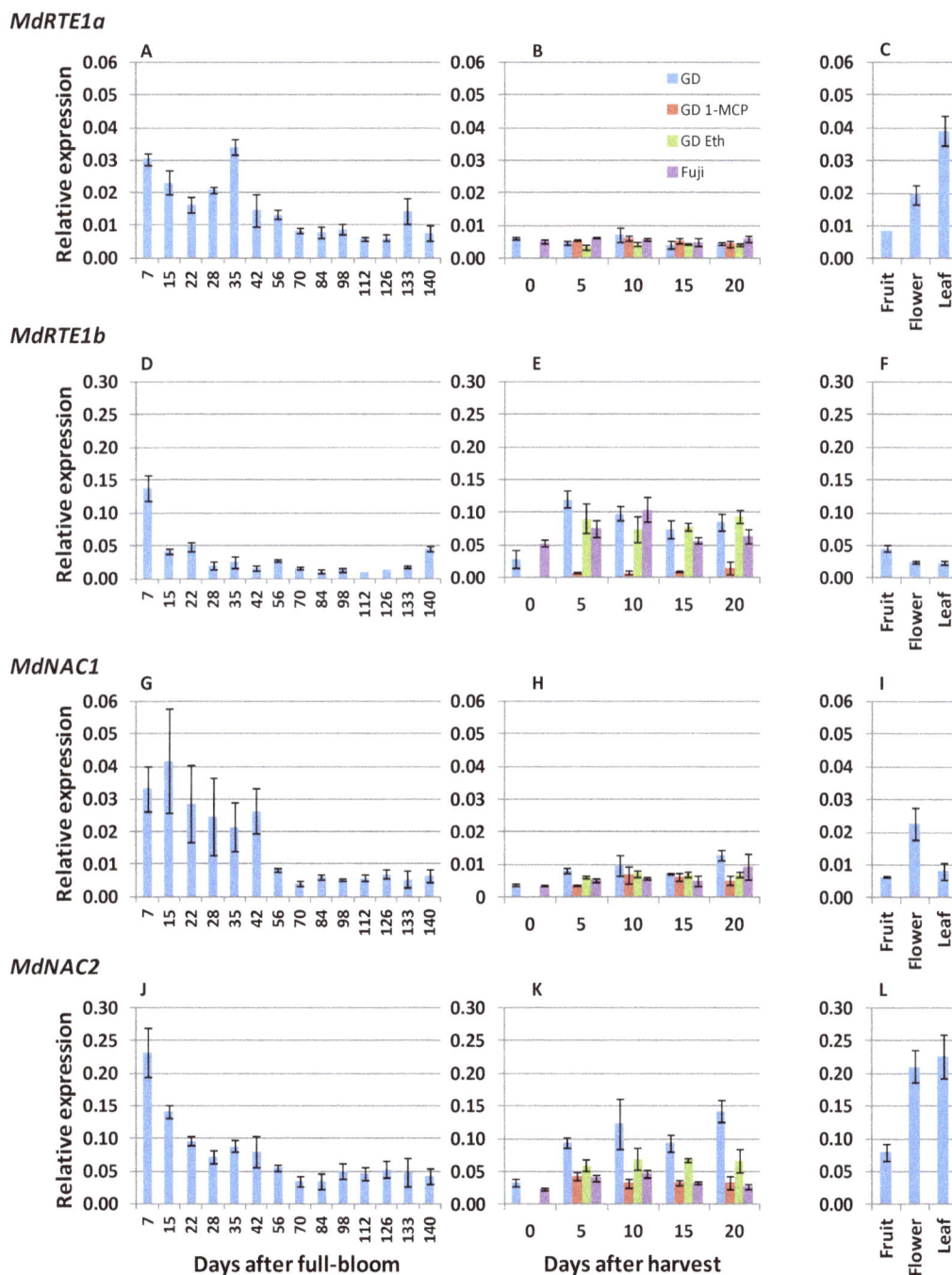

Figure 7. qRT-PCR assays of *MdRTE1a*, *MdRTE1b*, *MdNAC1* and *MdNAC2*

Expression of each gene is presented as relative fold change over the reference gene (actin). The bars represent SD measured from three replicates. **A**, **D**, **G** and **J**: Assays on the GD's fruit under growth and development (pre-harvest). **B**, **E**, **H** and **K**: Assays on the postharvest fruit of GD and 'Fuji'. GD (untreated), GD 1-MCP (treated with 1-MCP), GD Eth (treated with ethephon) and 'Fuji' (untreated). **C**, **F**, **I** and **L**: Assays on the fruit (at harvest), flowers (at full bloom) and young leaves of GD.

4.2 Expression of MdRTE1a, MdRTE1b, MdNAC1 and MdNAC2

RTE1 is a novel protein conserved in plants (Barry & Giovannoni, 2006; Resnick et al., 2006). The expression of *RTE1* was inducible by ethylene and over-expression of *RTE1* reduced ethylene sensitivity in *Arabidopsis* (Resnick et al., 2006). In tomato, over-expression of *GR*, a *RTE1* ortholog, blocked fruit from ripening (Barry & Giovannoni, 2006). These results indicated that RTE1 and GR function in ethylene signaling and/or fruit ripening. In this study, the expression of *MdRTE1b* was quickly enhanced in the untreated GD fruit during ripening (Figure 7E) and was highly inhibited by 1-MCP, indicating that *MdRTE1b* expression is under the regulation of ethylene and involved in apple fruit ripening. Thus, it is likely that MdRTE1b is the apple counterpart of RTE1 in *Arabidopsis*. However, the means by which MdRTE1b functions appeared to be in contrast with that of RTE1 and GR as higher expression of *MdRTE1b* is associated with higher ethylene production and greater fruit firmness loss in GD (i.e. increased ethylene sensitivity and accelerated ripening process) while lower expression of *MdRTE1b* is correlated with lower ethylene production and less fruit firmness loss (i.e. decreased ethylene sensitivity and prohibited ripening process) in 'Fuji' (Figure 7E).

In *Arabidopsis*, there is a related gene of *RTE1*, called *RTH* (*RTE1 HOMOLOG*) (Barry & Giovannoni, 2007; Resnick et al., 2006). RTE1 was reported to associate with one of the ethylene receptors, ETR1, while RTH does not have the same role as RTE1 in ethylene signaling (Dong et al., 2010). Based on our results, the expression of *MdRTE1a* is low and consistant in the untreated GD and 'Fuji' fruit during storage, and is not affected by 1-MCP or ethephon treatment (Figure 7B). These results suggested that *MdRTE1a* may not be involved in fruit ripening and ethylene regulation, similar to *RTH* in *Arabidopsis* (Dong et al., 2010). This similarity may partially explain why an apple ETR1 was not identified from the Y2H library when screened with MdRTE1a as bait in this study. *MdRTE1a* is expressed more in apple leaves and young fruit than in ripening fruit (Figure 7A-C), indicating that it may play roles in growth, but not in fruit ripening.

NAC proteins, characterized by a conserved NAC domain in N-terminus, are a large family of plant-specific transcription factors (Aida, Ishida, Fukaki, Fujisawa, & Tasaka, 1997; Mita, Henmi, & Ohno, 2006; Olsen, Ernst, Lo Leggio, & Skriver, 2005). There are 117 NAC genes in *Arabidopsis*, 151 in rice (Nuruzzaman et al., 2010) and 163 in *Populus* (Hu et al., 2010). NAC transcription factors regulate diverse biological processes, including stress response, growth and development, and signaling of major plant hormones such as auxin, ABA, jasminate and ethylene (Fan, Gao, Yang, Deng, & Li, 2007; He et al., 2005; Jiang, Li, Bu, & Li, 2009; Olsen et al., 2005). In *Arabidopsis*, *AtNAC2* is induced by ethylene precursor ACC treatment (He et al., 2005). A *Cymbidium* NAC gene is expressed abundantly when ethylene production burst occurred during necrosis of young inflorescences (Mita et al., 2006). All these results led to a conclusion that NAC genes are also involved in ethylene signaling and response. In this study, the expression of *MdNAC2* increased in the untreated GD fruit during ripening (Figure 7K), and was highly inhibited in the 1-MCP treated GD fruit, indicating that a modulation of ethylene on its action likely exists. More interestingly, the expression of *MdNAC2* in the ethephon treated GD fruit was even lower than that in the untreated GD fruit, suggesting that excessive ethylene might have a negative feedback regulation on the expression of *MdNAC2*. Such negative feedback regulation of excessive ethylene was also similarly observed for the expression of *MdRTE1b* (Figure 7E).

NOR (AY573802), another NAC gene in tomato, was isolated based on a tomato mutant *non-ripening* (*nor*), fruit of which neither produce climacteric ethylene nor ripen (Giovannoni, 2004). NOR regulates tomato fruit ripening, but it is not clear if it directly modulates ripening or via ethylene (Giovannoni, 2004; Lin, Zhong, & Grierson, 2009). According to the evidence in this study, MdNAC2 may regulate fruit ripening through ethylene signaling by interacting with MdRTE1b. Furthermore, the expression of *MdNAC2* was much higher in untreated GD fruit than in 'Fuji' fruit (Figure 7K), which showed much longer shelf life (less softening and lower ethylene production) than GD fruit. These results indicated that low expression of *MdNAC2* in 'Fuji' could serve an important part in explaining its high keeping quality.

In addition, the expression of *MdNAC1* and *MdNAC2* was strong in apple flowers (Figure 7I, L), suggesting their function in embryonic and floral development as addressed in Olsen et al. (2005). The high expression of *MdNAC2* in apple leaves (Figure 7L) is indicative of its role in vegetative growth (Mallory, Dugas, Bartel, & Bartel, 2004; Olsen et al., 2005). Both *MdNAC1* and *MdNAC2* were strongly expressed in early stage of fruit growth and development (Figure 7G, J), suggesting that the NAC proteins might be involved in fruit cell differentiation.

4.3 The Interactions between MdRTE1s and MdNACs

MdRTE1a and MdRTE1b were found to interact with MdNAC1 and MdNAC2 in yeast cells. Because of the similar expression patterns between *MdRTE1a* and *MdNAC1*, and between *MdRTE1b* and *MdNAC2*, it is possible that MdRTE1a and MdNAC1 might interact in actively growing organs and play roles in growth while MdRTE1b and MdNAC2 primarily in maturing/ripening fruit and function in ethylene signaling and/or fruit ripening.

The sequences of the *MdNAC1* and *MdNAC2* clones in the Y2H library only contain the C-terminal region (Figure 4); and neither has the N-terminal sequences in which the conserved NAC domain is located. In other words, the interactions between the MdRTE1s and MdNACs took place in the C-terminus of the two NAC proteins in yeast cells. However, we cannot exclude the possibility that the real world interactions between the MdRTE1s and MdNACs in apple cells, if they exist, could also involve the N-terminus of MdNACs although the NAC domain has been considered to have DNA-binding ability (Olsen et al., 2005).

The C-terminus of NAC proteins has been reported capable of activating gene expression (Duval, Hsieh, Kim, & Thomas, 2002; Meng, Cai, Zhang, & Guo, 2009; Ooka et al., 2003; Robertson, 2004). Duval et al. (2002) found that a C-terminal deletion construct of AtNAM was not able to activate a reporter gene in a yeast one-hybrid assay. It is possible that MdNAC1 and MdNAC2 may activate the transcription of MdRTE1s by their C-terminal interactions. Thus, MdNAC2 might be positioned at the top of MdRTE1b in ethylene signal transduction pathway. MdNAC2 might also function downstream of MdRTE1b through an unknown pathway or by bridging MdRTE1b and other components in the pathway. Further studies are needed to ensure the interaction between MdRTE1b and MdNAC2 and their functions in apple fruit ripening and ethylene signaling. It would be interesting to investigate whether MdRTE1b interacts with MdNAC2 in apple fruit or other plant cells using the bimolecular fluorescence complementation (BiFC) system. Detailed functional characterization of MdRTE1b and MdNAC2 would increase our understanding of apple fruit ripening.

In conclusion, apple MdRTE1a and MdRTE1b putatively interact with both MdNAC1 and MdNAC2 proteins in yeast cells. MdRTE1b is probably the apple counterpart of RTE1 in *Arabidopsis*. MdRTE1b and MdNAC2 may play an important role in fruit ripening and ethylene signaling, whereas MdRTE1a and MdNAC1 primarily function in the growth of young fruit, flowers and leaves.

Acknowledgements

We sincerely thank Drs Susan Brown and Christopher Watkins for their critical review of this manuscript. We also thank Dr. Brown for providing the apple trees for fruit sampling, Dr. Watkins for helping with 1-MCP treatment of the fruit, and Xia Xu, Yang Bai and Tuanhui Bai for technical assistance.

References

Adams, D. O., & Yang, S. F. (1979). Ethylene Biosynthesis - Identification of 1-Aminocyclopropane-1-Carboxylic Acid as an Intermediate in the Conversion of Methionine to Ethylene. *Proceedings of the National Academy of Sciences of the United States of America, 76*(1), 170-174. http://dx.doi.org/10.1073/pnas.76.1.170

Adams-Phillips, L., Barry, C., Kannan, P., Leclercq, J., Bouzayen, M., & Giovannoni, J. (2004). Evidence that CTR1-mediated ethylene signal transduction in tomato is encoded by a multigene family whose members display distinct regulatory features. *Plant Molecular Biology, 54*(3), 387-404. http://dx.doi.org/10.1023/B:PLAN.0000036371.30528.26

Aida, M., Ishida, T., Fukaki, H., Fujisawa, H., & Tasaka, M. (1997). Genes involved in organ separation in Arabidopsis: An analysis of the cup-shaped cotyledon mutant. *Plant Cell, 9*(6), 841-857. http://dx.doi.org/10.1105/tpc.9.6.841

Barry, C. S., & Giovannoni, J. J. (2006). Ripening in the tomato Green-ripe mutant is inhibited by ectopic expression of a protein that disrupts ethylene signaling. *Proceedings of the National Academy of Sciences of the United States of America, 103*(20), 7923-7928. http://dx.doi.org/10.1073/pnas.0602319103

Barry, C. S., & Giovannoni, J. J. (2007). Ethylene and fruit ripening. *Journal of Plant Growth Regulation, 26*(2), 143-159. http://dx.doi.org/10.1007/s00344-007-9002-y

Barry, C. S., McQuinn, R. P., Thompson, A. J., Seymour, G. B., Grierson, D., & Giovannoni, J. J. (2005). Ethylene insensitivity conferred by the Green-ripe and Never-ripe 2 ripening mutants of tomato. *Plant Physiology, 138*(1), 267-275. http://dx.doi.org/10.1104/pp.104.057745

Bruckner, A., Polge, C., Lentze, N., Auerbach, D., & Schlattner, U. (2009). Yeast Two-Hybrid, a Powerful Tool for Systems Biology. *International Journal of Molecular Sciences, 10*(6), 2763-2788. http://dx.doi.org/10.3390/ijms10062763

Dal Cin, V., Danesin, M., Boschetti, A., Dorigoni, A., & Ramina, A. (2005). Ethylene biosynthesis and perception in apple fruitlet abscission (Malus domestica L. Borck). *Journal of Experimental Botany, 56*(421), 2995-3005. http://dx.doi.org/10.1093/jxb/eri296

Dal Cin, V., Danesin, M., Botton, A., Boschetti, A., Dorigoni, A., & Ramina, A. (2007). Fruit load and elevation affect ethylene biosynthesis and action in apple fruit (Malus domestica L. Borkh) during development, maturation and ripening. *Plant Cell and Environment, 30*(11), 1480-1485. http://dx.doi.org/10.1111/j.1365-3040.2007.01723.x

Dal Cin, V., Rizzini, F. M., Botton, A., & Tonutti, P. (2006). The ethylene biosynthetic and signal transduction pathways are differently affected by 1-MCP in apple and peach fruit. *Postharvest Biology and Technology, 42*(2), 125-133. http://dx.doi.org/10.1016/j.postharvbio.2006.06.008

Dong, C. H., Jang, M., Scharein, B., Malach, A., Rivarola, M., Liesch, J., … Chang, C. (2010). Molecular association of the Arabidopsis ETR1 ethylene receptor and a regulator of ethylene signaling, RTE1. *Journal Biological Chemistry, 285*(52), 40706-40713. http://dx.doi.org/10.1074/jbc.M110.146605

Duval, M., Hsieh, T. F., Kim, S. Y., & Thomas, T. L. (2002). Molecular characterization of AtNAM: a member of the Arabidopsis NAC domain superfamily. *Plant Molecular Biology, 50*(2), 237-248. http://dx.doi.org/10.1023/A:1016028530943

Fan, J., Gao, X., Yang, Y. W., Deng, W., & Li, Z. G. (2007). Molecular cloning and characterization of a NAC-like gene in "navel" orange fruit response to postharvest stresses. *Plant Molecular Biology Reporter, 25*(3-4), 145-153. http://dx.doi.org/10.1007/s11105-007-0016-1

Giovannoni, J. J. (2004). Genetic regulation of fruit development and ripening. *Plant Cell, 16*, S170-S180. http://dx.doi.org/10.1105/tpc.019158

He, X. J., Mu, R. L., Cao, W. H., Zhang, Z. G., Zhang, J. S., & Chen, S. Y. (2005). AtNAC2, a transcription factor downstream of ethylene and auxin signaling pathways, is involved in salt stress response and lateral root development. *Plant Journal, 44*(6), 903-916. http://dx.doi.org/10.1111/j.1365-313X.2005.02575.x

Hu, R. B., Qi, G. A., Kong, Y. Z., Kong, D. J., Gao, Q. A., & Zhou, G. K. (2010). Comprehensive Analysis of NAC Domain Transcription Factor Gene Family in Populus trichocarpa. *Bmc Plant Biology, 10*, 145. http://dx.doi.org/10.1186/1471-2229-10-145

Itai, A., Tanabe, K., Tamura, F., & Tanaka, T. (2000). Isolation of cDNA clones corresponding to genes expressed during fruit ripening in Japanese pear (Pyrus pyrifolia Nakai): involvement of the ethylene signal transduction pathway in their expression. *Journal of Experimental Botany, 51*(347), 1163-1166. http://dx.doi.org/10.1093/jexbot/51.347.1163

Jiang, H., Li, H., Bu, Q., & Li, C. (2009). The RHA2a-interacting proteins ANAC019 and ANAC055 may play a dual role in regulating ABA response and jasmonate response. *Plant Signaling & Behavior, 4*(5), 464-466. http://dx.doi.org/10.4161/psb.4.5.8543

Kendrick, M. D., & Chang, C. (2008). Ethylene signaling: new levels of complexity and regulation. *Current Opinion in Plant Biology, 11*(5), 479-485. http://dx.doi.org/10.1016/j.pbi.2008.06.011

Kondo, S., Meemak, S., Ban, Y., Moriguchi, T., & Harada, T. (2009). Effects of auxin and jasmonates on 1-aminocyclopropane-1-carboxylate (ACC) synthase and ACC oxidase gene expression during ripening of apple fruit. *Postharvest Biology and Technology, 51*(2), 281-284. http://dx.doi.org/10.1016/j.postharvbio.2008.07.012

Lay-Yee, M., Dellapenna, D., & Ross, G. S. (1990). Changes in messenger RNA and protein during ripening in apple fruit (Malus domestica Borkh. cv Golden Delicious). *Plant Physiology, 94*(2), 850-853. http://dx.doi.org/10.1104/pp.94.2.850

Li, J. G., & Yuan, R. C. (2008). NAA and ethylene regulate expression of genes related to ethylene biosynthesis, perception, and cell wall degradation during fruit abscission and ripening in 'Delicious' apples. *Journal of Plant Growth Regulation, 27*(3), 283-295. http://dx.doi.org/10.1007/s00344-008-9055-6

Lin, Z. F., Alexander, L., Hackett, R., & Grierson, D. (2008). LeCTR2, a CTR1-like protein kinase from tomato, plays a role in ethylene signalling, development and defence. *Plant Journal, 54*(6), 1083-1093. http://dx.doi.org/10.1111/j.1365-313X.2008.03481.x

Lin, Z. F., Zhong, S. L., & Grierson, D. (2009). Recent advances in ethylene research. *Journal of Experimental Botany, 60*(12), 3311-3336. http://dx.doi.org/10.1093/jxb/erp204

Ma, J., & Ptashne, M. (1988). Converting a eukaryotic transcriptional inhibitor into an activator. *Cell, 55*(3), 443-446. http://dx.doi.org/10.1016/0092-8674(88)90030-X

Mallory, A. C., Dugas, D. V., Bartel, D. P., & Bartel, B. (2004). MicroRNA regulation of NAC-domain targets is required for proper formation and separation of adjacent embryonic, vegetative, and floral organs. *Current Biology, 14*(12), 1035-1046. http://dx.doi.org/10.1016/j.cub.2004.06.022

Meng, C. M., Cai, C. P., Zhang, T. Z., & Guo, W. Z. (2009). Characterization of six novel NAC genes and their responses to abiotic stresses in Gossypium hirsutum L. *Plant Science, 176*(3), 352-359. http://dx.doi.org/10.1016/j.plantsci.2008.12.003

Mita, S., Henmi, R., & Ohno, H. (2006). Enhanced expression of genes for ACC synthase, ACC oxidase, and NAC protein during high-temperature-induced necrosis of young inflorescences of Cymbidium. *Physiologia Plantarum, 128*(3), 476-486. http://dx.doi.org/10.1111/j.1399-3054.2006.00759.x

Moyle, R., Fairbairn, D. J., Ripi, J., Crowe, M., & Botella, J. R. (2005). Developing pineapple fruit has a small transcriptome dominated by metallothionein. *Journal of Experimental Botany, 56*(409), 101-112. http://dx.doi.org/10.1093/jxb/eri015

Nuruzzaman, M., Manimekalai, R., Sharoni, A. M., Satoh, K., Kondoh, H., Ooka, H., & Kikuchia, S. (2010). Genome-wide analysis of NAC transcription factor family in rice. *Gene, 465*(1-2), 30-44. http://dx.doi.org/10.1016/j.gene.2010.06.008

Olsen, A. N., Ernst, H. A., Lo Leggio, L., & Skriver, K. (2005). NAC transcription factors: structurally distinct, functionally diverse. *Trends in Plant Science, 10*(2), 79-87. http://dx.doi.org/10.1016/j.tplants.2004.12.010

Ooka, H., Satoh, K., Doi, K., Nagata, T., Otomo, Y., Murakami, K., … Kikuchi, S. (2003). Comprehensive analysis of NAC family genes in Oryza sativa and Arabidopsis thaliana. *DNA Research, 10*(6), 239-247. http://dx.doi.org/10.1093/dnares/10.6.239

Pfaffl, M. W. (2001). A new mathematical model for relative quantification in real-time RT-PCR. *Nucleic Acids Research, 29*(9), e45. http://dx.doi.org/10.1093/nar/29.9.e45

Reid, S. J., & Ross, G. S. (1997). Up-regulation of two cDNA clones encoding metallothionein-like proteins in apple fruit during cool storage. *Physiologia Plantarum, 100*(1), 183-189. http://dx.doi.org/10.1111/j.1399-3054.1997.tb03471.x

Resnick, J. S., Rivarola, M., & Chang, C. (2008). Involvement of RTE1 in conformational changes promoting ETR1 ethylene receptor signaling in Arabidopsis. *Plant Journal, 56*(3), 423-431. http://dx.doi.org/10.1111/j.1365-313X.2008.03615.x

Resnick, J. S., Wen, C. K., Shockey, J. A., & Chang, C. (2006). REVERSION-TO-ETHYLENE SENSITIVITY1, a conserved gene that regulates ethylene receptor function in Arabidopsis. *Proceedings of the National Academy of Sciences of the United States of America, 103*(20), 7917-7922. http://dx.doi.org/10.1073/pnas.0602239103

Rivarola, M., McClellan, C. A., Resnick, J. S., & Chang, C. (2009). ETR1-Specific Mutations Distinguish ETR1 from Other Arabidopsis Ethylene Receptors as Revealed by Genetic Interaction with RTE1. *Plant Physiology, 150*(2), 547-551. http://dx.doi.org/10.1104/pp.109.138461

Robertson, M. (2004). Two transcription factors are negative regulators of gibberellin response in the HvSPY-signaling pathway in barley aleurone. *Plant Physiology, 136*(1), 2747-2761. http://dx.doi.org/10.1104/pp.104.041665

Stepanova, A. N., & Alonso, J. M. (2009). Ethylene signaling and response: where different regulatory modules meet. *Current Opinion in Plant Biology, 12*(5), 548-555. http://dx.doi.org/10.1016/j.pbi.2009.07.009

Tacken, E., Ireland, H., Gunaseelan, K., Karunairetnam, S., Wang, D., Schultz, K., …Schaffer, R. J.. (2010). The Role of Ethylene and Cold Temperature in the Regulation of the Apple POLYGALACTURONASE1 Gene and Fruit Softening. *Plant Physiology, 153*(1), 294-305. http://dx.doi.org/10.1104/pp.109.151092

Tatsuki, M., & Endo, A. (2006). Analyses of expression patterns of ethylene receptor genes in apple (Malus domestica Borkh.) fruit treated with or without 1-methylcyclopropene (1-MCP). *Journal of the Japanese Society for Horticultural Science, 75*(6), 481-487. http://dx.doi.org/10.2503/jjshs.75.481

Tatsuki, M., Endo, A., & Ohkawa, H. (2007). Influence of time from harvest to 1-MCP treatment on apple fruit quality and expression of genes for ethylene biosynthesis enzymes and ethylene receptors. *Postharvest Biology and Technology, 43*(1), 28-35. http://dx.doi.org/10.1016/j.postharvbio.2006.08.010

Tatsuki, M., Hayama, H., & Nakamura, Y. (2009). Apple ethylene receptor protein concentrations are affected by ethylene, and differ in cultivars that have different storage life. *Planta, 230*(2), 407-417. http://dx.doi.org/10.1007/s00425-009-0953-z

Velasco, R., Zharkikh, A., Affourtit, J., Dhingra, A., Cestaro, A., Kalyanaraman, A., ... Malnoy, M. (2010). The genome of the domesticated apple (Malus x domestica Borkh.). *Nat Genet, 42*, 833-839. http://dx.doi.org/10.1038/ng.654

Wang, A. D., Tan, D. M., Tatsuki, M., Kasai, A., Li, T. Z., Saito, H., ... Haradaa, T. (2009). Molecular mechanism of distinct ripening profiles in 'Fuji' apple fruit and its early maturing sports. *Postharvest Biology and Technology, 52*(1), 38-43. http://dx.doi.org/10.1016/j.postharvbio.2008.09.001

Wang, A., Tan, D., Takahashi, A., Li, T. Z., & Harada, T. (2007). MdERFs, two ethylene-response factors involved in apple fruit ripening. *Journal of Experimental Botany, 58*(13), 3743-3748. http://dx.doi.org/10.1093/jxb/erm224

Whitelaw, C. A., LeHuquet, J. A., Thurman, D. A., & Tomsett, A. B. (1997). The isolation and characterisation of type II metallothionein-like genes from tomato (Lycopersicon esculentum L). *Plant Molecular Biology, 33*(3), 503-511. http://dx.doi.org/10.1023/A:1005769121822

Wiersma, P. A., Zhang, H., Lu, C., Quail, A., & Toivonen, P. M. A. (2007). Survey of the expression of genes for ethylene synthesis and perception during maturation and ripening of 'Sunrise' and 'Golden Delicious' apple fruit. *Postharvest Biology and Technology, 44*(3), 204-211. http://dx.doi.org/10.1016/j.postharvbio.2006.12.016

Yang, S. F., & Hoffman, N. E. (1984). Ethylene Biosynthesis and Its Regulation in Higher-Plants. *Annual Review of Plant Physiology and Plant Molecular Biology, 35*, 155-189. http://dx.doi.org/10.1146/annurev.arplant.35.1.155

Zegzouti, H., Jones, B., Frasse, P., Marty, C., Maitre, B., Latche, A., ... Bouzayen, M. (1999). Ethylene-regulated gene expression in tomato fruit: characterization of novel ethylene-responsive and ripening-related genes isolated by differential display. *Plant Journal, 18*(6), 589-600. http://dx.doi.org/10.1046/j.1365-313x.1999.00483.x

Zhou, X., Liu, Q., Xie, F., & Wen, C. K. (2007). RTE1 is a golgi-associated and ETR1-dependent negative regulator of ethylene responses. *Plant Physiology, 145*(1), 75-86. http://dx.doi.org/10.1104/pp.107.104299

2

Characterization of the TRPC3 Gene in Myotonic Goats: Further Insight Into *Myotonia congenita* and Muscular Dystrophy

M. M. Corley[1], & J. Caviness[1]

[1] Agriculture Research Station, Virginia State University, P.O. Box 9061, Virginia 23806, USA

Correspondence: M. M. Corley, Agriculture Research Station, Virginia State University, P.O. Box 9061, Virginia 23806, USA. E-mail: mcorley@vsu.edu

Abstract

Myotonia congenita (Mc) is a muscle disorder seen in both Myotonic goats and humans, caused by mutations in the chloride ion channel gene (CLCN1). Calcium signaling has been coupled to the function of the CLCN1, but this interaction is not well understood, as individuals with Mc do not experience muscular dystrophy (MD). Over expression of the Transient Receptor Cation Channel 3 (*TRPC3*), a protein responsible for calcium influx and muscle contraction causes an elevation in calcium which results in a phenotype of MD. Evaluation of the *TRPC3* gene in Myotonic goats has not been conducted. Therefore the objective of this experiment was to evaluate gene expression of the *TRPC3* gene in Myotonic vs. Spanish goats (control). Total RNA was isolated from whole blood samples. Cross species primers were designed from the human, bovine, and mouse *TRPC3* cDNA alignments. The goat partial *TRPC3* gene showed 98%, 92% and 91%, and 100%, 98%, and 98% nucleotide and amino acid sequence homology to the bovine, human and mouse *TRPC3* genes respectively. Quantitative Real Time PCR showed that gene expression of *TRPC3* was 77% higher (P<0.05) in Myotonic than Non-Myotonic (Spanish) goats. Male Myotonic goats expressed 67% higher levels (P<0.05) of TRPC3 than females. The *TRPC3* gene expression was 90% higher (P<0.05) in Myotonic goats older than 4 years of age. These data indicate that the *TRPC3* gene is a potential biomarker to further study *Myotonia congenita* in Myotonic goats and the interrelationship of the mechanism of calcium signaling in human Mc and MD.

Keywords: TRPC3, gene expression, *Myotonia congenita* Goat

1. Introduction

1.1 Significance of the Problem

The Myotonic goat ("falling goats", "stiff legged goats", epileptic goats, or fainting goats), was first known as the oldest recognized animal model for inherited muscular dystrophy (MD) (Bryant, 1979), since they exhibit an inherited disorder called *Myotonia congenita* (Mc). Myotonia is characteristic of hyperexcitability of the muscle cell membrane. Humans with this disorder often have prolonged muscle contractions and are unable to relax certain muscles after use (Curran & Keating, 1994; George, Crackower, Abdalla, Hudson, & Ebers, 1993; Koch et al., 1992). Myotonic disorders are classified either non-dystrophic or dystrophic myotonias. The non-dystrophic myotonias thus far have been shown to only involve the muscle system, whereas the dystrophic myotonias exhibit multisystem involvement concurrent with additional muscle weakness. Genes involved in Mc are the muscle voltage-gated sodium and chloride channel genes SCN4A ("Dinucleotide repeat polymorphisms at the SCN4A locus suggest allelic heterogeneity of hyperkalemic periodic paralysis and paramyotonia congenita," 1992; Du et al., 2012; Ruscák, 1997; Shirakawa, Sakai, Kitagawa, Hori, & Hirose, 2002) and CLCN1 (Bernard, Poulin, Puymirat, Sternberg, & Shevell, 2008; Grunnet et al., 2003; Wijnberg et al., 2012; Zhang, Bendahhou, Sanguinetti, & Ptácek, 2000), the Myotonic dystrophy protein kinase (DMPK) gene (Meola & Cardani, 2014), and the CCHC-type zinc finger, nucleic acid binding protein gene CNBP (Meola, 2013; Sun et al., 2011). In most forms of MD, membrane damage occurs after prolonged muscle contraction (Blake, Weir, Newey, & Davies, 2002; Petrof, Shrager, Stedman, Kelly, & Sweeney, 1993), which causes a delayed relaxation of the muscles (Beck, Fahlke, & George, 1996). Like humans (Curran & Keating, 1994; Skov, de Paoli, Lausten, Nielsen, & Pedersen, 2014), Mc in goats is attributed to a mutation in a chloride ion gene (clcn1) (Beck et al., 1996; Bryant & Conte-Camerino, 1991). Positively charged sodium ions signal the brain's message for the muscle cells to contract, while negatively charged chloride ions, tell the muscle cells to relax. *Myotonia congenita* results from an abnormal channel of chloride ions, which throws this relationship out of equilibrium. The muscle cells end up with more than enough sodium but not enough chloride, which causes abnormal repetitive electrical signals from the brain resulting in stiffness. Like the Myotonic goat, humans with Mc, do not exhibit obvious muscle wasting as seen with MD, and in most cases live a normal life span. This has led to a transition from viewing the Myotonic goat as an identical model to study human MD, to a renewed interest in what causes non-muscle wasting in this form of Mc. However, though there is a Myotonic phenotype, conflicting studies have shown that on one hand, Mc is not related to MD, as the characteristic muscle wasting does not occur, and on the other hand, loss of CLCN1 function is due to binding of the clc-1 mRNA to elevated levels of a CUG splicing regulator protein in skeletal muscle of individuals with Type 1 MD (Charlet-B et al., 2002). Other studies have found that mutations in both the CLCN1 and CNBP coexist in individuals with MD (Sun et al., 2011), indicating that there is a need for further understanding of the mechanism of Mc. Studies suggested that decreased calcium release or increased uptake may protect Myotonic goat muscle from destructive changes of calcium overload, which has been proposed to be a common factor for dystrophic change (Atkinson, Swift, & Lequire, 1981; Millay et al., 2009; Skov et al., 2014; Skov, Riisager, Fraser, Nielsen, & Pedersen, 2013; Swift, Atkinson, & LeQuire, 1979). It has been shown that disruption in calcium influx causes malfunction in muscle contraction (Allen, Gervasio, Yeung, & Whitehead, 2010). Furthermore, calcium alone is enough to induce dystrophy in skeletal muscle in mice (Millay et al., 2009). The Transient Receptor Potential Canonical proteins, particularly 3 (*TRPC3)* is involved in calcium cycling (Gonzalez-Cobos & Trebak, 2010) and muscle contraction (Ambudkar, 2009; Gailly, 2012; Tsvilovskyy et al., 2009). Over expression of TRPC3, coupled with an increase in calcium influx results in a phenotype of MD nearly identical to that observed in disease animal models lacking the cytoskeletal protein dystrophin (Millay et al., 2009). It was shown that inhibition of TRPC channels in mice considerably reduced calcium influx and dystrophic characteristics associated with a mutation in the dystrophin gene (mdx) (Vandebrouck, Martin, Colson-Van Schoor, Debaix, & Gailly, 2002; Whitehead, Yeung, & Allen, 2006). Experiments have shown that mouse mdx muscles are susceptible to stretch-induced damage caused by a series of stretched contractions resulting in a prolonged increase in resting intracellular calcium concentration. This is caused by calcium entry through stretch-activated channels via TRPC genes. This causes opening of the stretched channels thereby allowing calcium entry. The end result is significant muscle damage and consequently MD (Allen et al., 2010). The fact that TRPC3 is directly implicated in MD, and Mc is somewhat interconnected with calcium signaling, warrants further study.

1.2 Justification

The TRPC channels and muscular disease has resurrected interest in calcium signaling mechanisms relative to Mc (Beech, 2005; Guibert, Ducret, & Savineau, 2011; Nilius & Owsianik, 2011; Nilius, Voets, & Peters, 2005). We have isolated a partial sequence of the *TRPC3* gene in Myotonic goats (Accession: HQ847409). Gene expression of *TRPC3* in Myotonic goats has not been evaluated. Therefore this study was conducted to first isolate a partial sequence of the goat TRPC3 gene and evaluate gene expression of the *TRPC3* in Myotonic goats using quantitative Real Time PCR.

2. Method

2.1 Animals and Experimental Design

Animals used for the study were housed at Virginia State University Randolph farm in accordance with animal care and use guidelines. A total of 20 Spanish and Myotonic goats (5 males, 5 non- pregnant females of each breed), grazing pasture and supplemented with hay, cracked corn and ground soybean meal were selected.   Spanish goats were used to represent the non-Myotonic or non-Mc goats in the study.

2.2 Blood Collection and Serum Preparation

Goats were adequately restrained and blood collected via jugular venipuncture. In brief, the vein was identified by palpation and visual inspection. The area was swabbed with 70% alcohol. Gentle pressure was applied at the thoracic inlet to produce distension of the vein. Blood samples (3 ml) were collected in vials without coagulant using 16-20G needles. Blood samples were subsequently stored at -80°C for later molecular analysis.

2.3 Total RNA Isolation from Goat Blood

Total RNA was isolated from selected goat whole blood samples previously stored at -80°C using the modified RNA extraction protocol (Gauthier, Madison, & Michel, 1997). Approximately 500 µl of each sample was mixed with 3 ml of 1X Dulbecco's phosphate buffered saline (PBS) (Invitrogen; Grand Island, NY 14072) and centrifuged at 5,000 rpm for 20 minutes at 4°C in a Heraeus MegaFuge 16R centrifuge (ThermoFisher-Scientific). After discarding the supernatant, the addition of PBS and centrifugation was repeated until cells were clean. Once the cells were sufficiently washed, 1 ml of guanidinium thiocyanate solution (4M GTC, 25 mM Sodium citrate pH

7.0 and N-0.5% lauroylsarcosine, Sigma-Aldrich) was added and the cells were resuspended by pipetting. Under the fume hood, 10 microliters of β-mercaptoethanol (Omni-Pur), 2M sodium acetate (pH4) at one-tenth of the volume and 1 volume of phenol:chloroform:isoamyl alcohol (25:24:1) (Fisher Scientific; Fair Lawn, New Jersey 0740) were added and mixed before placing on ice for 15 minutes. The tubes were then centrifuged for 15 minutes at 4°C at 5,000 rpm. After centrifugation, the aqueous layer was removed and placed in a sterile tube. The samples were placed on ice for 10 minutes, 2 volumes of 95% ethanol added, and centrifuged for 15 minutes at 5,000 rpm, allowing RNA to precipitate. Two additional washes with 15 minutes of centrifugation at 5,000 rpm were completed using 200 μl of 70% ethanol before allowing samples to air dry. Samples were then resuspended in Diethylpyrocarbonate (DEPC) treated water and stored at -80°C. A NanoDrop 2000 Spectrophotometer (ThermoScientific) was used to measure RNA concentration and purity.

2.4 Isolation of Goat TRPC3

A comprehensive screening of the GenBank nucleotide databases was performed. The *TRPC3* sequences (bovine, human, mouse) were retrieved from the GenBank and sequence alignments generated using CLC Main WorkbenchBioinformatics software (clcbio.com). Oligonucleotide primers were designed from conserved regions of mRNA of the bovine human, and mouse *TRPC3* nucleotide sequence alignments. Primers and target regions used for isolation of the goat *TRPC3* gene are given in Table 1. The RT-PCR was conducted using the recommended protocol of the Verso 1-step RT-PCR kit (Thermo Scientific). Modified thermocycling conditions for 40 cycles were as follows: 50°C 15 minutes, 95°C, 2 minutes (initial denaturation), 95°C, 30 secs, 55°C, 1 minute, 72°C, 1 minute repeated 39 times and a final extension at 72°C for 5 minutes. The target *TRPC3* cDNA was visualized by 1.5% agarose gel electrophoresis and a UGenius UV gel documentation system (SynGene, Fredericksburg, MD) equipped with a high resolution CCD camera.

Table 1. Primers and Target Regions used for Isolation of the goat TRPC3 gene from goat whole blood

Accession No.	Primer name	Conserved Primer sequence	Target Region	Fragment Length (bp)
NM001104960	Forward	AGGATGACAGTGATGTAGA	2033-2057	213
	Reverse	ACCTGGACTTTGAGTTAC	2228-2245 (Rev-comp)	
NM001130698			2326-2538	
NM019510			2407-2619	

Table showing a summary of the GenBank Accession numbers of the TRPC3 genes used for nucleotide alignments and design of oligonucleotide primers from conserved regions among the species (*Bos taurus, Homo sapiens, Mus musculus*). The CLC Main Workbench Bioinformatics software was used to generate primers.

2.5 Nucleotide Sequencing of Goat TRPC3 cDNA

For *TRPC3* nucleotide sequencing, the cDNA (213bp) products were cut out and purified from agarose gels (Qiagen and Bio-Rad). Samples were sent for sequencing at GeneWiz (South Plainfield, New Jersey). Raw nucleotide sequences were analyzed using sequence analysis software (NCBI-BLAST, CLC Main Workbench). Subsequently, quantitative Real Time PCR (qRT-PCR) analysis was conducted to determine gene expression of goat *TRPC3*.

2.6 Measurement of Goat TRPC3 Gene Expression

Gene expression of *TRPC3* in whole blood was measured using (qRT-PCR). The analytical parameters were as previously published (Corley & Ward, 2013). In brief, the analysis was conducted using the iScript One Step RT-PCR kit with SYBR Green (BioRad). For each reaction 100ng of total RNA was used. The Actin gene was used as an internal standard for relative quantitative analysis of *TRPC3* gene expression.

2.5 Statistical Analysis

All data were analyzed using the General Linear Model procedure of SAS. Means were considered significant at the 5% level of probability using Duncan's Multiple Range Test.

3. Results and Discussion

3.1 Isolation and Analysis of Goat TRPC3

This study aimed to accomplish two objectives. Since the TRPC3 gene had not yet been identified in goats, our first aim was to isolate a partial sequence of this gene in the goats. After RNA isolation from goat blood, RT-PCR was performed. Agarose gel electrophoresis was performed to visualize the goat *TRPC3* cDNA. The cross species primers designed to target the Spanish and Myotonic goat *TRPC3* genes successfully amplified the expected 213 bp fragments (Figure 1). The goat *TRPC3* cDNA was sequenced and compared to the bovine, human, and mouse *TRPC3* genes to identify sequence similarities. The BLASTn (Altschul, Gish, Miller, Myers, & Lipman, 1990) of the goat *TRPC3* cDNA showed 98% and 92% and 91% sequence homology to the bovine and human and mouse *TRPC3* genes respectively (Figure 2). The BLASTp (Altschul et al., 1997) of the goat *TRPC3* amino acid alignment of the translated 213 bp nucleotide sequence was compared to the bovine, human, and mouse TRPC3 showing 100%, 98% and 98% homology respectively (Figure 3).

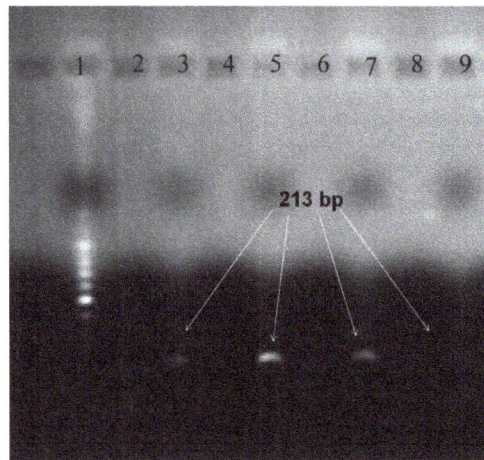

Lane 1= MW marker, Lanes 3,,5,7,9 = Goat TRPC3 cDNA of 213 bp

Figure1. Agarose Gel Electrophoresis of Goat TRPC3 cDNA after Reverse Transcriptase PCR

Gel electrophoresis of Goat TRPC3 cDNA. The gel is 1.5% agarose with 0.05mg/ml ethidium bromide.

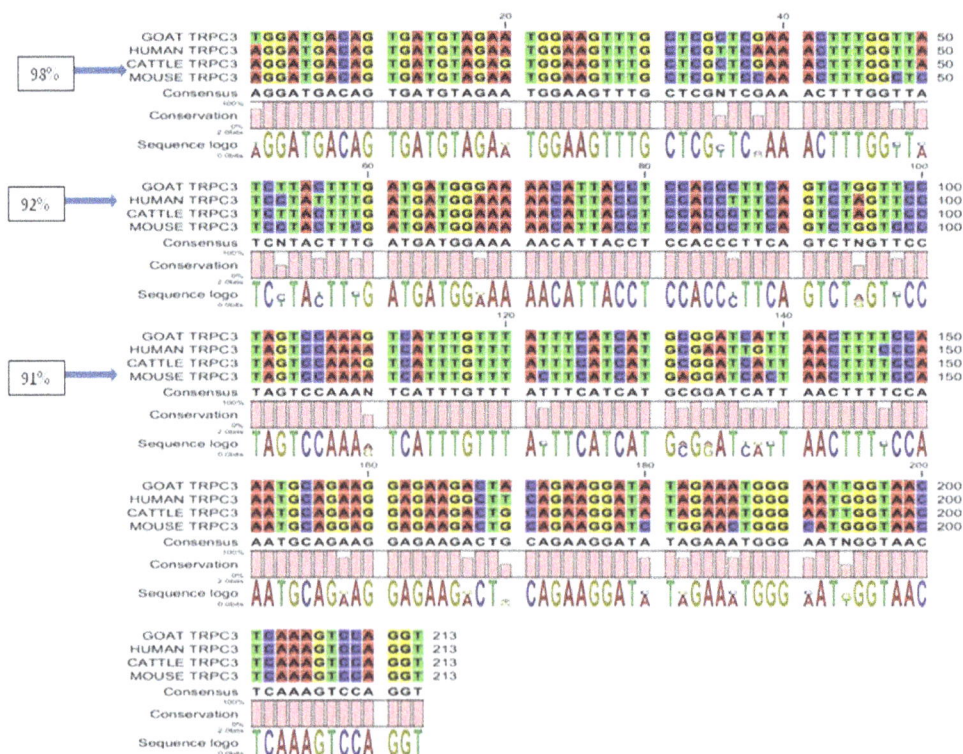

Figure 2. Nucleotide Sequence Alignment Showing Homology of Goat, Human, Bovine, and Mouse TRPC3 Genes

Nucleotide sequence alignment of partial goat TRPC 3 gene is shown. Alignment is shown to validate the conserved regions used to design the cross species oligonucleotide primers. Generated using CLC Main.

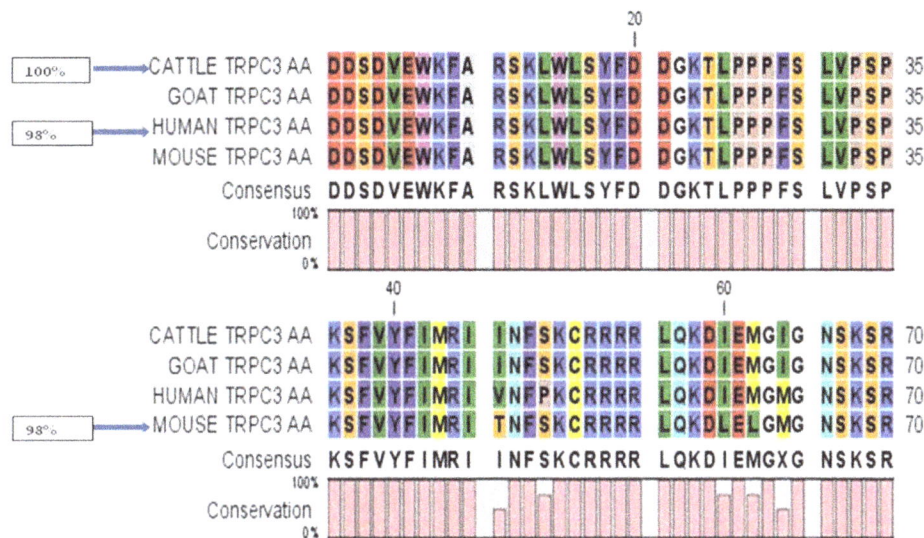

Figure 3. Amino Acid Alignment of Translated Goat, Human, Bovine, and Mouse TRPC3

Amino acid translated sequence alignment of partial goat TRPC3 gene is shown. Alignment is shown to validate the conserved regions used to design the cross species oligonucleotide primers. Degeneracy lends to a greater amino acid sequence homology of the partial TRPC3 genes. Generated w CLC Main Workbench Bioinformatics software Percentages represents the degree of homology among partial amino acid sequences of the TRPC3 gene.

3.2 TRPC3 Gene Expression

Our second aim was to measure gene expression of goat TRPC3 after nucleotide sequence verification of the target gene. Spanish goats (non-Myotonic) versus Myotonic goats were evaluated. Quantitative Real Time PCR successfully amplified the *TRPC3* gene in Spanish and Myotonic goats. The *TRPC3* gene was expressed 77% higher ($P<0.05$) in Myotonic goats when compared to non-Myotonic (Spanish) goats (Figure 4). This observation seems to be in line with previous studies in mice which demonstrated that an over expression of TRPC3 was correlated with characteristics of MD (Millay et al., 2009). Though Myotonic goats do not exhibit obvious muscle wasting (Bryant, 1979), there seems to be some involvement of the calcium signaling mechanism in goats that exhibit Mc as opposed to those that don't, as indicated by the TRPC3 gene expression data of this study. Myotonic males had 67% higher ($P<0.05$) gene expression of *TRPC3* than females (Figure 5). The Myotonic goats used in this study were sexually mature non-pregnant females and intact males. It has been shown that Mc is more severe in males than females, (Burge, Hanna, & Schorge, 2013), studies have shown that progesterone and testosterone inhibit CLCN1 channels containing the mutation F297S associated with dominantly inherited Mc, but is a non-genomic effect (Fialho, Kullmann, Hanna, & Schorge, 2008). Higher gene expression of the TRPC3 gene in males versus females in this study coincides with the severity of Mc in males with non dystrophic Mc. This further broadens the scope of insight into the mechanism of calcium signaling and muscle excitation or contraction in Mc. We evaluated gene expression of TRPC3 within two age groups of Myotonic goats; those younger and older than 4 yrs of age. These goats were exhibiting Mc, which is usually first manifested between 20 days to 6 months of age (Bryant, 1979). The *TRPC3* gene expression was 90% higher ($P<0.05$) in Myotonic goats older than 4 yrs old (Figure 6), indicating some form of progression of increased expression of TPRC3 over time. This could be indicative of the calcium cycling homeostasis depreciating with age in goats with Mc. It has already been documented that a high influx of calcium is enough to cause the onset of MD in mice (Millay et al., 2009), though Myotonic goats' Mc has been widely accepted as only the nondystrophic form. Molecular genetic studies have documented a coexistence of both CLCN1 mutations (characteristic of non dystrophic Mc), and CNBP (characteristic of MD) (Sun et al., 2011). Therefore further studies need to be conducted to determine the relationship between TRPC3, CLCN1 and CNBP in Myotonic goats. Such studies can then open the door for further study on the molecular characterization of these genes (TRPC3, CLCN1 and CNBP), and how they relate to Mc and MD. Older non-Myotonic goats (Spanish) also expressed higher levels ($P<0.05$) of TRPC3, but trended lower than that of Myotonic goats (Figure 6).

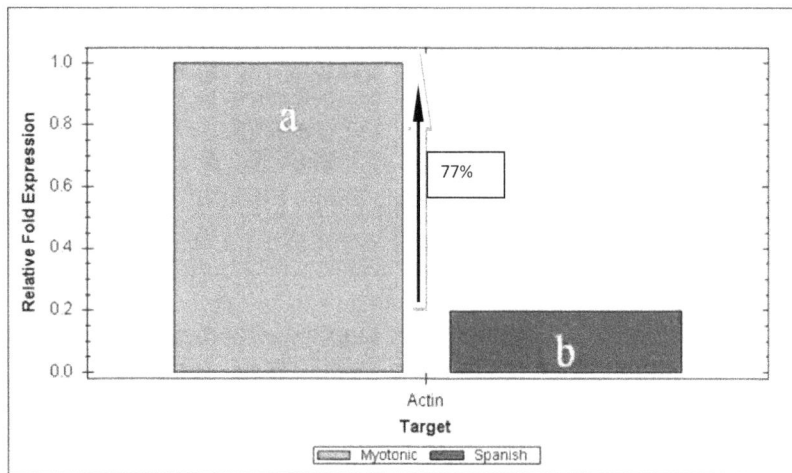

a, b means with different superscripts differ at P<0.05

Figure 4. Gene Expression of *TRPC3* in Myotonic and Non-Myotonic (Spanish) Goats

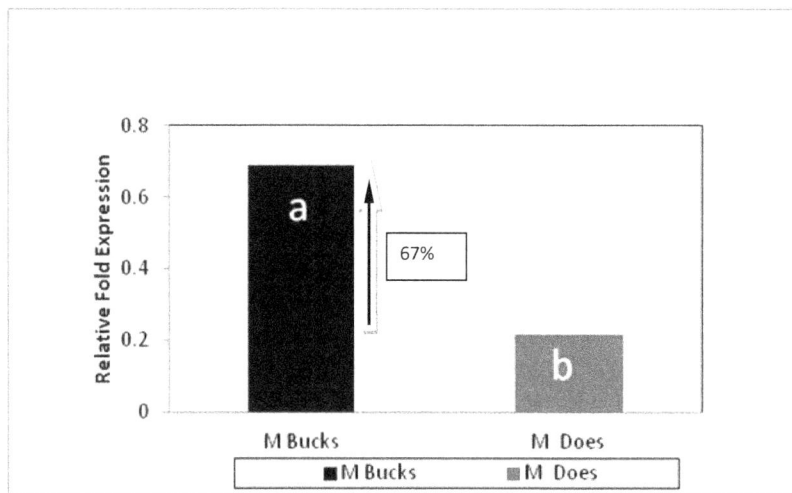

a, b means with different superscripts differ at P<0.05, n=10

Figure 5. Gender Effect on Gene Expression of *TRPC3* in Myotonic Goats

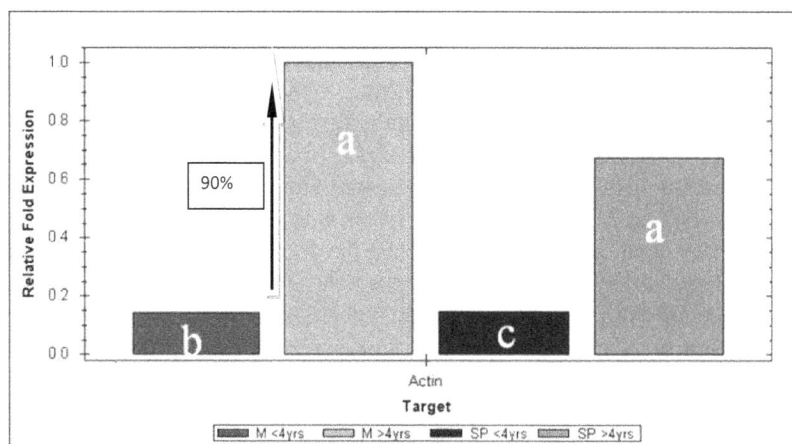

a, b means with different superscripts differ at P<0.05

Figure 6. Age Effect on Gene Expression of *TRPC3* in Myotonic Goats

4. Conclusion

The results of this study indicated that the cross species oligonucleotide primers designed from the conserved regions of the *TRPC3* genes used to amplify the goat *TRPC3* can be used on a blood sample derived from the cattle and mouse, but most importantly the human. It is clear that though TRPC3 is involved in MD, it may also be interrelated with the nondystrophic Mc in Myotonic goats. These data will serve as a basis for further study on calcium signaling in Mc in humans and other species, and also contribute to the beginning stages of the study of the goat *TRPC3* gene as a genetic marker for Mc. The Myotonic goat may then serve as a model not only for Mc, but a potential model to study other calcium signaling related neuromuscular disorders.

Acknowledgements

The authors would like to thank the Virginia State University animal care and laboratory staff at Randolph Farm and Animal Molecular Immunogenetics Laboratory respectively. This research was made possible by the Virginia State University Agricultural Research Station, Petersburg, Virginia, Journal Article Series 319.

References

Allen, D. G., Gervasio, O. L., Yeung, E. W., & Whitehead, N. P. (2010). Calcium and the damage pathways in muscular dystrophy. *Can J Physiol Pharmacol, 88*(2), 83-91. http://dx.doi.org/10.1139/Y09-058

Altschul, S. F., Gish, W., Miller, W., Myers, E. W., & Lipman, D. J. (1990). Basic local alignment search tool. *J Mol Biol, 215*(3), 403-410. http://dx.doi.org/10.1016/S0022-2836(05)80360-2

Altschul, S. F., Madden, T. L., Schäffer, A. A., Zhang, J., Zhang, Z., Miller, W., & Lipman, D. J. (1997). Gapped BLAST and PSI-BLAST: a new generation of protein database search programs. N*ucleic Acids Res, 25(*17), 3389-3402.

Ambudkar, I. S. (2009). Unraveling smooth muscle contraction: the TRP link. G*astroenterology, 137(*4), 1211-1214. http://dx.doi.org/10.1053/j.gastro.2009.08.025

Atkinson, J. B., Swift, L. L., & Lequire, V. S. (1981). Myotonia congenita. A histochemical and ultrastructural study in the goat: comparison with abnormalities found in human myotonia dystrophica. Am *J Pathol, 102(3*), 324-335.

Beck, C. L., Fahlke, C., & George, A. L. (1996). Molecular basis for decreased muscle chloride conductance in the myotonic goat. Pr*oc Natl Acad Sci U S A, 93(2*0), 11248-11252.

Beech, D. J. (2005). Emerging functions of 10 types of TRP cationic channel in vascular smooth muscle. Cl*in Exp Pharmacol Physiol, 32(8*), 597-603. http://dx.doi.org/10.1111/j.1440-1681.2005.04251.x

Bernard, G., Poulin, C., Puymirat, J., Sternberg, D., & Shevell, M. (2008). Dosage effect of a dominant CLCN1 mutation: a novel syndrome. J *Child Neurol, 23(2*), 163-166. http://dx.doi.org/10.1177/0883073807307974

Blake, D. J., Weir, A., Newey, S. E., & Davies, K. E. (2002). Function and genetics of dystrophin and dystrophin-related proteins in muscle. Ph*ysiol Rev, 82(2*), 291-329. http://dx.doi.org/10. 1152/physrev.00028.2001

Bryant, S. H. (1979). Myotonia in the goat. An*n N Y Acad Sci, 317,* 314-325.

Bryant, S. H., & Conte-Camerino, D. (1991). Chloride channel regulation in the skeletal muscle of normal and myotonic goats. Pf*lugers Arch, 417(6*), 605-610.

Burge, J. A., Hanna, M. G., & Schorge, S. (2013). Nongenomic actions of progesterone and 17β-estradiol on the chloride conductance of skeletal muscle. Mu*scle Nerve, 48(4*), 589-591. http://dx.doi.org/10.1002/mus.23887

Charlet-B, N., Savkur, R. S., Singh, G., Philips, A. V., Grice, E. A., & Cooper, T. A. (2002). Loss of the muscle-specific chloride channel in type 1 myotonic dystrophy due to misregulated alternative splicing. Mo*l Cell, 10(1*), 45-53.

Corley, M., & Ward, J. (2013). Expression of Fat and Cholesterol Biomarkers in Meat Goats. Jo*urnal of Agricultural Science, 3(1*), 78-90.

Curran, M., & Keating, M. (1994). A polymorphic dinucleotide repeat in the second intron of HUMCLC. Hu*m Mol Genet, 3(1*2), 2264.

Dinucleotide repeat polymorphisms at the SCN4A locus suggest allelic heterogeneity of hyperkalemic periodic paralysis and paramyotonia congenita. (1992). Am *J Hum Genet, 51(4*), 942.

Du, H., Grob, S. R., Zhao, L., Lee, J., El-Sahn, M., Hughes, G., ... Zhang, K. (2012). Myotonia congenita with strabismus in a large family with a mutation in the SCN4A gene. Eye (Lond), 26(8), 1039-1043. http://dx.doi.org/10.1038/eye.2012.80

Fialho, D., Kullmann, D. M., Hanna, M. G., & Schorge, S. (2008). Non-genomic effects of sex hormones on CLC-1 may contribute to gender differences in myotonia congenita. Neuromuscul Disord, 18(11), 869-872. http://dx.doi.org/10.1016/j.nmd.2008.07.004

Gailly, P. (2012). TRP channels in normal and dystrophic skeletal muscle. Curr Opin Pharmacol, 12(3), 326-334. http://dx.doi.org/10.1016/j.coph.2012.01.018

Gauthier, E. R., Madison, S. D., & Michel, R. N. (1997). Rapid RNA isolation without the use of commercial kits: application to small tissue samples. Pflugers Arch, 433(5), 664-668.

George, A. L., Crackower, M. A., Abdalla, J. A., Hudson, A. J., & Ebers, G. C. (1993). Molecular basis of Thomsen's disease (autosomal dominant myotonia congenita). Nat Genet, 3(4), 305-310. http://dx.doi.org/10.1038/ng0493-305

Gonzalez-Cobos, J. C., & Trebak, M. (2010). TRPC channels in smooth muscle cells. Front Biosci, 15, 1023-1039.

Grunnet, M., Jespersen, T., Colding-Jørgensen, E., Schwartz, M., Klaerke, D. A., Vissing, J., ... Dunø, M. (2003). Characterization of two new dominant ClC-1 channel mutations associated with myotonia. Muscle Nerve, 28(6), 722-732. http://dx.doi.org/10.1002/mus.10501

Guibert, C., Ducret, T., & Savineau, J. P. (2011). Expression and physiological roles of TRP channels in smooth muscle cells. Adv Exp Med Biol, 704, 687-706. http://dx.doi.org/10.1007/978-94-007-0265-3_36

Koch, M. C., Steinmeyer, K., Lorenz, C., Ricker, K., Wolf, F., Otto, M., ... Jentsch, T. J. (1992). The skeletal muscle chloride channel in dominant and recessive human myotonia. Science, 257(5071), 797-800.

Meola, G. (2013). Clinical aspects, molecular pathomechanisms and management of myotonic dystrophies. Acta Myol, 32(3), 154-165.

Meola, G., & Cardani, R. (2014). Myotonic dystrophies: An update on clinical aspects, genetic, pathology, and molecular pathomechanisms. Biochim Biophys Acta. http://dx.doi.org/10.1016/j.bbadis.2014.05.019

Millay, D. P., Goonasekera, S. A., Sargent, M. A., Maillet, M., Aronow, B. J., & Molkentin, J. D. (2009). Calcium influx is sufficient to induce muscular dystrophy through a TRPC-dependent mechanism. Proc Natl Acad Sci U S A, 106(45), 19023-19028. http://dx.doi.org/10.1073/pnas.0906591106

Nilius, B., & Owsianik, G. (2011). The transient receptor potential family of ion channels. Genome Biol, 12(3), 218. http://dx.doi.org/10.1186/gb-2011-12-3-218

Nilius, B., Voets, T., & Peters, J. (2005). TRP channels in disease. Sci STKE, 2005(295), re8. http://dx.doi.org/10.1126/stke.2952005re8

Petrof, B. J., Shrager, J. B., Stedman, H. H., Kelly, A. M., & Sweeney, H. L. (1993). Dystrophin protects the sarcolemma from stresses developed during muscle contraction. Proc Natl Acad Sci U S A, 90(8), 3710-3714.

Ruscák, J. (1997). [Molecular genetics of sodium channel myopathies]. Bratisl Lek Listy, 98(12), 701-707.

Shirakawa, T., Sakai, K., Kitagawa, Y., Hori, A., & Hirose, G. (2002). A novel murine myotonia congenita without molecular defects in the ClC-1 and the SCN4A. Neurology, 59(7), 1091-1094.

Skov, M., de Paoli, F. V., Lausten, J., Nielsen, O. B., & Pedersen, T. H. (2014). Extracellular magnesium and calcium reduce myotonia in isolated ClC-1 inhibited human muscle. Muscle Nerve. http://dx.doi.org/10.1002/mus.24260

Skov, M., Riisager, A., Fraser, J. A., Nielsen, O. B., & Pedersen, T. H. (2013). Extracellular magnesium and calcium reduce myotonia in ClC-1 inhibited rat muscle. Neuromuscul Disord, 23(6), 489-502. http://dx.doi.org/10.1016/j.nmd.2013.03.009

Sun, C., Van Ghelue, M., Tranebjærg, L., Thyssen, F., Nilssen, Ø., & Torbergsen, T. (2011). Myotonia congenita and myotonic dystrophy in the same family: coexistence of a CLCN1 mutation and expansion in the CNBP (ZNF9) gene. Clin Genet, 80(6), 574-580. http://dx.doi.org/10.1111/j.1399-0004.2010.01616.x

Swift, L. L., Atkinson, J. B., & LeQuire, V. S. (1979). The composition and calcium transport activity of the sarcoplasmic reticulum from goats with and without heritable myotonia. Lab Invest, 40(3), 384-390.

Tsvilovskyy, V. V., Zholos, A. V., Aberle, T., Philipp, S. E., Dietrich, A., Zhu, M. X., ... Flockerzi, V. (2009). Deletion of TRPC4 and TRPC6 in mice impairs smooth muscle contraction and intestinal motility in vivo. Gastroenterology, 137(4), 1415-1424. http://dx.doi.org/10.1053/j.gastro.2009.06.046

Vandebrouck, C., Martin, D., Colson-Van Schoor, M., Debaix, H., & Gailly, P. (2002). Involvement of TRPC in the abnormal calcium influx observed in dystrophic (mdx) mouse skeletal muscle fibers. J Cell Biol, 158(6), 1089-1096. http://dx.doi.org/10.1083/jcb.200203091

Whitehead, N. P., Yeung, E. W., & Allen, D. G. (2006). Muscle damage in mdx (dystrophic) mice: role of calcium and reactive oxygen species. Clin Exp Pharmacol Physiol, 33(7), 657-662. http://dx.doi.org/10.1111/j.1440-1681.2006.04394.x

Wijnberg, I. D., Owczarek-Lipska, M., Sacchetto, R., Mascarello, F., Pascoli, F., Grünberg, W., ... Drögemüller, C. (2012). A missense mutation in the skeletal muscle chloride channel 1 (CLCN1) as candidate causal mutation for congenital myotonia in a New Forest pony. Neuromuscul Disord, 22(4), 361-367. http://dx.doi.org/10.1016/j.nmd.2011.10.001

Zhang, J., Bendahhou, S., Sanguinetti, M. C., & Ptácek, L. J. (2000). Functional consequences of chloride channel gene (CLCN1) mutations causing myotonia congenita. Neurology, 54(4), 937-942.

3

Time-Resolved Förster Resonance Energy Transfer Analysis of Single-Nucleotide Polymorphisms: Towards Molecular Typing of Genes on Non-Purified and Non-PCR-Amplified DNA

Luca Nardo[1], Nicola Camera[1], Edoardo Totè[1], Maria Bondani[2], Roberto S. Accolla[3] & Giovanna Tosi[3]

[1] Department of Science and High Technology, University of Insubria, Italy

[2] Institute for Photonics and Nanotechnology, National Research Council, Italy

[3] Department of Surgical and Morphologic Sciences, University of Insubria, Italy

Correspondence: Luca Nardo, Department of Science and High Technology, University of Insubria, Via Valleggio, Como 11-22100, Italy. E-mail: luca.nardo@unimib.it

Abstract

Quantitative assessment of the fluorescence resonance energy transfer (FRET) efficiency between chromophores labeling the opposite ends of gene-specific oligonucleotide probes is a powerful tool to detect DNA polymorphisms with single-nucleotide resolution. The FRET efficiency can be most conveniently quantified by applying a time-resolved fluorescence analysis methodology, time-correlated single-photon counting. Recently, we probed by such technique the highly polymorphic DQB1 human gene. Namely, by using a single oligonucleotide probe and acting on non-amplified DNA samples contained in untreated cell extracts, we demonstrated the ability of pursuing unambiguous recognition of subjects bearing the homozygous DQB1-0201 genotype by exploiting the subtle, yet statistically significant, structural differences between the duplex formed by the probe with DQB1-0201 on the one end and duplexes formed with any of the other alleles, on the other end. The relevance of homozygous DQB1-0201 genotype recognition reseeds in the fact that the latter is overexpressed in subjects affected by insulin-dependent diabetes mellitus in north-eastern Italy.

In this article we review our preceding achievements and report on additional *in-vitro* experiments aimed at characterizing the duplexes obtained by annealing of the DQB1 allelic variants with a second oligonucleotide probe, with the final scope to achieve full genotyping of DQB1 on raw DNA samples by means of cross-combination of the FRET responses of both probes.

Keywords: polymorphism, molecular typing, oligonucleotide probe, time-correlated single-photon counting, time-resolved fluorescence resonance energy transfer

1. Introduction

Genetic diseases are, strictly speaking, pathologies resulting from site-specific mutations in the genome. However, in other cases the proneness of an individual to manifest non-genetic diseases can still be correlated to the presence of certain allelic variants of one or more genes. A particularly relevant instance is constituted by auto-immune diseases, which are notably correlated to the polymorphisms of the Human Leucocyte Antigen system (HLA). More specifically, insurgence of several auto-immune diseases is correlated with overexpression of certain genotypic variants of the highly polymorphic DQB1 human gene. For instance, in Caucasian population of north-eastern Italy the homozygous DQB1-0201 genotype is over-represented in subjects suffering from insulin-dependent diabetes mellitus (Tosi, 1993). The risk factor of a given subject with respect to contraction of multiple autoimmune diseases might thus be evaluated, and both preventive health care and early diagnosis be subsequently implemented, by means of wide-screen HLA genotyping.

Nowadays, molecular typing of polymorphic genes may be performed with single nucleotide sensitivity by applying several different techniques (Kim, 2007). However, as a rather common feature, virtually all these techniques require the Polymerase Chain Reaction (PCR) as initial step. Although it is out of doubt that PCR has revolutionized biology, it is not the ideal technique to be applied to the screening of vast populations, due to lack of both readiness and cost-effectiveness. During the last decade, molecular typing techniques based on exploitation of the differential fluorescence emission of fluorophore-labeled allele-specific oligonucleotide probes upon recognition of their genomic target have been given increasing interest because of their particularly neat response (i.e. high signal-to-noise ratio and low probability of false-positives detection), conferring to fluorescence-based assays notable readiness and ease of interpretation (Tyagi, 1998; Gaylord, 2005; Al Attar, 2008; Andreoni, 2009a, 2009b; Nardo, 2012). The common rationale of the majority of such techniques is that light emission by the fluorophore functionalizing the oligonucleotide probe (hereafter called the fluorescence donor, D) can be quenched by suitable dyes (hereafter referred to as fluorescence acceptors, A) in a strongly donor-acceptor-distance dependent way by virtue of a non-radiative decay mechanism of the D from the fluorescent excited state, called fluorescence resonance energy transfer (FRET). The latter occurs whenever the absorption spectrum of the A is significantly superimposed to the fluorescence emission spectrum of the D.

Molecular typing methods have been devised both making use of D-A dual-labeled probes (Tyagi, 1998; Andreoni, 2009a, 2009b; Nardo, 2012), which are virtually non-emitting while not annealed to their genomic target, and exploiting A selectively binding to the genomic DNA to be typed and probes single-labeled with a D whose fluorescence is "switched off" upon target recognition (Gaylord, 2005; Al Attar, 2008). In many instances (Tyagi, 1998), the response of a typing experiment is barely constituted by the presence/absence of a detectable fluorescence signal. Hence, this class of tests is frequently referred to as ON/OFF methods. However, further options are being explored in the last years, based on quantitative analysis of FRET-tuned fluorescence signals (Gaylord, 2005; Al Attar, 2008; Andreoni, 2009a, 2009b; Nardo, 2012). The basic advantage in doing so reseeds in the fact that quantitative evaluation allows at least partial release of the strict boundaries on signal-to-noise ratio required by ON/OFF strategies, thus making the implementation of assays on less pure DNA samples and/or on reduced DNA template concentrations conceivable, and ultimately opening the way to molecular typing of polymorphic genes directly on the unpurified and non-PCR-amplified DNA contained in crude cell lysates. The ability of determining even slight differences in the quantitative fluorescence response should also allow to perform typing with a single probe capable of hybridizing to any of the possible target sequences (i.e. to each possible polymorphic variant of the gene of interest) with sufficient selectivity with respect to other genomic regions. Allele-specific annealing would in the latter case occur through formation of duplexes whose structural details depend on the probe-target degree of complementarity, thus, in the end, from the target sequence. At least, a reduced number of probes with respect to ON/OFF experiments, in which a number of probes equal to the number of allelic variants displayed by the gene to be typed are intrinsically needed, are most likely necessary. This feature promises to substantially reduce the recourse to multiplexed analysis, thus speeding up and making easier the experimental and analytical procedures, as well as allowing working on much reduced template quantities. A first method to quantitatively evaluate FRET efficiencies is based on relative intensity measurements (Gaylord, 2005; Al Attar, 2008). Progresses in this direction have been made by Al Attar et al. (2008) who have provided as early as in 2008 a prove of principle of the feasibility of detecting the polymorphisms of the ABL portion of the BCR-ABL oncogene (Al Attar, 2008). However, this approach is affected by two major limitations: (i) quantitative assessments of the fluorescence intensity require detectors endowed with broad dynamic ranges, whose low quantum efficiency makes PCR amplification hardly avoidable; (ii) application to non-purified genomic material is prevented by the presence of endogenous fluorophores and, even more unavoidably, by the inherent difficulty in determining probe, target, and fluorescent impurities concentrations. Both the above-mentioned drawbacks can be overcome by quantitating the variations in FRET efficiency through measurements of the fluorescence decay time τ_D of D. Indeed, τ_D can be determined by applying a technique named time-correlated single-photon counting (TCSPC). TCSPC works by detecting and timing one photon per excitation pulse at most. The subsequent reconstruction of the statistical distribution of the detection times relative to the excitation pulses yields the fluorescence decay pattern, whose fitting provides the value of τ_D. For this reason, TCSPC is intrinsically independent from the D concentration and takes advantage of the usage of Geiger-mode light detectors, such as single-photon avalanche diodes, which are sensitive to even extremely deem fluorescence pulses, virtually reduced to few/single photons.

In 2009 we applied time-resolved FRET analysis to an in-vitro model system consisting in 25-bases oligonucleotides differing by a single A→T point mutation from each other (Andreoni, 2009a). The latter were made to anneal to a single oligonucleotide probe, dual-labeled at its 5' end with a D, namely 5-Dimethoxytrityloxy-5- [N-((tetramethylrhodaminyl)aminohexyl)-3-acrylimido]- 2'-deoxyUridine- 3'-[(2cyanoethyl)- (N,N-diisopropyl)] -phosphoramidite (TAMRA), and at its 3' end with a non-emitting A, 4'-(4-Nitrophenyldiazo) -2'- methoxy -5'- methoxy – azobenzene -4''-(N-ethyl)-N-ethyl-2-cyanoethyl-(N,N-diisopropyl) phosphoramidite (BHQ2). The choice of TAMRA as the fluorescent label was dictated by two considerations. Firstly, it is optimally excited with green-yellow light, which in turn is only slightly absorbed by endogenous fluorophores. Secondly, in aqueous environment its fluorescence decay is single-exponential, which considerably simplifies TCSPC data analysis. The probe was perfectly complementary to one of the oligonucleotides, and bore single-nucleotide mismatches with respect to the others. D fluorescence decay distributions were reconstructed for the different duplexes by means of a TCSPC system endowed with 30 ps resolution. The obtained τ_D values differed by at least 3 standard deviations from an oligonucleotide to the other. As sketched in Figure 1, the effect of single nucleotide mismatches at a duplex end (see S_1 and S_3 in the figure) was a systematic reduction of the D-A distance with respect to the perfectly matching duplex (S_0), which we interpreted as unzipping of the lateral bases. Conversely, single nucleotide mismatches at the center of a duplex (S_2) brought about an increase in the D-A distance, which we interpreted as due to a local unwinding of the double helical structure similar to that induced by intercalation (Nardo, 2008). In summary, time-resolved FRET proved to be sensitive to single-nucleotide mismatches and the response of the D fluorescence decay seemed to be related to the mismatch position.

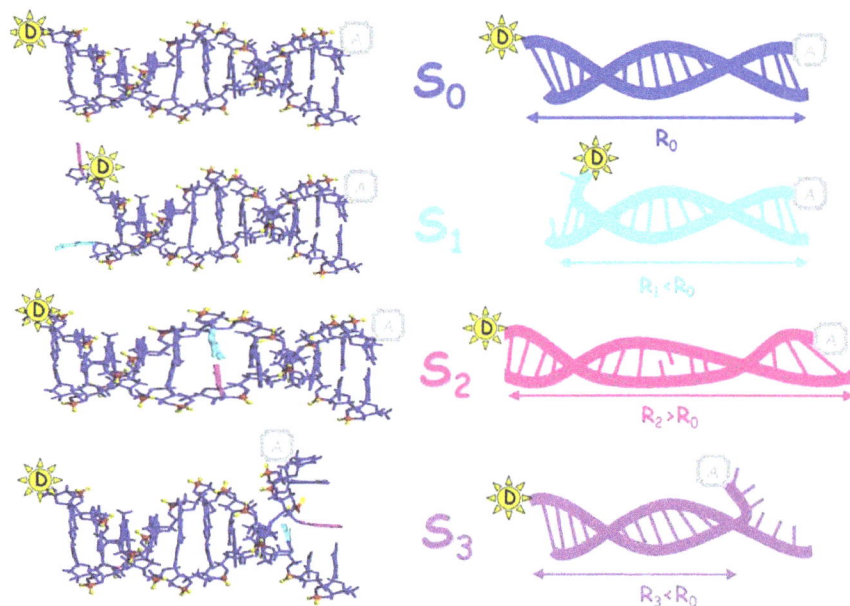

Figure 1. Effects of single-nucleotide mismatches in different positions of a double-stranded oligonucleotide: mismatches at a duplex end (see S_1 and S_3) induce a reduction of the D-A distance with respect to the perfectly matching duplex (S_0), while single nucleotide mismatches at the center of a duplex (S_2) increase the D-A distance

The subsequent step towards implementation of a DQB1-typing protocol based on time-resolved FRET was achieved shortly later (Andreoni, 2009b). Our aim was to demonstrate the feasibility of unambiguous recognition of the DQB1-0201 allele, whose presence in homozygous conditions confers, as discussed above, susceptibility to insulin-dependent diabetes mellitus in the population of north-eastern Italy (Tosi, 1993). The sequence of DQB1-0201 is different from that of any other allele in the highly polymorphic trait of the gene encompassing codons 52 to 57. Thus, recognition of that sequence implies recognition of the allele as a whole. For this reason, we decided to elect that region as our genomic target, and to pursue its specific recognition by an oligonucleotide probe with sequence corresponding to that of DQB1-0201 (P0), dual labeled with TAMRA and BHQ2. We used as templates 18-base oligonucleotides mimicking the eight distinct sequences exhibited in the above-mentioned trait by the DQB1 allelic variants. TCSPC measurements were performed for P0 hybridized to each target oligonucleotide. Significantly different τ_D values were determined for the eight duplexes. This result demonstrates that the subtle structural differences among duplexes formed by P0 with each allele may be sensed by applying time-resolved FRET analysis, and that the structures are sufficiently heterogeneous to allow full molecular typing of the target region with this unique probe. However, the selectivity of P0 for recognition of its target trait within the whole genome has proven to be not sufficient to permit application of the protocol on non-amplified genomic templates.

Nonetheless, we recently managed to pursue selective recognition of an extended trait of DQB1, also encompassing codon 51 and the first base of codon 58, by a 22-bases oligonucleotide probe (P1) in non-amplified and non-purified DNA samples obtained by bare cell lysis (Nardo, 2012). The results showed the possibility to reveal the presence of the DQB1-0201 allelic variant both in homozygous and heterozygous conditions, and to discriminate within the two cases. However, extension of the probe of a four-base long non-polymorphic trait reduced both the sensitivity of the FRET efficiency to equal D-A distance variations and the heterogeneity of the duplex structures. In conclusion, we failed to obtain a full discrimination of alleles. Namely, even in preliminary calibration measurements undergone on 22-bases oligonucleotide targets, P1 yielded hardly distinguishable τ_D values when annealed to DQB1-05031, DQB1-0602, and DQB1-0402-like sequences.

In this work we report τ_D data obtained by making to hybridize the same oligonucleotides used in (Nardo, 2012) with a different 22-base-pairs oligonucleotide probe (P2). The latter is dual-labeled with the same chromophores used in P0 and P1. However, to the aim of overcoming the ambiguities inherent to the P1 FRET response, the sequence of P2 is complementary to the DQB1-05031 sequence, while it obviously contains mismatches with respect to both the DQB1-0402-allele and the DQB1-0602-allele mimicking oligonucleotides. Significantly different τ_D values are measured with P2 in the three instances.

2. Method

2.1 Oligonucleotides and Sample Preparation

In Figure 2 the sequences of the eight target oligonucleotides reproducing the possible sequences manifested by the 22-bases polymorphic trait encompassing codons 51 to 57 and the first base of codon 58 are depicted. In the same figure, the TAMRA-BHQ2 dual-labelled probes P1 and P2, with sequences complementary to the DQB1-0201 and DQB1-05031 allele mimicking oligos, respectively, are also shown. Mismatch-sites of the target oligonucleotides with respect to P1 are black-shaded, while those with respect to P2 are light-gray shaded. Mismatching sites common to both probes are indicated by dark-grey shading. The unlabelled oligonucleotides were purchased by TIB Molbiol (Genova, Italy) and were desalted and further purified by double RP procedure.

0201	5'-	T	G	C	G	A	C	G	A	C	C	C	C	G	A	C	G	G	A	C	G	G	C	-3'
0501	5'-	T	G	C	G	G	C	G	T	C	C	C	C	G	C	C	G	G	A	C	A	A	C	-3'
0502	5'-	T	G	C	G	G	C	G	T	C	C	C	C	G	C	C	G	G	A	T	C	G	C	-3'
05031	5'-	T	G	C	G	G	C	G	T	C	C	C	C	G	C	C	G	G	A	C	T	G	C	-3'
0602	5'-	T	G	C	G	G	C	G	T	C	C	C	C	G	C	C	G	G	A	C	T	A	C	-3'
0301	5'-	T	G	C	G	G	C	G	A	C	C	C	C	G	G	C	G	G	A	C	T	G	C	-3'
0302	5'-	T	G	C	G	G	C	G	A	C	C	C	C	G	G	C	G	G	A	C	G	G	C	-3'
0402	5'-	T	G	C	G	G	C	G	A	C	C	C	C	G	C	C	G	A	A	C	T	G	C	-3'
Probe 1	5'-	A	C	G	C	T	G	C	T	G	G	G	G	C	T	G	C	C	T	G	C	C	G	-3'
Probe 2	5'-	A	C	G	C	C	G	C	A	G	G	G	G	C	G	G	C	C	T	G	A	C	G	-3'
Codons	5'-		51			52			53			54			55			56			57			-3'

Figure 2. Sequences of the eight target oligonucleotides and of the TAMRA-BHQ2 dual-labelled probes P1 and P2. Mismatch-sites of the target oligonucleotides with respect to P1 are black-shaded, those with respect to P2 are light-gray shaded. Mismatching sites common to both probes are indicated by dark-grey shading

The dual-labelled probes were purchased by Eurofins MWG Operon (Ebersberg, Germany) and purified by high-performance liquid chromatography. A nominal amount of twenty nanomoles of the lyophilized oligonucleotides was resuspended in appropriate volumes (indicated by the suppliers) of Tris-HCl EDTA buffer at pH 7.6 and10 mM ionic strength, to obtain 5 µM concentrated solutions. Equal amounts (100 µL) of probe and target stock solutions were mixed to obtain the solutions submitted to the annealing protocol. The latter, which is fully described elsewhere, consisted in heating the samples to 98°C in order to denaturate any residual secondary structures of the single-stranded oligonucleotides, cooling at the rate of 1°C/min to the optimal annealing temperature, derived from the melting temperature of the perfectly matching duplex as described in (Nardo, 2012), and letting the probe react with the (more or less) complementary oligonucleotide for 20 minutes. Finally, the solutions were cooled to 25°C at 3°C/min rate. The thermostat of a ThermoQuest-Finnigan gas chromatograph (San Jose, CA, USA) was used as the temperature controller.

The obtained solutions were diluted to a final probe concentration of 250 nM, to produce the sample for TCSPC analysis.

2.2 Time-Correlated Single-Photon Counting Apparatus

The fluorescence excitation/detection apparatus is extensively described elsewhere (Andreoni, 2009b). Briefly, a 113 MHz repetition rated train of excitation pulses (pulse duration 6.4 ps) at 532 nm is provided by a Nd: VAN cw-mode-locked laser (GE-100, Time Bandwidth Products, Zurich, CH) and suitably attenuated (typically by a factor of $> 10^3$) by means of neutral density filters, in order to achieve single-photon regime. Fluorescence is collected at 90° to the excitation beam through a long-wavelength-pass filter with cut at 550 nm, and focused on the sensitive area of a single-photon avalanche diode (PDM50, Micro-photon-devices, Bolzano, IT) by a 20X microscope objective. The avalanche pulses give the START signal to an integrated TCSPC board (which is a single card of a SPC 152 module, Becker & Hickl, Berlin, DE). The STOP signal is provided by the subsequent laser pulse, as detected by a fast photodiode internal to the laser. The START-STOP lag times occurring in a time window of 9 ns are digitized with 2.44 ps/channel resolution. The full-width at half-maximum duration of the detected excitation pulse is < 30 ps.

2.3 Data Analysis

Due to satisfactory yield of the annealing protocol, the data were treated as those obtained on the oligonucleotide samples of (Nardo, 2012). The data analysis procedure is explained in details in the quoted reference. Briefly, the absorbance of each sample at the TAMRA absorption peak (556 nm) was preliminarily measured. Each fluorescence decay histogram was acquired up to the fixed value of 65535 counts at the peak and the acquisition time measured. The absorption spectrum and decay distribution of a solution containing only the probe, at 250 nM nominal concentration, were also acquired. Each experimental decay pattern was normalized with respect to absorbance and acquisition time. The normalized decay obtained for the pure-probe sample was fitted to a three-exponential decay model, see Equation (1):

$$F(t) = y_0 + A_1 \exp[(t_0-t)/\tau_1] + A_2 \exp[(t_0-t)/\tau_2] + A_3 \exp[(t_0-t)/\tau_3] \qquad (1)$$

In the above equation, besides the decay times τ_i and pre-exponential factors A_i, two other parameters appear, namely y_0 and t_0. The former prepresents a constant offset added to the decay function, accounts for the both the dark-counts of the detectors (START signals provided by thermally induced avalances) and the detection of environmental non-time-correlated light, and was let free to vary during the fit. The latter represents the instant in which the excitation pulse impinges on the sample, and was fixed to the time-lag corresponding to the peak channel. The longest-lived component in the pure-probe decay is attributed to the emission of unquenched probe molecules and/or residual free TAMRA, whose fraction relative to the total TAMRA concentration is constant in all the samples. Thus, after synchronization of the histograms to the corresponding peak channels, we got rid of the unquenched TAMRA contribution to the decay of duplex samples by simply subtracting from each duplex histogram an exponential component with the same lifetime and pre-exponential factor as those derived for the longest-lived transient resolved in the pure-probe sample decay pattern. The background value y_0 was also subtracted to all the decay distributions. The τ_D values tagging the conformations of the different duplexes were derived by performing single-exponential fits of the so-obtained decay patterns according to Equation (2):

$$F(t) = A \exp[(t)/\tau_D] \qquad (2)$$

3. Results and Discussion

In a previous article (Nardo, 2012) the time-resolved FRET methodology was presented as a valid technique to recognize allelic-variants of a human gene (namely DQB1) by means of a D-A dual-labeled oligonucleotide probe (namely P1) capable of selectively hybridize to a 22-bases trait of DQB1 in unpurified and non-PCR amplified genomic DNA samples consisting in crude cell extracts. The allelic variants were tagged through determination of the decay time τ_D of the D fluorophore, which proved to be sensitive to the specificities of the probe-template duplex conformation. The τ_D values derived for P1 are reported in Table 1. As made apparent by straightforward comparison of the τ_D differences between different alleles with the experimental uncertainties in the τ_D values determination, most of the allelic sequences could be safely discriminated by using uniquely P1 as the probe. In particular, we specifically intended to recognize the IDDM-correlated DQB1-0201 allele. Hybridization of P1 to the latter yielded the τ_D value $\tau_D = 2725 \pm 3$ ps, while the least differing τ_D value was obtained in the case of the DQB1-0302 allele (which was the least different in sequence with respect to DQB1-0201). In the latter case we measured $\tau_D = 2599 \pm 3$ ps. The difference between the two τ_D values was as large as 42 standard deviations, corresponding to negligibly small probability of incurring in "false positive" recognition of DQB1-0201. More in general, the ability of discriminating among the possible homozygous genotypes by means of single τ_D measurements upon annealing with P1 can be evaluated under the very reasonable assumption of stochastic (*i.e.* Gaussian) distribution of the variable τ_D. If the latter assumption holds, the probability of getting an experimental value for τ_D different from the average values reported in Table 1 by more than 3σ is $\approx 0.3\%$. Similarly, the genotype DQB1-0YYY is erroneously attributed to a DQB1-0XXX homozygous DNA template whenever a value of τ_D more similar to $\tau_{D,YYY}$ than to $\tau_{D,XXX}$ is measured. The probability of such an event is determined by the value assumed by the "reliability parameter" $\Delta\tau_D/2\sigma$. Namely, it is $< 0.3\%$ if $\Delta\tau_D/2\sigma > 3$. Thus, in spite of the tiny τ_D differences between certain allelic variants, reliable typing is assured by the high repeatability of the τ_D measurements. Two exemplary instances of the above statement are offered by alleles DQB1-0301 and DQB1-0402 ($\Delta\tau_D = 19$ ps) on the one end, and alleles DQB1-0301 and DQB1-0501 ($\Delta\tau_D = 33$ ps). For the first pair of alleles, $\Delta\tau_D$ is comparable with the TCSPC system temporal resolution, whilst for the second one it is even lower. However, $\sigma(\tau_D) = 3$ ps for any of the above situations. Consequently, the probability of attributing erroneously the DQB1-0301 sequence to a DQB1-0402 template is $\approx 0.3\%$ ($\Delta\tau_D/2 \approx 3\sigma$), while we get an even lower probability of confusing DQB1-0301 with DQB1-0501 ($\Delta\tau_D/2 \approx 5.5\sigma$). On the contrary, it was impossible to distinguish between DQB1-05031 and DQB1-0602 decay times, as $\Delta\tau_D/2 < \sigma$. Moreover, discrimination between the latter alleles and DQB1-0502 was somewhat tricky, as $\Delta\tau_D/2 \approx \sigma$, meaning that the probability of mis-assignment was $\approx 33\%$.

Table 1. Values of τ_D determined for duplexes formed by the different DQB1-alleles-mimicking oligonucleotides with P1 (second column) and P2 (third column). The uncertainties are expressed in terms of the standard deviation among three parallels

DQB1-Allele-mimicking oligo	$\tau_D(P1) \pm \sigma(\tau_D)$ [ps]*	$\tau_D(P2) \pm \sigma(\tau_D)$ [ps]
DQB1-0201/0201	2725 ± 3	2563 ± 8
DQB1-0501/0501	2514 ± 3	2201 ± 13
DQB1-0502/0502	2404 ± 11	2030 ± 3
DQB1-05031/05031	2427 ± 2	2397 ± 2
DQB1-0602/0602	2430 ± 9	2194 ± 5
DQB1-0301/0301	2481 ± 3	2137 ± 4
DQB1-0302/0302	2599 ± 3	2489 ± 3
DQB1-0402/0402	2462 ± 3	1969 ± 1

*Data reproduced from (Nardo, 2012).

We now examine the results of similar experiments performed by using P2 as the oligonucleotide probe. The τ_D values obtained in the present case are listed in the last column of Table 1. They are obviously different from those obtained with P1, because of the different mismatches. By means of these values it is impossible to discriminate between allele DQB1-0501 and DQB1-0602. Thus, complete typing of DQB1 is not feasible also by using P2 as the unique probe. However, a cross-correlated analysis of the results obtained with both probes allows removal of any ambiguity in the attribution of the correct genotype to homozygous DQB1 samples. Namely, as shown in Figure 3a), the D fluorescence decay pattern is notably different for P2 hybridized to either DQB1-05031 (grey dots, the pertaining best single-exponential fitting curve is indicated by the black line) or DQB1-0602 (black circles, the pertaining best single-exponential fitting curve is indicated by the red line). We recall that the latter two alleles were indistinguishable with P1, as shown by the pertaining decay distributions reported in Figure 3 b), which, in spite of the zoomed scale, are almost superimposed. By using P2, the reliability parameter calculated in this instance takes the value $\Delta\tau_D/2\sigma \approx 20$ (by considering the higher standard deviation associated to determination of $\tau_{D,0602}$). Furthermore, the τ_D value obtained for P2 hybridized to DQB1-0502 differs from that measured in the case of DQB1-0602 by 33 standard deviations, and is even more different from that measured for DQB1-05031.

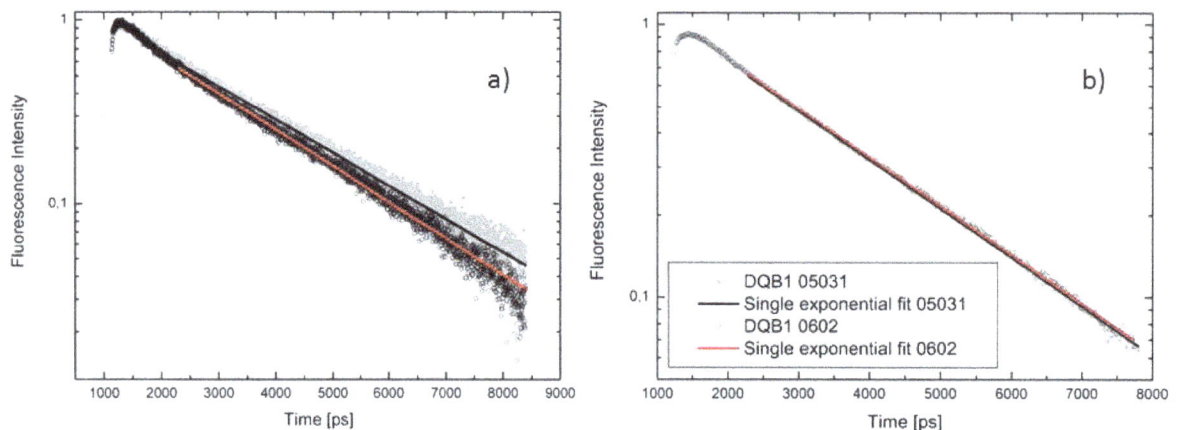

Figure 3. Fluorescence decay patterns of TAMRA labelling a) P2 and b) P1 hybridized to the oligos mimicking the DQB1-05031 (gray dots) and DQB1-0602 (black circles) allelic sequences in the trait encompassing codons 51-57 and the first base of codon 58. Solid lines indicate the corresponding single exponential decay best-fitting curve (see Equation 2 in the text)

In conclusion, we managed to pursue statistically reliable discrimination of each allelic variant of a highly polymorphic trait of the DQB1 human gene by combining the results of time-resolved FRET experiments performed by using two different D-A oligonucleotide probes. The above experiments have been successfully implemented on non-PCR amplified and non-purified DNA templates with one of the probes. The selectivity of the second probe in recognizing its genomic target in similarly raw samples is being evaluated. The performances of the method in typing heterozygous DQB1 genotypes are also currently under investigation.

Acknowledgements

The authors gratefully acknowledge the contribution of Prof. Alessandra Andreoni to fruitful discussions on the manuscript. We also thank Dr. Angelo Maspero for assistance during the annealing procedures. The postdoctoral fellowship of Dr. Luca Nardo has been funded by the project POR FSE Regione Lombardia, Ob. 2 Asse IV 2007-2013, through an FP7 cooperation programme.

References

Al Attar, H. A., Norden, J. O'Brien, S., & Monkman, A. P. (2008). Improved single nucleotide polymorphism detection using conjugated polumer/surfactant system and peptide nucleic acid. *Biosens. Bioelectron., 23*, 1466-1472. http://dx.doi.org/10.1016/j.bios.2008.01.005

Andreoni, A., Bondani, M., & Nardo, L. (2009a). Feasibility of single nucleotide polymorphism genotyping with a single probe by time-resolved Forster resonance energy transfer. *Mol. Cell. Probes, 23*, 119-121. http://dx.doi.org/10.1016/j.mcp.2008.12.008

Andreoni, A., Bondani, M., & Nardo, L. (2009b). Time-resolved FRET method for typing polymorphic alleles of the human leukocyte antigen system by using a single DNA probe. *Photochem. Photobiol. Sci., 8*, 1202-1206. http://dx.doi.org/10.1039/b906043j

Gaylord, B. S., Massie, M. R., Feinstein, S. C., & Bazan, G. C. (2005). SNP detection using peptide nucleic acid probes and conjugated polymers: applications in neurodegenerative disease identification. *Proc. Natl. Acad. Sci. USA, 102*, 34-39. http://dx.doi.org/10.1073/pnas.0407578101

Kim, S., & Misra, A. (2007). SNP genotyping: Technologies and biomedical applications. *Annu. Rev. Biomed. Eng., 9*, 289-320. http://dx.doi.org/10.1146/annurev.bioeng.9.060906.152037

Nardo, L., Bondani, M., & Andreoni, A. (2008). DNA-ligand binding-mode discrimination by characterizing fluorescence resonant energy transfer trough lifetime measurements with picosecond resolution. *Photochem. Photobiol., 84*, 101-110. http://dx.doi.org/10.1111/j.1751-1097.2007.00204.x

Nardo, L., Tosi, G., Bondani, M., Accolla, R. S., & Andreoni, A. (2012). Typing of a polymorphic human gene conferring susceptibility to insulin-dependent diabetes mellitus by picosecond-resolved FRET on non-purified/non-amplified genomic DNA. *DNA Res., 19*, 347-355. http://dx.doi.org/10.1093/dnares/dss017

Tosi, G., Mantero, G., Magalini, A. R., Primi, D., Soffiati, M., Pinelli, L., ... Accolla, R. S. (1993). HLA-DQB1 typing of north-east Italian IDDM patients using amplified DNA, oligonucleotide probes, and a rapid DNA-enzyme immunoassay (DEIA). *Mol. Immunol., 30,* 69-76. http://dx.doi.org/10.1016/0161-5890(93)90427-D

Tyagi, S., Bratu, D. P., & Russel Kramer, F. (1998). Multicolor molecular beacons for allele discrimination. *Nat. Biotechnol., 16,* 49-53. http://dx.doi.org/10.1038/nbt0198-49

4

Discovery of a Choline-Responsive Transcriptional Regulator in *Burkholderia xenovorans*

Ricardo Martí-Arbona[1], Tuhin S. Maity[1], John M. Dunbar[1], Clifford J. Unkefer[1] & Pat J. Unkefer[1]

[1] Los Alamos National Laboratory, Los Alamos, NM 87545, United States

Correspondence: Ricardo Martí-Arbona, Los Alamos National Laboratory, Los Alamos, NM 87545, United States. E-mail: rm-a@lanl.gov

Abstract

The search for effectors of novel transcriptional regulators is a challenging task. Here, we present the prediction and validation of an effector for a novel transcriptional regulator (TR). The clustering of genes around the gene coding for Bxe_A0425, a TR in *Burkholderia xenovorans* LB400 and its closest orthologs, suggests the conservation of a functional operon composed a several open reading frames from which a TR, a transporter, and two oxidoreductases can be easily identified. A search of operons containing these functional components revealed a remarkable resemblance of this system to the evolutionarily convergent and functionally conserved operons found in *Escherichia coli*, *Bacillus subtilis*, *Staphylococcus xylosus* and *Pseudomonas aeruginosa*. These operons are involved in the uptake and catabolism of choline to create the potent osmo-protectant molecule glycine betaine. We used frontal affinity chromatography coupled to mass spectrometry to screen for the binding of choline and other intermediates of the glycine biosynthesis pathway to the TR Bxe_0425. We then used electrophoretic mobility shift assays to confirm our results. We found that choline was the sole metabolite binding to this TR and identified choline as an effector molecule for Bxe_A0425. These findings suggest that this operon in *B. xenovorans* is involved in the uptake and catabolism of choline to protect the organism from osmotic stress.

Keywords: FAC-MS, uptake and catabolism of choline, EMSA, osmo-protectant, *Burkholderia xenovorans*, osmoregulator, Bxe_A0425, transcriptional regulator

1. Introduction

As a group, *Burkholderiales* are well-adapted organisms capable of surviving in environments ranging from animal hosts to contaminated soils that subject them to wide array of stresses. Among the stresses that they are adapted to overcome is osmotic stress, which they can encounter in animal cells and fluids as well as soils. One of the most effective mechanisms that bacteria have developed to overcome osmotic stress is the accumulation of glycine betaine, an osmo-protectant, either through choline uptake or *de novo* synthesis from choline (Csonka & Hanson, 1991). Many organisms contain evolutionarily convergent operons that are involved in the uptake and conversion of choline to glycine betaine (Figure 1). These operons maintain remarkable functional similarity although in many cases they exhibit very little sequence homology. For example, in *Escherichia coli*, the operon encodes for a choline transporter (BetT), two oxidoreductases, a NADH-dependent glycine betaine aldehyde dehydrogenase (BetB), and a FADH-dependent choline dehydrogenase (BetA) (Lamark et al., 1991). The *Bacillus subtilis* operon exhibits the choline dehydrogenating function but the enzyme catalyzing the choline oxidation is different; the choline dehydrogenase (BetA) is replaced by a cholineoxidase (GbsB) (Boch, Kempf, Schmid, & Bremer, 1996) that belongs to a family of alcohol dehydrogenases commonly found in higher organisms.

We sought to identify the effector metabolite for a TR in *Burkholderia xenovorans*. This TR, Bxe_A0425, and the operon in which it resides have remained largely uncharacterized. We present data showing that Bxe_A0425 binds choline and that the operon in which Bxe_A0425 resides is of suitable composition to be involved in choline uptake and catabolism.

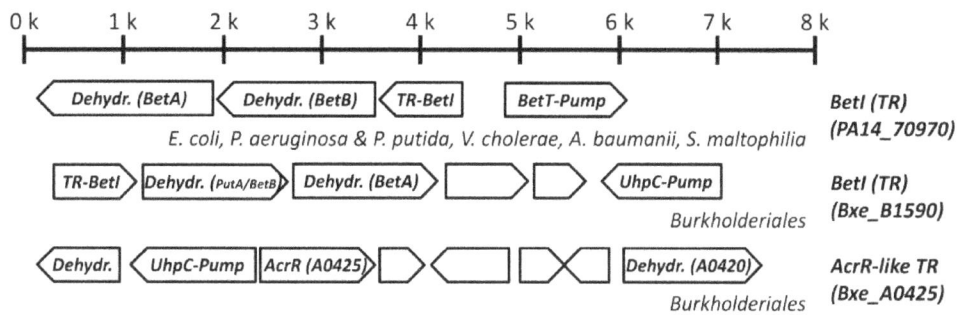

Figure 1. Genomic clustering for the known operon for choline uptake and catabolism in various organisms (row one) and in the *Burkholderiales* order (row two)

Row three presents the suggested secondary operon for choline uptake and catabolism in *Burkholderiales*. The organisms in row one are *Escherichia coli, Pseudomonas aeruginosa, Pseudomonas putida, Vibrio cholera, Acinetobacterbaumanii, Stenotrophomonasmaltophilia.*

2. Methods

2.1 Gene Cluster Analysis

Physical gene clustering around the gene coding for Bxe_A0425 in *B. xenovorans* and its closest analogs was analyzedby utilizing an online integrated portal for comparative and functional genomics [Microbes Online (Dehal et al., 2010)]. We utilized a homolog approach similar to that used by Martí-Arbona et al. (2003), Hall et al. (2013) and Nguyen et al. (2013). Literature searches were performed using Google Scholar.

2.2 Cloning, Expression and Purification of the GST-Bxe_A0425 Protein

2.2.1 Cloning

The gene Bxe_A0425 was cloned from *B. xenovorans* as described in Nguyen and co-workers for Bxe_B2842 (Nguyen et al. 2013) with the following primer modifications: forward 5'-GATCCATGGATGTCCGACAGTGAAACCCC-3' and reverse 5'-GATCGGATCCTCAACCGTCCGAAGGCGCCG-3'.

2.2.2 Protein Expression and Purification

The affinity chromatography-based purification of the GST-Bxe_A0425 was performed as described by Nguyen and co-workers for GST-Bxe_B2842 (Nguyen et al. 2013).

2.3 Frontal Affinity Chromatography-Mass Spectrometry (FAC-MS)

2.3.1 Effector Library Design

The functional assignments of the genes clustering around the gene coding for the TR Bxe_A0425 suggest that this operon could be involved in the uptake and dehydrogenation of choline. Possible effectors were identified based on their occurrence inthe metabolism of glycine, serine and threonine in the KEGG database (http://www.genome.jp/kegg/). Choline and several other compounds in the metabolism of glycine, serine and threonine were used to investigate the effector-binding specificity of Bxe_A0425. Ten µM solutions of commercially available compounds were prepared in buffer C and adjusted to pH 7.4. The compounds used in this screening library were glycerate-3-phosphate, AMP, phosphocholine, creatine, 5-aminolevulinate, hypotaurine, taurine, sarcosine, dihydroxy fumarate, L-threonine, L-serine, glycine, 2,3-diphospho-D-glycerate and choline.

2.3.2 FAC-MS Experiments

The FAC-MS approach to TR discovery used here was previously described by our group (Hall et al., 2013; Martí-Arbona et al., 2012; Nguyen et al., 2013). The preparation of the FAC-MS column was performed as described by Hall et al. (2013) for the protein GST-Bxe_A0736.

2.4 Screening for TR/DNA/Effector Interactions.

The binding of the effector molecule to the TR/DOS complex was monitored by electrophoretic mobility shift assays (EMSAs) as described by our group (Maity et al., 2012). dsDNA (1 µM) was incubated with varying amounts of GST-Bxe_A0425 (0 to 20 µM). Free and protein-bound DNAs were separated on a 1% TBE-agarose gel and then visualized by staining with an ethidium bromide solution. The DNA sequence of the DOS utilized in the EMSA experiments was 5'-GATGTTAGTTACTGATCGGTAACATG-3', which is located between 35 and 76 bp up-stream of the Bxe_A0425 gene.

3. Results and Discussion

3.1 Gene Cluster Analysis

The genes clustered around the gene coding for Bxe_A0425 in *B. xenovorans* and its closest orthologs suggest the conservation of a functional operon composed a several open reading frames (ORFs) from which a TR, a transporter/pump and two short-chain dehydrogenating enzymes can be easily identified. A literature search of functional operons containing these functional components revealed a remarkable resemblance of this system to the evolutionarily convergent and functionally conserved operons found in *E. coli* (Lamark et al., 1991), *B. subtilis* (Boch, Kempf, Schmid, & Bremer, 1996) and *S. xylosus* (Rosenstein, Futter-Bryniok, & Gotz, 1999). These reported functional operons have been linked to the uptake and dehydrogenation of choline to generate glycine betaine, a potent osmo-protectant molecule. Importantly, homology indicated that the genes named *betB*, *gbsA*, *putA*, *cudA* and possibly others all encode for BetB, a NAD-dependent dehydrogenase.

Choline uptake and catabolism operons are examples of convergent evolution, meaning that some organisms developed to utilize a set of bacterial NAD-dependent dehydrogenases while others developed to use FADH-dependent oxidases. Moreover, a comparative analysis of the genes in the Bet-operon in *Pseudomonales* (Figure 2) showed that divergence in gene homology is observed not only in the oxidoreductases in the operon but also in transporters. For example, *Pseudomonas aeruginosa* utilizes the BetT transporter; *Pseudomonas mendocina* uses a major facilitator superfamily transporter (MSF); *Pseudomonas fluorescens* uses an OpuD transporter (BCCT family transporter) and in *Pseudomonas syringae* the transporter is missing from the operon. *B. xenovorans* does not contain a BetT-like transporter; instead, it contains transporters in the UhpC family (MSF-type) of transporters (Bxe_A0426 and Bxe_B1595) like *P. mendocina*.

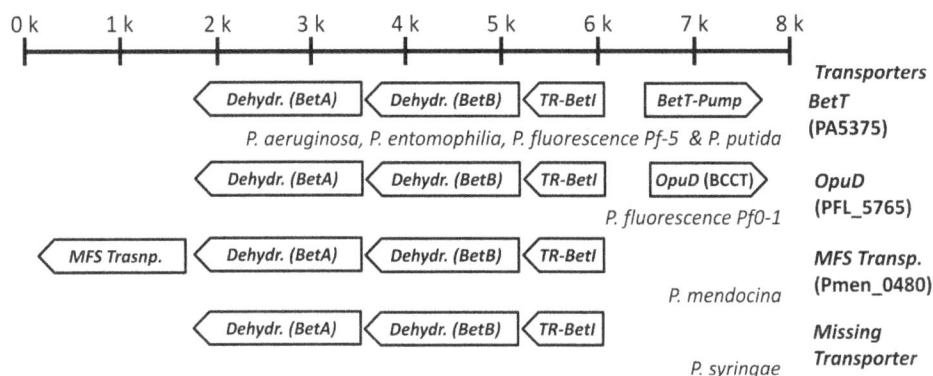

Figure 2. Genomic clustering for choline uptake and catabolism (Bet-operon) of choline in representative *Pseudomonales*

Each row represents a member(s) of the *Pseudomonales* with a variation in the transporter gene within the Bet-operon.

Figure 1 presents a comparison of the known operons encoding for the choline uptake systems in other bacteria with two operons in *B. xenovorans* of very similar composition, each apparently under the control of a single AcrR-like TR. The traditional Bet-operon is depicted in row two and shows the conservation of both dehydrogenating enzymes (BetA: Bxe_B1592 and PutA/BetB: Bxe_B1591) and introduces a new type of MSF-type transporter (UhpC family) similar to transporters observed in *P. mendocina*. The operon containing the TR Bxe_A0425 (Figure 1: row three) also contains a MFS transporter annotated as UhpC (Bxe_A0426), a putative oxidoreductase (Bxe_A0427) and a putative monooxygenase/dehydrogenase (Bxe_A0420). These observations suggest the possible involvement of this operon in the uptake and catabolism of choline.

3.2 Screening for the Effector Molecule Binding to the TR

Based on the implication of choline uptake and metabolism by the gene cluster analysis and the involvement of the general metabolic area of glycine, serine and threonine metabolism in glycine betaine generation, we generated a library of possible effectors from commercially available substrates and products from these pathways. This compound library contained glycerate-3-phosphate, AMP, phosphocholine, creatine, 5-aminolevulinate, hypotaurine, taurine, sarcosine, dihydroxyfumarate, L-threonine, L-serine, glycine, 2,3-diphospho-D-glycerate and choline. Analysis of this library by FAC-MS of the immobilized Bxe_A0425 TR showed a wide distribution of the elution time of the infused compounds. Compounds that quickly eluted from the column were cataloged as no-binders of Bxe_A0425 while those that were retained and eluted at a later time were be cataloged as binders. Large retention time implies high the affinity of Bxe_A0425 for the ligand. Each compound's ion intensity was normalized and an elution profile was created using Sigma Plot 12.3 (Figure 3). The affinity of Bxe_A0425 for these compounds was choline (black) >>> 2,3-diphospho-D-glycerate (dark grey) > glycine (yellow) > L-serine (green) > L-threonine (Brown) > dihydroxyfumarate (red) > sarcosine (dark red) > taurine (purple) > Hypotaurine (green) > 5-aminolevulinate (dark blue) > creatine (grey) > phosphocholine (orange) > AMP (cian) > glycerate-3-phosphate (blue). Compared with all the other compounds tested, the retention of choline was significant, suggesting that under the arbitrary conditions of our FAC-MS screen assay, cholinewas the best effector for Bxe_A0425.

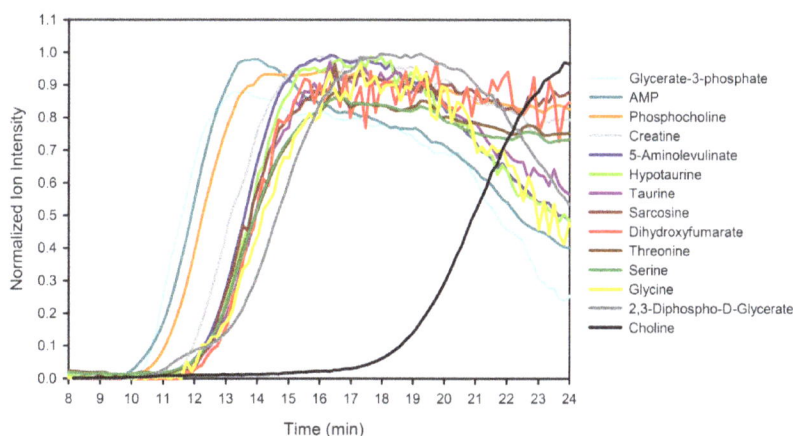

Figure 3. Screening for effectors of Bxe_A0425

The elution profiles of the library of potential effectors (10 μM each) after flowing past immobilized GST-Bxe-A00425are shown. The retention of choline (black line) in the column was much greater, suggesting that it is the best effector for Bxe_A0425.

3.3 Effector Binding to the TR/DOS Complex

Bxe_A0425 is a member of the TetR family of TRs. Maity et al. (2012) demonstrated that GST-Bxe_A0425 is capable of binding its DOS to create a TR/DOS complex in the absence of an effector molecule. It is therefore possible that it functions as a transcriptional repressor (Maity et al., 2012). We used EMSA experiments to test the ability of choline to disrupt the TR/DOS complex. Figure 4 shows a set EMSA gels. The gel in the left panel was run in the absence of the effector molecule (choline) and the gel in the right panel was run with 10 mM choline. The concentration of GST-Bxe_A0425 was varied from 0 (lane 1, the electrophoretic mobility of free DNA) to 20 μM (lane 8, the electrophoretic mobility of the fully formed TR/DOS complex). Under the experimental conditions studied, an increase in the intensity of the free DNA band in lanes 3, 4 and 5 is observed when choline is added to the assay. These changes suggest that choline indeed affects the TR/DOS complex and causes the protein to release the DNA. This observation supports the results obtained from the FAC-MS experiments and indicates that Bxe_A0425 in involved in the regulation of this newly identified operon.

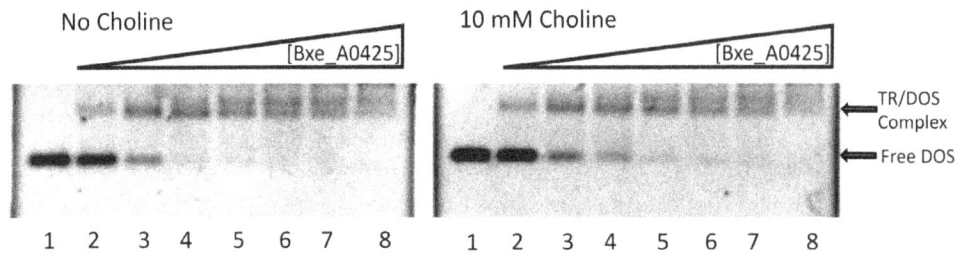

Figure 4. EMSA assays to test the ability of choline to disrupt the TR/DOS complex

Lane 1 shows the electrophoretic mobility of free DNA while lane 8 shows the electrophoretic mobility of the fully formed TR/DOS complex. Lanes 2 to 8 contain increasing concentrations of the GST-Bxe_A0425 TR, the concentration were 0.3, 0.6, 1.2, 2.5, 5.0, 10.0 and 20.0 µM, respectively. Lanes 3 to 5 show a change in the free DOS band intensity, suggesting that choline disrupts the TR/DOS binary complex.

3.4 Conclusions

We found Bxe_A0425 in a cluster of genes in *B. xenovorans* and other *Burkholderiales* containing close homologues of the TR and in an apparent operon. The general functional annotation of the genes in the clusters was consistent with the possibility that the operon and the TR are involved in choline uptake and metabolism to produce glycinebetaine. Bxe_A0425 bound choline when screened against a library of metabolites characteristic of this general metabolic area. Results from an EMSA assay were consistent withcholine impacting the association of this TR with its DOS. Taken together, these findings suggest that TR Bxe_A0425 and the operon in which it resides are involved in choline uptake and conversion to glycine betaine.

Acknowledgements

This work was supported by the Los Alamos National Laboratory's Laboratory Directed Research Program (20090107DR). We are grateful for the editorial improvements made by Dr. Virginia A. Unkefer.

References

Boch, J., Kempf, B., Schmid, R., & Bremer, E. (1996). Synthesis of the osmoprotectant glycine betaine in *Bacillus subtilis*: Characterization of the *GbsAB* genes. *Journal Bacteriology, 178*(17), 5121-5129.

Csonka, L. N., & Hanson, A. D. (1991). Prokaryotic osmoregulation: Genetics and physiology. *Annual Reviews of Microbiology, 45*, 569-606. http://dx.doi.org/10.1146/annurev.mi.45.100191.003033

Dehal, P. S., Joachimiak, M. P., Price, M. N., Bates, J. T., Baumohl, J. K., Chivian, D., ... Arkin, A. P. (2010). Microbesonline: An integrated portal for comparative and functional genomics. *Nucleic Acids Research, 38*, 396-400. http://dx.doi.org/10.1093/nar/gkp919

Hall, R. S., Martí-Arbona, R., Hennelly, S. P., Maity, T. S., Mu, F., Dunbar, J. M., ... Unkefer, P. J. (2013). *In-vitro* characterization of an L-kynurenine-responsive transcription regulator of the oxidative tryptophan degradation pathway in *Burkholderia xenovorans*. *Journal of Molecular Biology Research, 3*(1), 55-65. http://dx/doi/org/10.5539/jmbr.v3n1p55

Lamark, T., Kaasen, I., Eshoo, M. W., Falkenberg, P., McDougall, J., & Strom, A. R. (1991). DNA sequence and analysis of the Bet genes encoding the osmoregulatory choline-glycine betaine pathway of *Escherichia coli*. *Molecular Microbiology, 5*(5), 1049-1064. http://dx.doi.org/10.1111/j.1365-2958.1991.tb01877.x

Maity, T. S., Close, D. W., Valdez, Y. E., Nowak-Lovato, K., Martí-Arbona, R., Nguyen, T. T., ... Dunbar, J. (2012). Discovery of DNA operators for TetR and MarR family transcription factors from *Burkholderia xenovorans*. *Microbiology, 158*(2), 571-582. http://dx.doi.org/10.1099/mic.0.055129-0

Martí-Arbona, R., Teshima, M., Anderson, P. S., Nowak-Lovato, K. L., Hong-Geller, E., Unkefer, C. J., & Unkefer, P. J. (2012). Identification of new ligands for the methionine biosynthesis transcriptional regulator (MetJ) by FAC-MS. *Journal of Molecular Microbiology and Biotechnology, 22*(4), 205-214. http://dx.doi.org/10.1159/000339717

Martí-Arbona, R., Xu, C., Steele, S., Weeks, A., Kuty, G. F., Seibert, C. M., & Raushel, F. M. (2006). Annotating enzymes of unknown function: N-formimino-L-glutamate deiminase is a member of the Amidohydrolase Superfamily. *Biochemistry, 45*(7), 1997-2005. http://dx.doi.org/10.1021/bi0525425

Nguyen, T. T., Martí-Arbona, R., Hall, R. S., Maity, T. S., Valdez, Y. E., Dunbar, J. M., ... Unkefer, P. J. (2013). Identification and *in-vitro* characterization of a novel OhrR transcriptional regulator in *Burkholderia xenovorans* LB400. *Journal of Molecular Biology Research, 3*(1), 37-46. http://dx.doi.org/10.5539/jmbr.v3n1p37

Rosenstein, R., Futter-Bryniok, D., & Gotz, F. (1999). The choline-converting pathway in *Staphylococcus xylosus* C2A: Genetic and physiological characterization. *Journal of Bacteriology, 181*(7), 2273-2278.

5

Chromosomal Q-Heterochromatin and Age in Human Population

Ibraimov A. I.[1,2], Akanov A. A.[1], Baygazieva G. D.[1] & Meimanaliev T. S.[1]

[1] Kazakh National Medical University after name S. D. Asfendyarov, Almaty, Kazakhstan

[2] Laboratory of Human Genetics, National Center of Cardiology and Internal Medicine, Bishkek, Kyrgyzstan

Correspondence: Ibraimov A. I., Laboratory of Human Genetics, National Center of Cardiology and Internal Medicine, Bishkek, Kyrgyzstan. E-mail: ibraimov_abyt@mail.ru

Abstract

Individuals in the population differ from each other in number, location, size and intensity of fluorescence of chromosomal Q-heterochromatin regions (Q-HRs). It has been shown that human populations differ significantly as concerns this feature. The question remains open as to whether there exist differences between individuals belonging to different age groups. This fact is very important for understanding the possible biological role of the broad quantitative variability of chromosomal Q-HRs in human genome. The quantitative content of chromosomal Q-HRs was studied in individuals of Kazakh and Russian nationality from different age groups. It was shown that chromosomal Q-HRs are most numerous in the genome of neonates, while they are the least numerous in the genome of elderly subjects (aged 60 years and older). It is supposed that the lesser amount of Q-HRs in the genome of elderly subjects is due to the selective advantage in their survival to old age. The possible selective value of chromosomal Q-HRs is discussed.

Keywords: chromosomal Q-heterochromatin, Q-heterochromatin polymorphism, Q-heterochromatin selection, age groups, human adaptation

1. Introduction

A fundamental feature of chromosomes of higher eukaryotes, including man, is the presence of two evolutionary consolidated types of genetic material: euchromatin and heterochromatin. Euchromatin – the conservative portion of the genome – contains transcribed structural genes, while heterochromatin – the variable portion of the genome – predominantly consists of nontranscribed repeated DNA sequences.

Heterochromatin is universally distributed in the chromosomes of all the higher eukaryotes, amounting to 10% - 60% of their genome. About 15% - 20% of the human genome is composed of heterochromatin regions (HRs) (John, 1988). Chromosomal HRs does not change during ontogenesis and are clearly inherited as discrete features.

To-date, two types of heterochromatin are known: C- and Q-heterochromatin. There are several significant differences between them, including the fact that C-heterochromatin is encountered in chromosomes of all higher eukaryotes, while Q-heterochromatin is only present in man, the chimpanzee and gorilla. In man C-heterochromatin is present in all his chromosomes, varying mainly in size, while Q-heterochromatin can only be detected on seven autosomes and the Y-chromosome (Paris Conference, 1971, 1975).

Individuals in a population differ from each other in the number, location, size and intensity of staining (fluorescence) of chromosomal Q-heterochromatin regions (Q-HRs). It has been shown that certain human populations differ significantly as concerns this feature (Geraedts & Pearson, 1974; Müller et al., 1975; Buckton et al., 1976; Lubs et al., 1977; Yamada & Hasegawa, 1978; Al-Nassar, 1981; Ibraimov & Mirrakhimov, 1982a, 1982b, 1982c; Ibraimov et al., 1982, 1986, 1990, 1991, 2000, 2013; Stanyon et al., 1988; Kalz et al., 2005; Decsey et al., 2006). The question remains open as to whether there exist differences between individuals belonging to different age groups in population. This fact is very important for understanding the possible biological role of the broad quantitative variability of human chromosomal Q-HRs.

2. Materials and Methods

2.1 Sample Characteristics

Using identical methods we studied neonates, as well as students from Kazakh National Medical University in

Almaty. The group of elderly subjects (aged 60 years and over) consisted of individuals living in the Old People's Homes of Almaty. In the studied samples only individuals whose parents were from the same ethnic group are included.

2.2 Cytogenetic Methods

Chromosomal preparations were made using short-term cultures of peripheral blood lymphocytes, with the exception of newborn infants where umbilical blood was used. The cell cultures were processed according to slightly modified (Ibraimov, 1983a) conventional methods (Hungerford, 1965). The dye used was quinacrine mustard. All the chromosomal preparations were analyzed by one and the same cytogeneticist (A.I.I.) to investigate chromosomal Q-HR variability. Calculation and registration of chromosomal Q-HRs was performed using the criteria and methods described in detail elsewhere (Ibraimov et al., 1982, 1990).

2.3 Quantitative Characteristics of Q-HR Variability and Methods Used for Comparisons

Q-HR variability of autosomes in populations is usually described in the form of three main quantitative characteristics:

1) The distribution of the number of Q-HRs in a population, i.e., distribution of individuals having different numbers of Q-HRs in the karyotype regardless of their location on seven Q-polymorphic autosomes, which also reflected the range of Q-HRs variability in the population genome;

2) The derivative of this distribution, an important population characteristic, is the mean number of Q-HRs per individual;

3) The frequency of Q-HRs on seven Q-polymorphic autosomes (3, 4, 13-15, 21 and 22) in the population. Despite the fact that in human autosomes there are twelve loci in which Q-HRs can be detected (3 cen, 4 cen, 13 p11, 13 p13, 14 p11, 14 p13, 15 p11, 15 p13, 21 p11, 21 p13, 22 p11, 22 p13), individuals with 24 Q-HRs in their genome could exist, but such cases have not as yet been reported. In individuals of a population the number of Q-HRs on the autosomes usually ranges from zero to ten (Yamada & Hasegawa, 1978; Al-Nassar et al., 1981; Ibraimov & Mirrakhimov, 1985; Ibraimov, 2010).

The distribution of the numbers and mean number of Q-HR per individual in samples were compared using the Student *t*-test.

3. Results

Table 1 and Table 2 show the distribution of the numbers and mean number of Q-HRs on autosomes in studied age groups of Kazakh and Russian nationality. As can be seen from these Tables, in every case neonates are characterized by the highest mean number value, and by a broad range of variability in the distribution of Q-HRs as compared to individuals from older groups and especially from groups of elderly subjects.

Table 1. Distribution of the numbers and mean number of chromosomal Q-HRs per individual in Kazakh samples

Number of Q-HRs	Newborns (n = 389)	18 - 25 years (n = 239)	60 years and older (n = 33)
	I	II	III
0	4 (1.0%)		2 (6.0%)
1	9 (2.3%)	16 (6.7%)	5 (15.1%)
2	60 (15.4%)	36 (15.1%)	7 (21.2%)
3	85 (21.9%)	58 (24.3%)	12 (36.4%)
4	97 (24.9%)	60 (25.1%)	6 (18.2%)
5	76 (19.5%)	44 (18.4%)	1 (3.0%)
6	38 (9.8%)	17 (7.1%)	
7	20 (5.1%)	7 (2.9%)	
8		1 (0.4%)	
Total	1520 (99.9%)	881 (100.0%)	84 (99.9%)
Mean number of Q-HRs	3.91 ± 0.07	3.69 ± 0.09	2.55 ± 0.213
Statistics	$t_{I, II}$ = 1.808;	df = 626;	P = 0.071;
	$t_{I, III}$ = 5.068;	df = 420;	P = <0.001;*
	$t_{II, III}$ = 4.259;	df = 270;	P = <0.001;*

* - these differences are statistically significant.

Table 2. Distribution of the numbers and mean number of chromosomal Q-HRs per individual in Russian samples

Number of Q-HRs	Newborns (n = 83)	18 - 25 years (n = 60)	60 years and older (n = 80)
	I	II	III
0	1 (1.2%)		4 (5.0%)
1	4 (4.8%)	4 (6.6%)	6 (7.5%)
2	9 (10.8%)	7 (11.7%)	28 (35.0%)
3	14 (16.9%)	14 (23.3%)	23 (28.7%)
4	30 (36.1%)	17 (28.3%)	14 (17.5%)
5	16 (19.3%)	13 (21.6%)	5 (6.2%)
6	6 (7.2%)	4 (6.6%)	
7	3 (3.6%)	1 (1.6%)	
Total	321 (99.9%)	224 (99.9%)	212 (99.9%)
Mean number of Q-HRs	3.87 ± 0.156	3.73 ± 0.177	2.65 ± 0.133
Statistics	$t_{I, II}$ = 0.502;	df = 141;	P = 0.575
	$t_{I, III}$ = 5.895;	df = 161;	P = <0.001*
	$t_{II, III}$ = 4.980;	df = 138;	P = <0.001*

* - these differences are statistically significant.

As seen in Tables 1 and 2, there are no statistically significant differences between newborns and samples of young adults of the same nationality (18-25 – years old). However, a closer analysis showed that, in fact, these two age groups significantly differ from each other. It turned out that the main reason for the lack of differences between the two age groups was combining individuals of both sexes into one group. It has long been noted that at the population level, due to the largest block of Q-heterochromatin on Y chromosome in males, amount of Q-HRs on autosomes is less than in females (Ibraimov et al., 2000). When we compared newborns and students based on their gender, we found that these two age groups were significantly different from each other (Table 3).

Table 3 shows the distribution of the numbers and mean number of Q-HRs on autosomes in males and females in two age groups of Kazakh nationality. As can be seen from these Tables, in every case females are characterized by the highest mean number value, and by a broad range of variability in the distribution of the numbers of chromosomal Q-HRs as compared to males. These differences are statistically significant. Such regularities as in Kazakhs was also found in Russian samples (data not shown).

Table 3. Distribution of the numbers and mean number of Q-HRs on autosomes in males and females in newborn and 18 - 25 years individuals of Kazakh nationality

Number of Q-HRs	Newborns Boys (n = 207)	Newborns Girls (n = 182)	Males 18 - 25 years (n = 49)	Females 18 - 25 years (n = 190)
	I	II	III	IV
0	3	1		
1	5	4	9	7
2	39	21	12	24
3	47	38	13	45
4	51	46	11	49
5	37	39	4	40
6	18	20		17
7	7	13		7
Total	770	750	136	745
Mean number	3.72 ± 0.102	4.12 ± 0.111	2.78 ± 0.176	3.92 ± 0.104
Statistics	$t_{I, II} = 2.649$;	df = 387;	P = 0.008*	
	$t_{II, III} = 5.775$;	df = 229;	P =<0.001*	
	$t_{III, IV} = 5.119$;	df = 237;	P =<0.001*	
	$t_{I, III} = 4.137$;	df = 254;	P =<0.001*	
	$t_{II, IV} = 1.313$;	df = 370;	P = 0.190	

* - these differences are statistically significant.

Results of the comparative analyses of the frequency of Q-HRs on autosomes in individuals from different age groups are presented in Table 4. As it can be seen from this Table, a tendency is noted towards a decrease in the absolute frequency of Q-HRs with age in all the chromosomes. Of interest is the fact that this difference was pronounced in chromosome 3 and 13 containing more than half the Q-HRs of the human population genome. At the same time, individuals from all the age groups did not differ significantly from each other in the portion of Q-HRs in seven Q-polymorphic autosomes, in keeping with our previous observations (Ibraimov, 1993, 2010; Ibraimov et al., 1986, 1990, 1991, 2000, 2013).

Table 4. Q-HRs frequencies in seven Q-polymorphic autosomes of Kazakh and Russian samples

Location of Q-HRs	Kazakhs			Russians		
	Newborns (n = 389)	18 - 25 years (n = 239)	60 years and older (n = 33)	Newborns (n = 83)	18 - 25 years (n = 60)	60 years and older (n = 80)
3	369 (47.4)*	213 (44.5)	21 (31.8)	98 (59.0)	71 (59.2)	68 (42.5)
	24.3**	24.2	25.0	30.5	31.9	32.1
4	74 (9.5)	48 (10.0)	3 (4.5)	10 (6.0)	8 (6.7)	5 (3.1)
	4.9	5.4	3.5	3.1	3.7	2.4
13	480 (61.7)	268 (56.1)	26 (39.4)	104 (62.5)	72 (60.0)	73 (45.6)
	31.6	30.4	30.7	32.4	31.5	34.4
14	117 (15.0)	68 (14.2)	8 (12.1)	25 (15.0)	17 (14.1)	17 (10.6)
	7.7	7.7	9.3	7.8	7.4	8.0
15	161 (20.7)	77 (16.1)	9 (13.6)	33 (19.9)	22 (18.3)	22 (13.7)
	10.6	8.7	10.3	10.3	10.0	10.4
21	196 (25.2)	108 (22.6)	10 (15.1)	30 (18.1)	16 (13.3)	18 (11.2)
	12.9	12.3	12.4	9.3	7.3	8.5
22	123 (15.8)	99 (20.7)	7 (10.6)	21 (12.6)	18 (15.0)	9 (5.6)
	8.0	11.2	8.8	6.5	8.2	4.2
Всего	1520 (195.4)	881 (184.3)	84 (126.6)	321 (193.4)	224 (186.6)	212 (132.3)
	100.0	99.9	100.0	99.9	100.0	100.0
Mean number of Q-HRs	3.91	3.69	2.55	3.86	3.16	2.65

* -Q-HRs frequency from the number of chromosomes analyzed;

** -Q-HRs frequency as percentage of the overall number of chromosomal Q-HR.

4. Discussion

Despite the fact that human chromosomal Q-HRs has been studied for more than forty years, their biological role and nature remain unclear. A remarkable feature of human chromosomal Q-HRs is their wide quantitative variability characterized by the fact that individuals in a population differ in the number, location, size and intensity of fluorescence of these specific fluorescent areas (Paris Conference, 1971, 1975). The existence of population Q-HRs variability in twelve polymorphic loci of seven autosomes is a well-established fact (Lubs et al., 1977; Al-Nassar et al., 1981; Ibraimov & Mirrakhimov, 1982a, 1982b, 1982c; Ibraimov et al., 1982, 1986, 1990, 1991, 2000, 2013; Kalz et al., 2005).

We have studied over 20 human populations of Eurasia and Africa, representing all the three major racial groups. These studies showed that certain human populations differ considerably from each other and these differences are mainly related to the environment they live in and not to racial or ethnic features. In particular, the mean number of Q-HRs per individual is considerably lower in the genome of populations living permanently in the North and at high altitude, as well as in newcomers well adapted to extreme environmental conditions of high altitudes (mountaineers) and the Far North (borers - oil industry workers of West Siberia), as compared to populations living in temperate zones of Eurasia and in low-altitude subequatorial Africa (Ibraimov & Mirrakhimov, 1982a, 1982b, 1982c; Ibraimov et al., 1982, 1986, 1990, 1991, 2000, 2013).

Based on these observations, we have at one time put forward the hypothesis on the possible selective value of chromosomal Q-HRs in the adaptation of human populations to certain extreme environmental factors – in particular to cold and hypoxia. These same data are the basis of our cytogenetic model of the origin of contemporary man. We feel that a conclusive role in this unique process could be played by features of the chromosomal Q-HR system, only inherent to our nearest ancestors and inherited by us, and not by the appearance of new specific structural genes or gene complexes (Ibraimov, 1993, 2010, 2011; Ibraimov et al., 1986).

If our assumption on the possible selective value of the amount of chromosomal Q-HRs in human adaptation and the origin of man is correct, than validation of this hypothesis requires, among other things, further reliable data on the amount of Q-HRs in the genome of individuals from different age groups, because we have sufficient reasons to believe that individuals with a lesser amount of Q-HRs in their genome are not only able to adapt themselves better to certain extreme environmental factors, but also have more chances to live to old age (Ibraimov & Karagulova, 2006a, 2006b).

In the literature we have found only two studies directly or indirectly devoted to chromosomal Q-HR variability in individuals from different age groups. Thus, Buckton et al. (1976) have examined three Scottish populations. Two of the populations, newborn babies and 14 year old cohort, were from the Scottish mainland (Edinburgh area). The third population (65 years and over) was from Barra, a relatively isolated island in the Outer Hebrides. The authors came to the conclusion that "rather more variation is found in the Q-band intensity polymorphisms: the island population appears to have fewer Brilliant and intense variants than do the other two groups, 2.9 per person as compared to 4.2 and 3.9 for the newborn and 14 years old subjects, respectively; this may be an age differences rather than a population differences". However, the authors remarked that "these figures may reflect a real phenomenon in that the Barra population shows less polymorphic variation than the mainland populations; but equally well they may reflect a possible age difference, perhaps in the way that chromosomes from aged individuals respond to technical treatment? These possibilities could be evaluated by studying Barra and mainland population samples with a vertical age distribution". Unfortunately, as far as we know, such a study has not yet been carried out in either Scotland or elsewhere.

In Siberia, Nazarenko (1987) tried to reveal "age-related changes in the frequency of chromosomal Q-polymorphism in an ethnically homogeneous normal human population". The age of the subjects studied ranged from 7 to 101 years. The author divided them into 3 age groups: 7 to 19 years old, 20 to 29 and 50 years old and over. He found the following changes in the mean number of Q-variants per individual in these 3 age groups: 3.67 ± 0.10; 3.45 ± 0.14 and 3.12 ± 0.18, respectively.

Two explanations are proposed (Geraedts & Pearson, 1974; Nazarenko, 1987) for the somewhat decreased mean number of chromosomal Q-HRs in the older individuals as compared to those under 20 years of age who were similar in other respects: (1) Q-HRs instability during ontogenesis; and (2) the dependence of adaptability and survival upon the location of Q-HRs.

The first hypothesis is to be confirmed by continuous follow-up in a large sample of individuals (from the onset of zygote formation to death). As far as we know, such studies have not been performed. The second hypothesis is not confirmed by our data showed that as a rule, increases in the mean number of Q-HRs are accompanied by increases in absolute Q-HR frequencies on all the autosomes, and vice versa. Populations with a low mean number of O-HRs per individual also have a narrow range of variability in the amount of chromosomal Q-HRs (Ibraimov, 1993, 2010, 2011; Ibraimov et al., 2000). This is also documented by data presented here.

Earlier, examining individuals of different age groups, representing two racial-ethnic groups (Kyrgyz and Russians, living in Kyrgyzstan), we confirmed the observations of British and Russian researchers (Ibraimov & Karagulova, 2006a). The reasons for returning to this problem were as following: a) the above quotes of British researchers imply that the authors were not sure of their findings because of their choice of older people, who were from the island of Barra, whereas the newborn and 14 years old subjects were taken from mainland Scotland (Buckton et al., 1976); b) differences in the mean number of Q-HRs on the individual in the population between the three age groups, obtained in Siberia (Nazarenko, 1987), were not statistically significant; c) in the previous study (Ibraimov & Karagulova, 2006a, 2006b), we used the data obtained in different years with the help of chemical reagents which were not always of high quality (all produced in the former USSR). In this study, we tried to eliminate all these problems by using high-quality chemical reagents and culture medium (Sigma Chemical Co., USA), and the collection of blood samples was carried out for one year in Almaty. As shown above, our new data have fully confirmed all previous observations on the existence of differences in the number of chromosomal Q-HRs in the genome between individuals of different age groups in the human population.

Thus, the existence of considerable differences in the amount of chromosomal Q-HRs in individuals of different age groups in the population is an established fact. However, there is no agreement as to the nature of Q-HRs variability, although arguments are based on the 'selectionist' hypothesis. One approach, implying that Q-HRs with different locations were basically similar in structural and functional features, were defined by Ibraimov et al. (1986) in the following manner: "of primary importance to an individual is the dose and not the location of

Q-HR". The term "dose" is defined as the amount of Q-heterochromatin material in the genome regardless of its location in any chromosome. In other words, this approach is based on the assumption that Q-HR lack locus-specificity. Those favoring the alternative approach believe that deviations from expected Q-HR frequencies, observed in any loci, reflect some structural and functional features of these loci and are due either to selection or to nonfortuitous segregation of chromosomes bearing the given Q-HR (Geraedts & Pearson, 1974; Mikelsaar et al., 1975; Nazarenko et al., 1987).

The hypothesis on which the project is based on assumption that older individuals probably have a lower number of Q-HRs in their genome than younger ones from the same population. If this proves to be true, then we have the right to assume that individuals with a smaller amount of Q-HRs in their genome are not only able to adapt themselves better to certain extreme environmental factors, but also have more chances, other things being equal, to live to old age. And indeed, our data indicate that: 1) in a population there is a clear-cut tendency towards a decrease in the number of chromosomal Q-HRs with age, regardless of racial and ethnic features on the individuals (Tables 1 and 2); 2) of all the age groups the genome of neonates contains the greatest amount of Q-HRs; 3) decreases in the number of Q-HRs with age are not due to the "loss" of Q-heterochromatin on individual loci or chromosomes, but occur simultaneously in all the seven Q-polymorphic autosomes (Table 4), in keeping with all our previous observations (Ibraimov, 1993, 2010, 2011; Ibraimov et al., 1986, 1990, 1991, 1997, 2013).

We believe that the decrease in the amount of Q-HRs with age in a population is not an ontogenetic process, but rather the result of natural selection where individuals with a greater number of chromosomal Q-HRs in their genome than on the average in a population "fall out". Therefore, we once again suggest that the cause of the observed differences lies in differential adaptability of individuals with different amounts of Q-heterochromatin material in their genome regardless of its location in any chromosome.

References

Al-Nassar, K. E., Palmer, C. G., Connealy, P. M., & Pao-Lo, Y. (1981). The genetic structure of the Kuwaiti population. II. The distribution of Q-band chromosomal heteromorphisms. *Hum. Genet., 57*, 423-427. http://dx.doi.org/10.1007/BF00282021

Buckton, K. E., O'Riordan, M. L., Jacobs, P. A., Robinson, J. A., Hill, R., & Evans, H. J. (1976). C- and Q-band polymorphisms in the chromosomes of three human populations. *Ann. Hum. Genet., 40*, 90-112. http://dx.doi.org/10.1111/j.1469-1809.1976.tb00168.x

Decsey, K., Bellovits, O., & Bujdoso, G. M. (2006). Human chromosomal polymorphism in a Hungarian sample. *Int. J. Hum. Genet., 6*(3), 177-183.

Geraedts, J. P. M., & Pearson, P. L. (1974). Fluorescent chromosome polymorphisms: frequencies and segregation in a Dutch population. *Clin. Genet., 6*, 247-257. http://dx.doi.org/10.1111/j.1399-0004.1974.tb02086.x

Hungerford, D. A. (1965). Leucocytes cultured from small inocula of whole blood and preparation of metaphase chromosomes by treatment with hypotonic KCl. *Stain Technol., 40*, 333-338.

Ibraimov, A. I. (1983). Chromosome preparations of human whole lymphocytes – an improved technique. *Clin. Genet., 24*, 240-242. http://dx.doi.org/10.1111/j.1399-0004.1983.tb00077.x

Ibraimov, A. I. (1993). The origin of modern humans: a cytogenetic model. *Hum. Evol., 8*(2), 81-91. http://dx.doi.org/10.1007/BF02436607

Ibraimov, A. I. (2010). Chromosomal Q-heterochromatin regions in populations and human adaptation. In M. K. Bhasin & C. Susanne (Eds.), *Anthropology Today: Trends and Scope of Human Biology* (pp. 225-250). Delhi: Kamla- Raj Enterprises.

Ibraimov, A. I. (2011). Origin of modern humans: a cytogenetic model. *Hum. Evol., 26*(1-2), 33-47.

Ibraimov, A. I., Akanov, A. A., Meymanaliev, T. S., Karakushukova, A. S., Kudrina, N. O., Sharipov, K. O., & Smailova, R. D. (2013). Chromosomal Q-heterochromatin polymorphisms in 3 ethnic groups (Kazakhs, Russians and Uygurs) of Kazakhstan. *Int. J. Genet., 5*(1), 121-124.

Ibraimov, A. I., & Karagulova, G. O. (2006a). Chromosomal Q-heterochromatin regions in individuals of various age groups. *Int. J. Hum. Genet., 6*(3), 219-228.

Ibraimov, A. I., & Karagulova, G. O. (2006b). Chromosomal Q-heterochromatin variability in neonates deceased during first year of age. *Int. J. Hum. Genet., 6*(4), 281-285.

Ibraimov, A. I., & Mirrakhimov, M. M. (1982a). Human chromosomal polymorphism. III. Chromosomal Q-polymorphism in Mongoloids of northern Asia. *Hum. Genet., 62,* 252-257. http://dx.doi.org/10.1007/BF00333531

Ibraimov, A. I., & Mirrakhimov, M. M. (1982b). Human chromosomal polymorphism. IV. Chromosomal Q-polymorphism in Russians living in Kyrghyzia. *Hum. Genet., 62,* 258-260. http://dx.doi.org/10.1007/BF00333532

Ibraimov, A. I., & Mirrakhimov, M. M. (1982c). Human chromosomal polymorphism. V. Chromosomal Q-polymorphism in African populations. *Hum. Genet., 62,* 261-265. http://dx.doi.org/10.1007/BF00333533

Ibraimov, A. I., & Mirrakhimov, M. M. (1985). Q-band polymorphism in the autosomes and the Y chromosome in human populations. In A. A. Sandberg & R. Alan (Eds.), *Progress and Topics in Cytogenetics. The Y chromosome. Part A. Basic Characteristics of the Y chromosome* (pp 213-87). New York: Liss Inc.

Ibraimov, A. I., Axenrod, E. I., Kurmanova, G. U., & Turapov, O. A. (1991). Chromosomal Q-heterochromatin regions in the indigenous population of the northern part of West Siberia and new migrants. *Cytobios, 67,* 95-100.

Ibraimov, A. I., Karagulova, G. O., & Kim, E. Y. (2000). The relationship between the Y chromosome size and the amount of autosomal Q-heterochromatin in human populations. *Cytobios, 102,* 35-53.

Ibraimov, A. I., Kurmanova, G. U., Ginsburg, E. Kh., Aksenovich, T. I., & Axenrod, E. I. (1990). Chromosomal Q-heterochromatin regions in native highlanders of Pamir and Tien-Shan and in newcomers. *Cytobios, 63,* 71-82.

Ibraimov, A. I., Mirrakhimov, M. M., Axenrod, E. I., & Kurmanova, G. U. (1986). Human chromosomal polymorphism. IX. Further data on the possible selective value of chromosomal Q-heterochromatin material. *Hum. Genet., 73,* 151-156. http://dx.doi.org/10.1007/BF00291606

Ibraimov, A. I., Mirrakhimov, M. M., Nazarenko, S. A., Axenrod, E. I., & Akbanova, G. A. (1982). Human chromosomal polymorphism. I. Chromosomal Q-polymorphism in Mongoloid populations of Central Asia. *Hum. Genet., 60,* 1-7. http://dx.doi.org/10.1007/BF00281254

John, B. (1988). The biology of heterochromatin. In R. S. Verma (Ed.), *Heterochromatin: molecular and structural aspects* (pp. 1-147). Cambridge University Press.

Kalz, L., Kalz-Fuller, B., Hedge, S., & Schwanitz, G. (2005). Polymorphism of Q-band heterochromatin; qualitative and quantitative analyses of features in 3 ethnic groups (Europeans, Indians, and Turks). *Int. J. Hum. Genet., 5*(2), 153-163.

Lubs, H. A., Patil, S. R., Kimberling, W. J., Brown, J., Hecht, F., Gerald, P., & Summitt, R. L. (1977). Racial differences in the frequency of Q- and C-chromosomal heteromorphism. *Nature, 268,* 631-632. http://dx.doi.org/10.1038/268631a0

Mikelsaar, A. V. N., Kaosaar, M. E., Tuur, S. J., Viikmao, M. H., Talvik, T. A., & Laats, J. (1975). Human karyotype polymorphism. III. Routine and fluorescence microscopic investigation of chromosomes in normal adults and mentally retarded children. *Humangenetik, 26,* 1-8.

Müller, H. J., Klinger, H. P., & Glasser, M. (1975). Chromosome polymorphism in a human newborn population. II. Potentials of polymorphic chromosome variants for characterizing the idiogram of an individual. *Cytogenet Cell Genet., 15,* 239-255. http://dx.doi.org/10.1159/000130522

Nazarenko, S. A. (1987). Age dynamics of fluorescent polymorphism in human chromosomes. *Cytol. Genet. (Russian), 21,* 183-186.

Paris Conference and Supplement. (1971, 1975). Standartization in human cytogenetics. *Birth Defects, 6,* 1-84.

Stanyon, R., Studer, M., Dragone, A., De Benedictis, G., & Brancati, C. (1988). Population cytogenetics of Albanians in Cosenza (Italy): frequency of Q- and C-band variants. *Int. J. Anthropol., 3,* 19-29.

Yamada, K., & Hasegawa, T. (1978). Types and frequencies of Q-variant chromosomes in a Japanese population. *Hum. Genet., 44,* 89-98. http://dx.doi.org/10.1007/BF00283578

6

Inference of Specific Gene Regulation by Environmental Chemicals in Human Embryonic Stem Cells

Sachiyo Aburatani[1] & Wataru Fujibuchi[2]

[1] Computational Biology Research Center (CBRC), National Institute of Advanced Industrial Science and Technology, Tokyo, Japan

[2] Center for iPS Research and Application, Kyoto University, Kyoto, Japan

Correspondence: Sachiyo Aburatani, Computational Biology Research Center (CBRC), National Institute of Advanced Industrial Science and Technology, AIST Tokyo Waterfront BIO-IT Research Building, 2-4-7 Aomi, Koto-ku, Tokyo 135-0064, Japan. E-mail: s.aburatani@aist.go.jp

Abstract

We are exposed to many environmental chemicals in our daily life. Certain chemicals threaten our health, especially that of embryos and can cause serious developmental problems. To prevent abnormal development and diseases caused by chemicals, it is important to clarify the mechanisms of chemical toxicity in embryonic cells. The gene regulatory network is one of the useful methods for clarifying functional mechanisms in living cells, so we applied a statistical method to infer the gene regulatory network in human embryonic stem cells. In this study, we improved our previously developed SEM approach for inferring a network model from 9 gene expression profiles in human embryonic stem cells, which were exposed to various chemicals. The estimated regulatory models clarified the differences between chemicals, and the shapes of the inferred models reflected the features of the chemical toxicities. The toxicity of acrylamide affected neuronal cell-related genes, while that of diethylnitrosamine disturbed cell differentiation-related genes. On the other hand, the TCDD network reflected feedback regulation, and finally disturbed neuronal cell-related genes. In the Thalidomide network, cell differentiation genes related to axis formation in embryonic cells were affected by thalidomide toxicity.

Keywords: structural equation modeling, environmental chemical, gene regulatory network, embryonic stem cell

1. Introduction

1.1 Introduction of the Problem

Environmental pollution is a byproduct of our usual life activities. Vehicle exhaust contains gases, including many noxious chemicals. Factories discharge industrial waste in the air, ground, and water. Many rivers are polluted by domestic sewage and wastewater. The emitted chemicals are sometimes trapped in clouds and then contaminate the ground in rainfall. Thus, we are exposed to many chemicals in our daily life, and some environmental chemicals can cause serious developmental toxicity effects. Developmental toxicity is either a structural or functional alteration, and these alterations interfere with the normal developmental programming in early embryos. These interferences can cause abnormal development and diseases (Baccarelli & Bollati, 2009; Hou et al., 2012). One of the most infamous environmental chemicals is methylmercury, which is known to affect fetal development (Yuan, 2012; Tatsuta et al., 2012). Furthermore, other chemicals are also considered to be toxic, since they can cause abnormal cell differentiation in embryos (Rappolee et al., 2012; He et al., 2012; Harrill et al., 2011).

To prevent chemically-induced developmental abnormalities and diseases, it is important to clarify the mechanisms of chemical toxicity in embryonic cells (Gündel et al., 2007; Thompson & Bannigan, 2008). The gene regulatory network is one of the useful methods to clarify the regulatory mechanisms. To infer the networks among the genes from the mRNA levels, various algorithms, including Boolean and Bayesian networks, have been developed (Akutsu et al., 2000; Friedman et al., 2000). In our previous investigation, we developed an approach based on graphical Gaussian modeling (GGM) in combination with hierarchical clustering, and we could infer the huge network among all of the genes by this approach. (Aburatani et al., 2003; Aburatani &

Horimoto, 2005). However, GGM infers only the undirected graph, whereas the Boolean and Bayesian models infer the directed graph, which shows causality. Although all of these approaches are suitable for establishing the relationships among the genes, they cannot reveal the relationships between un-observed factors and genes, because of insufficient information in the gene expression profiles. To clarify the mechanisms of biological processes in living cells, un-observed factors, which affect the target gene's expression, should also be considered. Thus, an alternative approach that includes un-observed factors should be applied.

Recently, we developed a new statistical approach based on Structural Equation Modeling (SEM), to infer the protein-DNA interactions for gene transcriptional control from only the gene expression profiles, in the absence of protein information (Aburatani, 2011; 2012). We applied this approach to reveal the causalities within the well-studied transcriptional regulation system in yeast (Aburatani, 2011). The significant features of SEM are the inclusion of latent variables within the constructed model and the ability to infer the network, including the cycle structure. Furthermore, the SEM approach allows us to strictly evaluate the inferred model, by using fitting scores. The linear relationships between variables are assumed to minimize the differences between the model's covariance matrix and the calculated sample covariance matrix. Some fitting indices are defined for evaluating the model adaptability, and thus the most suitable model can be selected by SEM (Bollen, 1989; Duncan, 1975; Pearl, 2001).

Here, we applied the SEM approach to infer the regulatory relationships among 9 neurodevelopmentally-related genes. The expression profiles of these genes were measured in human embryonic stem cells exposed to four environmental chemicals. The chemicals are known to have harmful toxicities that affect the developmental process in human embryos. Thus, inferring the regulatory network among the developmentally-related genes will help us to reveal the mechanisms of toxicity-dependent responses in the embryo. Furthermore, we improved our SEM approach for assuming preliminary initial models from the time-series data. By using this new approach, we can construct an initial model for the SEM calculation in the absence of known regulatory interactions. The resulting gene expression data clarified the chemical-specific interactions among the developmentally-related genes.

2. Methods

2.1 Expression Data

We utilized the expression data that were measured to clarify the effects of environmental chemical exposure on neuronal differentiation (He et al., 2012; Fujibuchi et al., 2011). In these expression data, nine genes considered to be affected by chemicals were measured in human embryonic stem cells: GATA2, Lmx1A, MAP2, Nanog, Nestin, Nodal, Oct3/4, Pax6 and Tuj1 (He et al., 2012; Fujibuchi et al., 2011). The expression of beta-actin was also measured, as an internal control. The expression levels of these 10 genes were measured in human embryonic stem cells exposed to four chemicals: acrylamide, diethylnitrosamine, TCDD and thalidomide (He et al., 2012; Fujibuchi et al., 2011). The toxicities of these chemicals are different: acrylamide is neurotoxic, diethylnitrosamine is genotoxic, TCDD is carcinogenic, and thalidomide has other toxicity. The human embryonic cells were exposed to each chemical for several time periods: 24 hours, 48 hours, 72 hours and 96 hours. Each chemical was also tested at 5 concentrations: very low, low, medium, high and very high. The expression of the selected genes was measured twice under each condition by RT-PCR, and thus 160 (4 time periods x 5 concentrations x 2 repeats x 4 chemicals) expression patterns per gene were measured (Fujibuchi et al., 2011).

First, the expression level of each gene was normalized to the internal beta-actin control and averaged, as follows:

$$E_g = \frac{1}{N} \sum_{i=1}^{N} \log_2 \left(\frac{e_g^i}{e_{bActin}^i} \right) \tag{1}$$

Here, N is the number of repeated experiments, e_g^i is the measured expression level of gene g under one set of

conditions, and e_{bActin}^i is the beta-actin expression level measured under the same conditions. By dividing by

the expression level of beta-actin, the intracellular expression level of each gene was normalized. To minimize the experimental error, the logarithms of the normalized expression data were obtained and averaged.

2.2 Extraction of Causalities from Expression Data

Usually, we assume an initial model from previous knowledge for the SEM calculation, but there are no defined regulations among the selected genes in this study. Thus, we had to construct an initial model of each chemical from the regulatory relationships between the gene pairs. To detect the regulatory relationships from the measured time series expression data, cross correlation coefficients were applied to the expression profiles. These cross correlation coefficients were calculated for each chemical and each concentration. Cross correlation is utilized as a measure of similarity between two waves in signal processing by a time-lag application, and it is also applicable to pattern recognition (Li & Caldwell, 1999). In a time series analysis, the cross correlation between two time series describes the normalized cross covariance function. Therefore, the range of cross correlation values is from -1 to +1. If we let $X_t = \{x_1, \cdots, x_N\}$, $Y_t = \{y_1, \cdots, y_N\}$ represent two time series datasets including N time points, then the cross correlation is given by

$$r_{xy} = \frac{\sum_{t=1}^{N}\{x_i - \overline{x}\}\{y_{i+d} - \overline{y}\}}{\sqrt{\sum_{t=1}^{N}\{x_i - \overline{x}\}^2}\sqrt{\sum_{t=1}^{N}\{y_{i+d} - \overline{y}\}^2}} \qquad (2)$$

where d is the time-lag between variables X and Y. In this case, the expression profiles were measured at four time points, and thus three cross correlations of each gene pair were calculated with d=-1, 0, and 1.

2.3 Construction of the Initial Models

To infer the chemical-dependent regulatory networks, the differences between times and concentrations should be merged. In this study, we developed a new method for constructing an initial model of each chemical, with the merging of time and concentration conditions. Figure 1 shows the newly developed method. First, we constructed lag matrices to simplify the information from the time series data. The elements of the lag matrices were the time lags, which were defined for the calculation of the cross correlation. In this study, cross correlations were calculated with three lags, -1, 0, and +1. The absolute values of these three cross correlations were compared, and the lag value d with the highest absolute value was arranged as a matrix element. Lag matrices were constructed for each concentration, and thus five lag matrices were obtained for each chemical (Figure 1a).

In the next step, we merged the difference in the concentrations of each chemical. Binomial relationships were extracted from each lag matrix. For each chemical, there are five lag matrices according to the chemical concentration, and we considered that the chemical-specific relationships among the genes will be conserved in several lag matrices. If the same relationships existed in several lag matrices, then the binomial relationships were duplicated (Figure 1b).

We subsequently constructed one frequency matrix of binary relationships for each chemical. We counted the frequency of the appearance of relationships in binomial relationships. The number representing the frequency of each gene pair was arranged in this matrix, and thus the range was from 0 to 5 (Figure 1c). In the frequency matrix, we can merge the differences in the concentrations, since the elements of the frequency matrix indicate the information for the different concentrations. We selected the possible relationships from the frequency matrix. It is considered that a possible relationship would be indicated by its frequency of appearance. Thus, we selected the relationships with two or more values in the frequency matrix (Figure 1d). At the final step, an initial model was constructed with the selected possible relationships. By this approach, an initial model can include cyclic structures (Figure 1e).

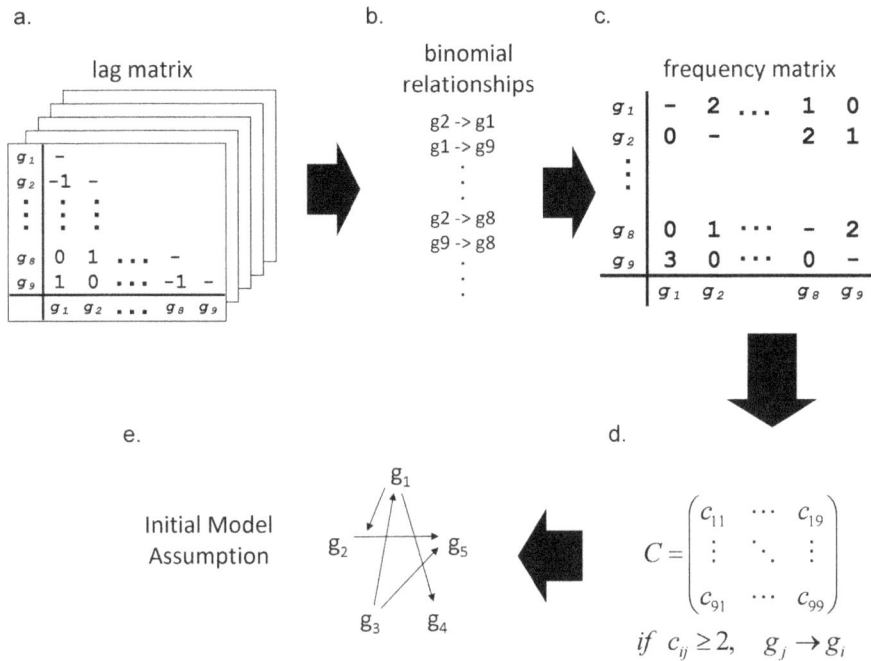

Figure 1. Developed procedure for initial model construction

The procedure for constructing an initial model from the time-lag information of the cross correlation coefficients. (a) Time-lag matrices for each chemical. In this study, three time-lags were selected for the calculation of the cross correlation coefficients. Thus, three cross correlation coefficient values were obtained between all gene pairs. The time-lag value with the highest absolute value among the cross correlation coefficients was selected. Time-lag matrices were constructed for each concentration, so five time-lag matrices were obtained for each chemical. (b) Binomial relationships. These relationships were extracted from the five time-lag matrices. If the same relationships exist in several concentration matrices, then the extracted binomial relationships are duplicated in this step. (c) Frequency matrix of causal relationships between all gene pairs. From the binomial relationship, we can count the frequency of relationships between gene pairs. (d) Selection of possible causal relationships from the frequency matrix. The possible relationships between genes are considered to persist at several chemical concentrations. Thus, we selected the relationships with two or more values in the frequency matrix. (e) Construction of an initial model with selected causal relationships. By this approach, an initial model can include cyclic structures.

2.4 Structural Equation Modeling without Latent Variables (SEM without LV)

After the construction of an initial model for each chemical, we applied the SEM calculation to infer the network model that fit the measured expression data. Usually, two types of variables can be included in the SEM model: observed and latent. These variables constitute the structural models that consider the relationships between the latent variables and the measurement models that consider the relationships between the observed variables and the latent variables. These relationships can be presented both algebraically, as a system of equations, and graphically, as path diagrams.

In this study, the nine developmentally-related genes (GATA2, Lmx1A, MAP2, Nanog, Nestin, Nodal, Oct3/4, Pax6 and Tuj1) were defined as the observed variables. Meanwhile, none were defined as latent variables, which were common regulators of several genes. The un-observed factor, which affected each gene's expression, was displayed as an error. The observed variables were classified as one of two types: exogenous variables and endogenous variables. Exogenous variables are not regulated by other variables in the system, as opposed to endogenous variables, which are regulated by other variables in the system. In the initial model, the starting genes are defined as exogenous variables without errors, while all other genes are defined as endogenous variables with errors. We inferred the regulatory relationships that exist between the observed variables in the network model. The model is defined as follows:

$$y = \Lambda y + \varepsilon \qquad (3)$$

Here, y is a vector of p observed variables (measured gene expression patterns), and Λ is a $p \times p$ matrix representing the regulatory relationships between the observed variables. Errors that affect the observed endogenous variables are denoted by ε. The above equation can be represented in the SEM matrix format as:

$$\begin{bmatrix} O \\ y \end{bmatrix} = \begin{bmatrix} O & O \\ O & \Pi \end{bmatrix} \begin{bmatrix} O \\ y \end{bmatrix} + \begin{bmatrix} O \\ \varepsilon \end{bmatrix} \tag{4}$$

In this study, we did not define the latent variables, and thus Os were arranged as zero partial matrices, which denote no relationships with q latent variables. The SEM is based on a covariance analysis defined as $S = \Sigma(\theta)$, where S is the covariance matrix calculated from the observed data and $\Sigma(\theta)$ is the matrix-valued function of the parameter θ. Let Φ denote the covariance matrix of the error terms ε, and G denote the $p \times (p + q)$ combined matrix of the $p \times q$ zero matrix and the $p \times p$ identity matrix. The covariance matrix of model is given by

$$\Sigma(\theta) = G \begin{bmatrix} I - O & O \\ O & I - \Pi \end{bmatrix}^{-1} \Phi \begin{bmatrix} I - O & O \\ O & I - \Pi \end{bmatrix}^{-1'} G' \tag{5}$$

Each element of the covariance matrix model is expressed as a function of the parameters that appear in the model. The unknown parameters were estimated, in order to minimize the difference between the model covariance matrix and the sample covariance.

The SEM software package SPSS AMOS 17.0 (IBM, USA) was used to fit the model to the data. The quality of the fit was estimated by the goodness-of-fit index (GFI), which measures the relative discrepancy between the empirical data and the inferred model, and the adjusted GFI (AGFI), which is the GFI modified according to the degrees of freedom. Furthermore, we used CFI and RMSEA as fitting scores, to evaluate the model fitting. Since these indices have threshold values, as criteria to decide whether the model is suitable to obtain data independent of a huge sample number, they are considered to be useful to clarify the degree of model fitting in this study.

2.5 Parameter Estimation

Parameter estimation was performed by comparing the actual covariance matrix S, calculated from the measured data, and the estimated covariance matrices $\Sigma(\theta)$ of the constructed model. To minimize the difference between S and $\Sigma(\theta)$, the Maximum Likelihood (ML) method is commonly used as a fitting function to estimate the SEM parameters:

$$F_{ML}(S, \Sigma(\theta)) = \log|\Sigma(\theta)| - \log|S| + tr(\Sigma(\theta)^{-1} S) - p \tag{6}$$

Here, $\Sigma(\theta)$ is the estimated covariance matrix, S is the sample covariance matrix, $|\Sigma|$ is the determinant of matrix Σ, $tr(\Sigma)$ is the trace of matrix Σ, and p is the number of observed variables. The principal objective of SEM is to minimize $F_{ML}(S, \Sigma(\theta))$, which is the objective function and is used to obtain the maximum likelihood. Generally, $F_{ML}(S, \Sigma(\theta))$ is a nonlinear function. Therefore, iterative optimization is required to minimize $F_{ML}(S, \Sigma(\theta))$ and to find the solutions (Joreskog & Sorbom, 1984).

2.6 Iteration for Optimal Model

In the SEM analysis, both the parameters and network structures are fitted to the measured data. The parameters are estimated by maximum likelihood, and the network structures are evaluated by the scores of goodness of fit indices. The goodness of fit scores indicate the similarity between the constructed model and the measured data. Through the acceptance or rejection of the models, the optimal model that describes the measured data can be selected.

By using the estimated parameters, the variance-covariance matrix between the variables could be calculated in the network model. This model's variance-covariance matrix is compared with the actual variance-covariance matrix between observed variables, which is calculated from the measured data. The similarity between a constructed model and the actual data is defined in a quantitative manner by the fitting scores. In this study, four different fitting scores were utilized: GFI, AGFI, CFI and RMSEA. Values of GFI, AGFI and CFI above 0.90 are required for a good model fit. RMSEA is one of the most popular parsimony indexes displayed in the table, and RMSEA values below 0.05 represent a good model fit (Spirtes et al., 2001). Furthermore, RMSEA values of 0.10 or more are considered to indicate that the constructed model is far from the actual data. To optimize the model, we developed an iteration algorithm as follows:

Step 1: Deletion of a non-significant edge from the constructed network model

Use 0.05 as the significance level for the determination of the significant regulation among the variables. After the parameters are estimated, the inverse matrix of the Fisher information matrix of parameters is calculated. The inverse matrix of Fisher information represents the asymptotic parameters' covariance matrix. The probability of each parameter is calculated by using this asymptotic parameters' matrix, since all of the parameters are usually normally distributed.

Step 2: Reconstruction of the network model

The structure of the network model without the non-significant edge is completely different from that of the former model. Thus, all parameters should be re-calculated from the reconstructed model, and the similarity of the network structure should also be re-calculated.

Step 3: Iteration of Steps 1 and 2 until all edges become significant

Since the probabilities of all of the edges in the reconstructed models have also changed, the deletion of the non-significant edges is executed step-by-step.

Step 4: Addition of a possible causal edge to the reconstructed model

According to the Modification Index (MI), we add a new causal edge between the observed variables. The MI measures how much the chi-square statistic is expected to decrease if a particular parameter setting is constrained (Joreskog & Sorbom, 1984). The MI value indicates the possibility of new causality between the variables, and thus we add a new edge according to the highest MI score.

Step 5: Iteration from Steps 1 to 3

The addition of a new edge to a constructed model changes the structure of the network model again. In other words, all parameters, including the probabilities of all edges, have also changed. Thus, we execute the iteration from Step 1 to Step 3 again.

Step 6: Determination of significant relationships among error terms

After all of the edges are significant and all of the MI scores are lower than 10.0 in the constructed model, the significant relationships between the error terms are estimated by the MI scores. The relationships among the error terms have no direction, and thus they are a correlation between error terms. The relationships between the error terms were considered to be other regulatory systems in the living cell. Thus, these relationships among the error terms were used for the calculations, but were not incorporated into the network, and thus they have been excluded from the figures.

3. Results

3.1 Initial Model Assumption

To construct the initial network model of each chemical, we utilized our newly developed method. One of the distinguishing features of our new method is its ability to include the cyclic structure in the network model. Cyclic regulation, such as feedback regulation, is considered to be important for living cells to control normal gene expression, and the new method is useful to detect the cyclic regulation from the gene expression data. The initially constructed models are shown in Figure 2. The initial model of TCDD was the most complex structure. The components of the constructed models were 9 genes with 19 relationships in Acrylamide, 8 genes with 12 relationships in Diethylnitrosamine, 9 genes with 23 relationships in TCDD, and 8 genes with 10 relationships in Thalidomide.

There are some obvious features in the network diagram of each initial model. The numbers of exogenous and endogenous genes are different from each other. In the initial Acrylamide model, four genes were arranged as exogenous variables, but only Oct3/4 was arranged as the last endogenous variable. Thus, it is considered that acrylamide quickly affected the expression of many genes, and only one gene was affected later. In contrast, only one gene was arranged as an exogenous variable and many genes were arranged as the last endogenous variables in the initial Thalidomide model. These differences between the initial chemical models summarized the distinctive gene expression profiles for each chemical. The initial TCDD model involved some cyclic regulation, even though the other models had only hierarchical regulation.

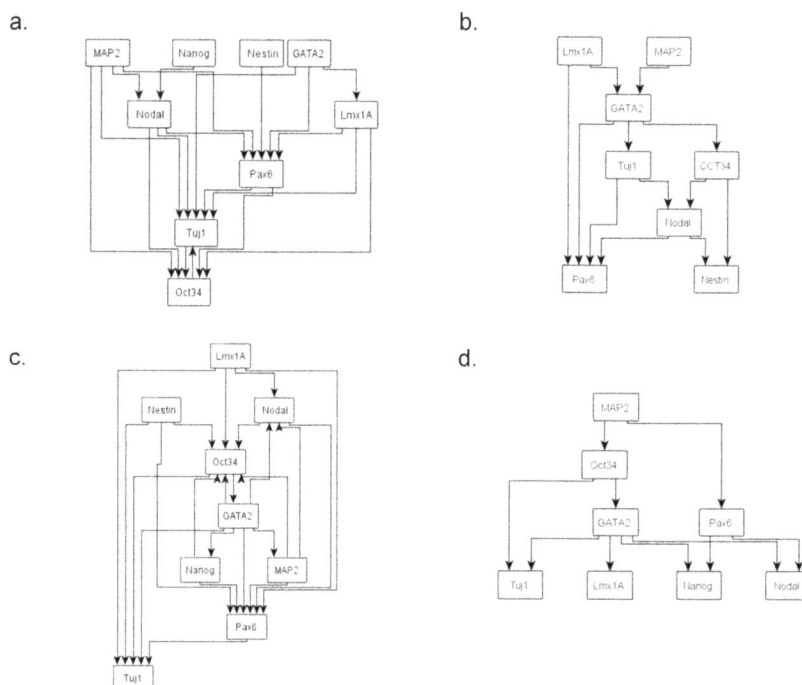

Figure 2. Initial network models

The initial models of the selected chemicals were constructed by the developed approach. One initial model was constructed for each chemical, since the initial model included summarized time-series information and concentration information. (a) Initial model constructed from all gene expression profiles with all Acrylamide exposure. (b) Initial model of Diethylnitrosamine. (c) Initial model of TCDD. (d) Initial model of Thalidomide. The numbers of genes in the initial models were 9 in Acrylamide, 8 in Diethylnitrosamine, 9 TCDD, and 8 in Thalidomide.

Before the calculation of SEM, all of the initial models were simplified, since the initial models included some duplicated interactions among the genes, such as direct interactions between two genes and indirect interactions between them. In the simplification process for the initial models, the longest path between two genes was retained, since the arrows indicated only time precedence, not causalities in the initial model. Therefore, the difference between direct and indirect interactions is not important. By retaining the longest paths, all of the preceding information was included, as the simplest diagram.

3.2 Inferred Networks by SEM

The final inferred networks for each chemical and the goodness of fit scores are depicted in Figure 3, and the estimated regression weights of the edges are displayed in Table 1. The inferred networks of the chemicals revealed distinct structures. The differences between the gene regulation by chemicals were clarified by the shapes of the inferred network models. The Acrylamide network was a centralized model, the Diethylnitrosamine network was a ladder-like model, the TCDD network was a closed circular structure, and the Thalidomide network was a diffusion type.

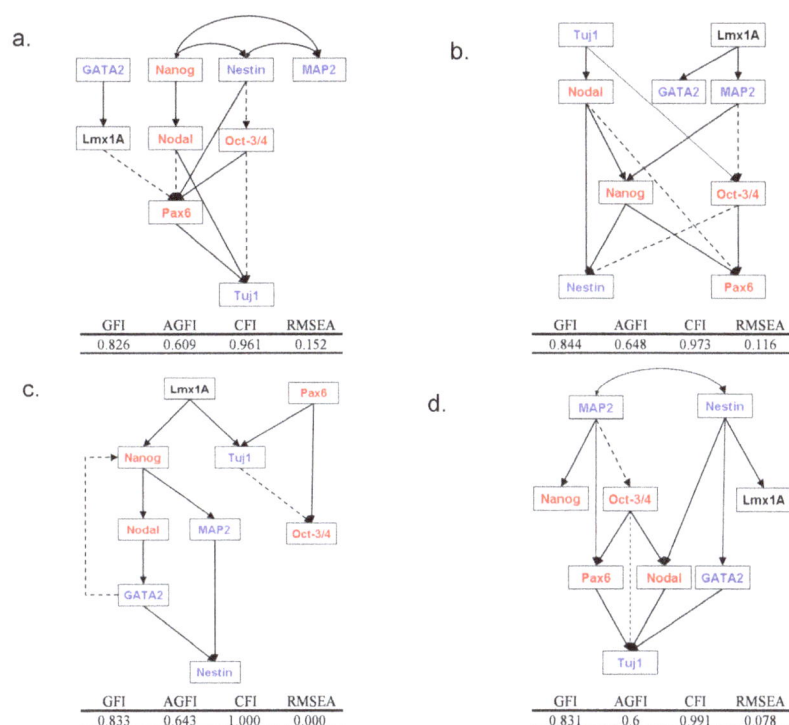

Figure 3. Inferred Toxic-dependent Networks

The optimal model for each chemical, obtained by the developed SEM iteration procedure. A positive relationship between genes is displayed with a solid arrow. A negative relationship between genes is displayed with a dashed arrow. Gene names with blue characters indicate "neurodevelopment related genes", genes with red characters indicate "cell differentiation-related genes" and genes with black characters indicate "related to transcription of insulin". (a) Acrylamide model; (b) Diethylnitrosamine model; (c) TCDD model and (d) Thalidomide model. The fitting scores are displayed under each model.

One of the unique features of the inferred Acrylamide network was that many genes were arranged at the top phase in the regulatory network, and only one gene was arranged as the final result of all regulation in the network. On the other hand, the shape of the Diethylnitrosamine network looked like a ladder, and two serial regulations interacted with each other. One serial regulation started from Lmx1A, and the other started from Tuj1. These top phase genes were considered as signal input genes, and they were different from those in the Acrylamide and Thalidomide networks. For example, Tuj1 was arranged as a signal input gene in the Diethylnitrosamine network, but it was arranged as an output object in the Acrylamide and Thalidomide networks. The unique feature of the TCDD network is the involvement of some closed circular structures in the inferred model. Among the parts of the circular structure, the regulatory direction from GATA2 to Nodal was different from the other relationships. Furthermore, the regression weight between GATA2 and Nodal was estimated as a negative value. Thus, it was considered that the inferred regulation from GATA2 to Nodal reflected feedback control by GATA2. In the Thalidomide network, the shape of the network model was reversed, as compared to that of the Acrylamide network. Only two genes were arranged at the top phase in the regulatory network, but many genes were arranged at the middle phase in the model. This means that only a few genes are directly affected by thalidomide, but finally many genes are affected throughout the gene regulatory network.

Table 1. Regression weight and probability of each edge

Acrylamide				Diethylnitrosamine			
Parent	Child	Regression Weight	P	Parent	Child	Regression Weight	P
GATA2	Lmx1A	0.921	***	Tuj1	Nodal	0.702	***
Nanog	Nodal	0.522	0.003	Lmx1A	MAP2	0.378	0.003
Nestin	Oct-34	-0.437	0.01	Tuj1	Oct-34	0.63	***
Nestin	Pax6	0.64	***	MAP2	Oct-34	-0.475	***
Nodal	Pax6	-0.803	***	Nodal	Nanog	0.295	***
Lmx1A	Pax6	-0.232	***	MAP2	Nanog	0.754	***
Oct-34	Pax6	0.592	***	Lmx1A	GATA2	0.636	***
Nodal	Tuj1	0.843	***	Nodal	Nestin	0.33	***
Pax6	Tuj1	1.09	***	Nodal	Pax6	-0.209	0.005
Oct-34	Tuj1	-0.702	***	Nanog	Pax6	0.418	***
				Nanog	Nestin	0.902	***
				Oct-34	Pax6	1.11	***
				Oct-34	Nestin	-0.193	***

TCDD				Thalidomide			
Parent	Child	Regression Weight	P	Parent	Child	Regression Weight	P
GATA2	Nanog	-0.787	***	MAP2	Oct-34	-0.443	0.023
GATA2	Nestin	0.22	***	MAP2	Pax6	0.349	0.005
Lmx1A	Nanog	1.374	***	Nestin	Nodal	1.03	***
Lmx1A	Tuj1	0.476	0.004	Nestin	GATA2	0.664	***
MAP2	Nestin	0.906	***	Oct-34	Pax6	0.932	***
Nanog	MAP2	1.024	***	Oct-34	Nodal	0.258	***
Nanog	Nodal	0.967	***	Oct-34	Tuj1	-0.597	***
Nodal	GATA2	0.931	***	Pax6	Tuj1	1.12	***
Pax6	Oct-34	0.988	***	Nodal	Tuj1	0.349	***
Pax6	Tuj1	0.5	0.003	GATA2	Tuj1	0.167	0.015
Tuj1	Oct-34	-0.324	***	MAP2	Nanog	0.84	***
				Nestin	Lmx1A	0.842	***
				Tuj1	Nanog	0.196	0.002

4. Discussion

Our inferred model revealed the differences between the gene regulation by environmental chemicals. Furthermore, the shapes of the network models reflected the different features of the chemical toxicities well. In the Acrylamide network, the effects of acrylamide toxicity finally aggregated to Tuj1, which is known to contribute to microtubule stability in neuronal cells (Rosenstein et al., 2003). Acrylamide is neurotoxic, and thus it is reasonable that its effect finally aggregated to a neuronal cell-related gene.

In the Diethylnitrosamine network, the cell differentiation genes were arranged from the middle to lower steps. This means that diethylnitrosamine disturbed normal cell differentiation in the embryonic stem cell. These harmful effects were considered to be caused by the carcinogenic genotoxicity of diethylnitrosamine (Ito et al., 1992; Puatanachokchai et al., 2006; Iatropoulos et al., 2006). On the other hand, the neuronal-related genes were arranged at a later phase in the TCDD network model. Although both diethylnitrosamine and TCDD have the same carcinogenic toxicities, their regulatory mechanisms were different.

From the Thalidomide network, it was considered that the receptors of thalidomide toxicity may be rarer than those of other chemicals. However, several types of genes are finally affected by thalidomide chemical toxicity. Among the cell differentiation genes, Nodal and Nanog are important for normal early embryonic development. Nodal is related to the development of the left-right axial structure (Hamada et al., 2002; Grandel & Patel, 2009), and its signaling pathway is important very early in development, for cell fate determination and many other developmental processes (Grandel & Patel, 2009). Nanog is a key factor for maintaining pluripotency in embryonic stem cells (Mitsui, 2003; Chambers et al., 2003). According to the abnormal expression of these cell differentiation-related genes, the thalidomide phenotype, with its harmful side effects, may occur in the human embryo.

We applied an improved SEM approach to reconstruct a gene regulatory model from the gene expression data in human embryonic stem cells, and we have shown that SEM is a powerful approach to estimate the gene regulation caused by chemical toxicity. The inferred networks clarified the differences between the gene regulation by chemicals, and the features of the chemical toxicities were well reflected in the network structures. Thus, the network construction by SEM is one of the useful approaches for inferring the regulatory relationships among genes. Furthermore, the inferred network among genes can be utilized for the estimation of a chemical's effect, from experimentally obtained expression profiles. The ability to identify expression profiles and the corresponding biological functions is expected to provide further possibilities for SEM in the inference of the effects of chemical toxicity on regulatory mechanisms.

Acknowledgements

We would like to express our gratitude to Dr. Yamane (Kyoto Univ.) and Dr. Ohsako (Univ. of Tokyo) for providing the expression profiles in human embryonic stem cells exposed to 15 chemicals. Their support was quite valuable for this investigation.

References

Aburatani, S. (2012). Network Inference of pal-1 Lineage-Specific Regulation in the C. elegans Embryo by Structural Equation Modeling. *Bioinformation, 8*(14), 652-657. http://dx.doi.org/10.6026/97320630008652

Aburatani, S., & Horimoto, K. (2005). Elucidation of the Relationships between LexA-Regulated Genes in the SOS response. *Genome Informatics, 16*(1), 95-105.

Aburatani, S., Kuhara, S., Toh, H., & Horimoto, K., (2003). Deduction of a gene regulatory relationship framework from gene expression data by the application of graphical Gaussian modeling. *Signal Processing, 83*, 777-788. http://dx.doi.org/10.1016/S0165-1684(02)00476-0

Aburatani. S. (2011). Application of structure equation modeling for inferring a serial transcriptional regulation in yeast. *Gene. Regul. Syst. Bio., 5*, 75-88. http://dx.doi.org/10.4137/GRSB.S7569

Akutsu, T., Miyano, S., & Kuhara, S. (2000). Algorithms for identifying Boolean networks and related biological networks based on matrix multiplication and fingerprint function. *J. Comput. Biol., 7*, 331-343. http://dx.doi.org/ 10.1145/332306.332317

Baccarelli, A., & Bollati, V. (2009). Epigenetics and environmental chemicals. *Curr. Opin. Pediatr., 21*(2), 243-251. http://dx.doi.org/10.1097/MOP.0b013e32832925cc

Bollen, K. A. (1989). *Structural Equations with Latent Variables*. New York: Wiley-Interscience.

Chambers, I., Colby, D., Robertson, M., Nichols, J., Lee, S., Tweedie, S., & Smith, A. (2003) Functional expression cloning of Nanog, a pluripotency sustaining factor in embryonic stem cells. *Cell, 113*(5), 643-655. http://dx.doi.org/10.1016/S0092-8674(03)00392-1

Duncan, O. D. (1975). *Introduction to Structural Equation Models*. New York: Academic Press.

Friedman, N., Linial, M., Nachman, I., & Pe'er, D. (2000). Using Bayesian networks to analyze expression data. *J. Comput. Biol., 7*, 601-620. http://dx.doi.org/10.1145/332306.332355

Fujibuchi, W. (2011). Prediction of Chemical Toxicity by Network-based SVM on ES-cell Validation System. *The Proc. of the 2011 Joint Conference of CBI-Society and JSBi.*

Grandel, C., & Patel, N. H. (2009). Nodal signaling is involved in left-right asymmetry in snails. *Nature, 457*(7232), 1007-1011. http://dx.doi.org/10.1038/nature07603

Gündel, U., Benndorf, D., Bergen, M., Altenburger, R., & Küster, E. (2007). Vitellogenin cleavage products as indicators for toxic stress in zebra fish embryos: A proteomic approach. *Proteomics, 7*(24), 4541-4554. http://dx.doi.org/10.1002/pmic.200700381

Hamada, H., Meno, C., Watanabe, D., & Saijoh, Y. (2002). Establishment of vertebrate left-right asymmetry. *Nat. Rev. Genet., 3*(2), 103-113. http://dx.doi.org/10.1038/nrg732

Harrill, J. A., Robinette, B. L., & Mundy, W. R. (2011). Use of high content image analysis to detect chemical-induced changes in synaptogenesis in vitro. *Toxicol. In. Vitro., 25*(1), 368-387. http://dx.doi.org/10.1016/j.tiv.2010.10.011

He, X., Imanishi, S., Sone, H., Nagano, R., Qin, X-Y., Yoshinaga, J., ... Ohsako, S. (2012). Effects of methylmercury exposure on neuronal differentiation of mouse and human embryonic stem cells. *Toxicol. Lett., 212*, 1-10. http://dx.doi.org/10.1016/j.toxlet.2012.04.011

Hou, L., Zhang, X., Wang, D., & Baccarelli, A. (2012). Environmental chemical exposures and human epigenetics. *Int. J. Epidemiol., 41*(1), 79-105. http://dx.doi.org/10.1093/ije/dyr154

Iatropoulos, M. J., Wang, C. X., Keutz, K. E., & Williams, G. M. (2006). Assessment of chronic toxicity and carcinogenicity in an accelerated cancer bioassay in rats of Nifurtimox, an antitrypanosomiasis drug. *Exp. Toxicol. Pathol., 57*(5-6), 397-404. http://dx.doi.org/10.1016/j.etp.2006.01.005

Ito, N., Hasegawa, R., Imaida, K., Masui, T., Takahashi, S., & Shirai, T. (1992). Pathological markers for non-genotoxic agent-associated carcinogenesis. *Toxicol. Lett., 64-65*, 613-620. http://dx.doi.org/10.1016/0378-4274(92)90239-G

Joreskog, K. G., & Sorbom, D. (1984). LISREL-VI: Analysis of Linear Structural Relationships By the Method of Maximum Likelihood. Redondo Beach: Doss-Haus Books.

Li, L., & Caldwell, G. E. (1999). Coefficient of cross correlation and the time domain correspondence. *J. of Electromyography and Kinesiology, 9*, 385-389. http://dx.doi.org/10.1016/S1050-6411(99)00012-7

Mitsui, K. (2003). The homeoprotein Nanog is required for maintenance of pluripotency in mouse epiblast and ES cells. *Cell, 113*(5), 631-642. http://dx.doi.org/10.1016/S0092-8674(03)00393-3

Pearl, J. (2001). *Causality: Models, Reasoning, and Inference* (2nd ed.). Cambridge: Cambridge University Press.

Puatanachokchai, R., Kakuni, M., Wanibuchi, H., Kinoshita, A., Kang, J. S., Salim, E. I., … Fukushima, S. (2006). Lack of promoting effects of phenobarbital at low dose on diethylnitrosamine-induced hepatocarcinogenesis in TGF-alpha transgenic mice. *Asian Pac. J. Cancer Prev., 7*(2), 274-278.

Rappolee, D. A., Xie, Y., Slater, J. A., Zhou, S., & Puscheck, E. E. (2012). Toxic stress prioritizes and imbalances stem cell differentiation: implications for new biomarkers and in vitro toxicology tests. *Syst. Biol. Reprod. Med., 58*(1), 33-40. http://dx.doi.org/10.3109/19396368.2011.647381

Rosenstein, J. M., Mani, N., Khaibullina, A., & Krum, J. M. (2003). Neurotrophic effects of vascular endothelial growth factor on organotypic cortical explants and primary cortical neurons. *J. Neurosci., 23*(35), 11036-11044.

Spirtes, P., Glymour, C., & Scheines, R. (2001). *Causation, Prediction, and Search* (2nd ed.). Cambridge: The MIT Press.

Tatsuta, N., Nakai, K., Murata, K., Suzuki, K., Iwai-Shimada, M., Yaginuma-Sakurai, K., … Satoh, H. (2012). Prenatal exposures to environmental chemicals and birth order as risk factors for child behavior problems. *Environ. Res., 114*, 47-52. http://dx.doi.org/10.1016/j.envres.2012.02.001

Thompson, J., & Bannigan, J. (2008). Cadmium: toxic effects on the reproductive system and the embryo. *Reprod. Toxicol., 25*(3), 304-315. http://dx.doi.org/10.1016/j.reprotox.2008.02.001

Yuan, Y. (2012). Methylmercury: a potential environmental risk factor contributing to epileptogenesis. *Neurotoxicology, 33*(1), 119-126. http://dx.doi.org/10.1016/j.neuro.2011.12.014

7

Identifying Significant Biological Markers in Klotho Gene Variants Across Wide Ranging Taxonomy

Tommy Rodriguez[1]

[1] Pangaea Biosciences, Department of Research & Development, Miami, FL, USA

Correspondence: Tommy Rodriguez, Pangaea Biosciences, 1020 North Lakemont Ave., Winter Park, FL 32792, USA. E-mail: trodriguez@pangaeabio.com

Abstract

Biological aging is marked by progressively degenerative physiological change that causes damage to tissues and organs. Errors in biopolymers accumulate over time; mitochondrial dysfunction, telomere attrition, and wider genomic instability lead to an altered state of intercellular communication. In this investigation, my focus will be aimed at examining and identifying specifically critical biomarkers in genetic variants of KLOTHO (a transmembrane protein involved in the genetic regulation of age-related disease) among organisms with varied life spans that range across wide taxonomical rankings. Here, I investigate the correlation between lower and higher frequency α-amino acid compositions in Klotho protein factors within a grouped methodology; as to also include several demonstrative techniques in comparative sequence analysis for inferring relatedness in evolutionary context.

Keywords: Klotho, KL gene, protein, protein evolution, comparative sequence analysis, multiple sequence alignment, Kalign, phylogenetc analysis, amino acid composition analysis

1. Introduction

Promising new research in life extension has led to remarkable advancements in our basic understanding of the molecular mechanisms associated with aging. Within the last two decades, much of this research has focused on identifying potential genomic candidates for longevity, while seeking to explain how these individual genes can affect the biological process of aging. One of the proteins in particular (KLOTHO) has been the subject of several recently published papers. A handful of studies now suggest that genetic variants of KLOTHO [encoded by the KL gene] are associated with human aging and tumor suppression, and trials on model organisms involving KLOTHO variants has shown improvement in cognition and deceleration in age-related development (Dubal et al., 2014). In one such case, Dubal et al. (2014) demonstrated that systemic overexpression of Klotho variants in transgenic mice enhanced cognition and increased longevity by an average ratio of ±25 percent; whereas Klotho-deficient mice manifested a syndrome resembling accelerated human aging and displayed extensive and accelerated arteriosclerosis (Dubal et al., 2014).

For reasons that are not yet fully understood, Klotho-associated mechanisms, "change cellular calcium homeostasis, by both increasing the expression and activity of TRPV5 and decreasing that of TRPC6 (Kurosu et al., 2005)." Moreover, altered mineral-ion homeostasis could be a cause of premature aging‑like phenotypes (Kurosu et al., 2005). In order to gain a more comprehensive understanding of the underlying functions in molecular components of KLOTHO, we should begin by examining KL-derivative patterns across a wide evolutionary spectrum, without limiting ourselves to one individual taxa or another. Because the lifespan of organisms vary widely among species, a comparative approach could help us identify a set of unique signatures in the molecular variation patterns of KLOTHO. In turn, this may help provide meaningful reference toward a fully systematic investigation. Throughout the course of this paper, I seek to address those areas thoroughly.

It should then be noted: This paper does not presume on a solution to longevity; nor does it seek to draw parallels between the many collective biological features that allow organisms to persist in semi-optimal states. Instead, my focus will be aimed at identifying and examining specifically critical biomarkers, or sequence motifs, in genetic variants of KLOTHO among organisms with varied life spans that range across wide taxonomical rankings. Here, I investigate the correlation between lower and higher frequency α-amino acid compositions in Klotho protein factors within a grouped methodology; as to also include several demonstrative techniques in comparative sequence analysis for inferring relatedness in evolutionary context.

2. Methods

2.1 Sequence Selection and Group Categorization

Biological aging is marked by progressively degenerative physiological change that causes [irreversible] damage to tissues and organs. Errors in biopolymers accumulate over time; mitochondrial dysfunction, telomere attrition, and wider genomic instability lead to an altered state of intercellular communication. Within a broader spectrum, degrees of longevity vary widely among organisms. Only a tiny percentage of species exhibit characteristics that is counter-intuitive to universal mechanisms of biological aging. *Turritopsis dohrnii* and *Polycelis feline* are two such examples of eukaryotic organisms that appear to exhibit, "limitless telomere regenerative capacity fueled by a population of highly proliferative adult stem cells (Tan et al., 2012)." Several other species of marine Crustaceans share similar traits that prolong biological aging.

Individual-scale longevity is largely determined by inheritance; genes can explain up to 35 percent (American Federation for Aging Research, 2012). As noted earlier, recent studies indicate that Klotho gene variants may be potential candidates for the genetic regulation of age-related disease. For this purpose, I selected the KL gene as a prime candidate for analysis and evaluation. Furthermore, using homologous variants of Klotho protein sequences as a primary source for comparative studies, I grouped and categorized six distinct genomic datasets of KLOTHO [in FASTA format] by the following criteria:

i.   Group *type a* { *Heterocephalus glaber*; *Chelonia mydas*; *Danio rerio*; } – consisting of organisms that are recognized as having favorable biological attributes in one or more of the following areas: life cycle duration, tumor-suppression, cancer-resistance, immunity, tissue regeneration.

ii.   Group *type b* { *Homo sapiens*; *Mus musculus*; *Mesocricetus auratus*; } – consisting of organisms that reside within a moderate range of natural life expectancy due to age-related decline, disease, and predispositions.

The six protein sequences used for curating the experimental dataset were obtained via NCBI Protein database. The accession references for each individual sequence are detailed in the sections below.

2.1.1 Regarding group *type a*

Two of three *type a* organisms exhibit extraordinarily prolonged life cycles. *Heterocephalus glaber* are highly resistant to cancer, and can maintain a youthful vascular function and cellular oxidant-antioxidant phenotype relatively longer and are better protected against aging-induced oxidative stress than shorter-living rodents (Buffenstein, 2009; Csiszar et al., 2007; Pérez et al., 2009). *Danio rerio* have shown the ability to regenerate their fins, skin, and heart (Jopling, 2010; Sun et al., 2009). Jopling et al. (2010) demonstrated that [zebrafish] can fully regenerate its heart after amputation of up to 20 percent of the ventricle. Although very little is known about the average lifespan of green sea turtles, *Chelonia mydas* reach sexual maturity between 20 to 50 years of age (United States Fish and Wildlife Service, 2005).

2.2 Arriving at Reliable Models for Comparative Analysis

2.2.1 Performing Pairwise Sequence Alignment

Following group categorization, I deployed UGENE's pairwise sequence alignment (PSA) tool in order to identify similarity probabilities among my six protein sequences [length: 1,037]. Initially, I produced five pairwise sequence alignments by pairing each Klotho protein sequence against *Homo sapiens* [type b] and then grouping the results in numerical sequence [see below]. In this operation, Hirschberg algorithm was selected for optimal matching. A reliable algorithmic selection, Hirschberg increases efficiency by maximizing the sum of pairwise scores with quasi-gap penalties (Chao & Zhang, 2008). The Hirschberg algorithm can be derived from the Needleman-Wunsch algorithm by observing the following (Hirschberg, 1975):

if $(Z, W) = NW[\text{score}]\ (X, Y)$ is the optimal alignment of (X, Y), and $X = X^l + X^r$ is an arbitrary partition of X, there exists a partition $Y^l + Y^r$ of Y such that $NW[\text{score}]\ (X, Y) = NW[\text{score}]\ (X^l, Y^l) + NW[\text{score}]\ (X^r, Y^r)$

When calculating the highest number of consistent patterns in a local alignment, gap penalty scores are often disregarded (Lassmann & Sonnhammer, 2005). However, because I am interested in obtaining a global alignment rather than calculating common subsequences, I applied a gap open penalty score of ten; including a gap extension penalty of two. The resulting PSA provided a first glimpse into similarity probabilities. As noted below, my numerically grouped PSA outputs infer high degree homology between three of the six sequences and two of six sequences; now categorized as group *type a* and group *type b*. PSA validates sequence placement among *type a* organisms and *type b* organisms.

Table 1. Similarity probabilities among *type a* and *type b* (PSA)

Group	Identification	Similarity appx.	FASTA description
type *b*	Homo sapiens	100%	>gi|2618596|dbj|BAA23382.1
type *b*	Mus musculus	85%	>gi|2618594|dbj|BAA23381.1
type *b*	Mesocricetus auratus	85%	>gi|524966649|ref|XP_005083219.1
type *a*	Heterocephalus glaber	54%	>gi|351702474|gb|EHB05393.1
type *a*	Chelonia mydas	52%	>gi|465963328|gb|EMP30051.1
type *a*	Danio rerio	50%	>gi|528492619|ref|XP_690797.4

2.2.2 Performing Multiple Sequence Alignment

A series of multiple sequence alignment (MSA) operations were then performed in preparation for phylogenetic analysis. In contrast to pairwise alignment, multiple sequence alignment can reveal subtle similarities among *groups* of proteins (Chao & Zhang, 2008). Here, I selected Kalign for multiple sequence alignment; an accurate and fast MSA algorithm (Lassmann & Sonnhammer, 2005). Kalign is an extension of Wu-Manber approximate pattern-matching algorithm, which is based on Levenshtein distances. This strategy enables Kalign to estimate sequence distances faster and more accurately than other popular iterative methods. Comparisons done by Lassmann and Sonnhammer (2005) show that Kalign is about 10 times faster than ClustalW and, depending on the alignment size, up to 50 times faster than other iterative methods; Kalign also delivers better overall resolution (Lassmann & Sonnhammer, 2005).

Kalign is noted for producing optimal execution times, and this procedure would require minimal computational resources. First, I initiated UGENE's multiple sequence alignment tool by importing and processing the six protein sequences in FASTA format. Due to parameter setting sensitivity in protein data types, Kalign for MSA gap penalty scores were modified slightly during successive intervals until an optimal global alignment was obtained. Each interval resulted in a 3,092 base-pair alignment, followed by a phylogenetic diagram. These operations were then repeated in successive fashion, upon conducting protein to nucleotide conversions.

2.3 Reverse KLOTHO Protein-DNA Translation for Phylogenetic Reconstruction

Generally speaking, protein sequences are intolerant of change in evolutionary context. Over the span of evolutionary time, protein sequences undergo selective constraints for protein function and protein structure, and these are conserved over much longer periods than individual codons (Martin & Palumbi, 1993). The most direct evolutionary changes to protein occur at the amino and carboxyl termini in the form of domain insertions, repetitions, and deletions (Marsh & Teichmann, 2010). Likewise, we must also consider the possibility that convergent evolution can occur to produce apparent similarity between proteins that are evolutionarily unrelated, but perform similar functions and have similar structures (Bastien, 2008). Thus, multiple substitutions at a single DNA base more accurately reflect mutational history (Martin & Palumbi, 1993).

The challenges in utilizing protein sequences to infer divergence events led me to consider a secondary option to protein-protein comparison in the context of phylogenetic reconstruction. Perhaps it would be necessary to cross-check the original diagram(s) produced by protein-protein comparison with a nucleotide derivative. Using my custom-based API translator (Protein to DNA Bio Translator), I reverse translated each sequence from its original data type, namely protein, into a workable [and theoretical] dataset made exclusively of KL-derivative nucleotide sequences (DNA).

Following conversion, the six DNA sequences were imported to UGENE's bioinformatics software, and upon file import and MSA execution, an alternate radial phylogenetic diagram was generated that would allow me to cross-confirm the original(s) obtained by protein-protein comparison. I implemented PHYLIP neighbor-joining method coupled with distance matrix model F84 on the 3,092 base-pair alignment; this procedure would also require additional bootstrapping compilers to help evaluate the strength of the nodes. Lastly, it is worth noting that the resulting phylogenetic tree(s) do not assume an evolutionary clock; it is in effect an unrooted tree.

The PHYLIP neighbor-joining method is capable of generating highly probable diagrams in scenarios involving low degrees of variance, regardless of dataset size. An accurate and statically consistent polynomial-time algorithm, PHYLIP neighbor-joining does not assume that all lineages evolve at the same rate (as proteins evolve at different rates), and it constructs a tree by successive clustering of lineages, setting branch lengths as the lineages join [where a set of n taxa requires $n - 3$ iterations; each step is repeated by $(n - 1) \times (n - 1)$] (Felsenstein, 1981). For illustration purposes, the following formulas demonstrate a standard neighbor-joining Q-matrix algorithm:

$$Q(i,j) = (n-2)\, d(i,j) - \sum \{n, k = 1\}\, d(i,k) - \sum \{n, k = 1\}\, d(j,k) \tag{1}$$

Pair to node (distances):

$$(f,u) = \tfrac{1}{2}\, d(f,g) + \tfrac{1}{2}(n-2)\, [\, \sum \{n, k = 1\}\, d(f,k) - \sum \{n, k = 1\}\, d(g,k)\,] \tag{2}$$

Taxa to node (distances):

$$d(u,k) = \tfrac{1}{2}\, [\, d(f,k) + d(g,k) - d(f,g)\,] \tag{3}$$

2.4 Analyzing α-Amino Acid Compositions

Given its mass and length, the molecular weight measurements of protein structures are fundamentally important to its biochemical characterization and function. Moreover, the α-amino acid composition of each protein structure may contribute to the overall quality of a protein. If certain α-amino acids are optimal for protein structure, natural selection should have acted over evolutionary time to increase the frequency of these α-amino acids (Anthis et al., 2013). As noted by Mannakee and Gutenkunst (2012), catalytic domains in protein evolve faster, while non-catalytic domains in protein evolve more slowly. This may also suggest that networks typically evolve under stabilizing selection (Mannakee & Gutenkunst, 2012).

In addition to a phylogenetic reconstruction, this investigation puts a strong emphasis on discerning meaningful patterns, or sequence motifs, in lower and higher frequency α-amino acid compositions. By comparing the results of composition frequency percentages among group *type a* and group *type b*, I hope to identify any potentially significant molecular markers in evolutionary conserved chemical properties of Klotho that may have increased or decreased over evolutionary time, within a particular group and across a wider evolutionary spectrum. This procedure incorporates a few methods and techniques, such as α-amino acid residue calculations for determining molecular weight and frequency percentage.

A number of highly efficient web applications are well suited for α-amino acid composition analysis. In this phase of my investigation, I use Composition/Molecular Weight Calculation tool (University of Delaware, 2014) for obtaining the sum ratio of α-amino acid residues and approximate residue charge, and Protein Calculator (Anthis et al., 2013) to determine atomic compositions. I then apply the resulting figures and datasets in comparative analysis in order to evaluate the overall α-amino acid distributions.

The sum ratios in α-amino acid counts vary slightly due to irregularities in protein sequence length. This would also explain discrepancies in PSA similarity approximations. Amino acid counts are quantified by the total number of residues in an individual protein sequence. These figures are later arranged and disbursed as to determine frequency percentage within a protein sequence arrangement. The total estimate in molecular weight per arrangement is also deciphered by accounting for the total sum of individual residue weights. Lastly, my molecular weight formula integrates the following two variables: (x) the individual sum residue weight(s); (y) subtracted by the molecular weight of H_2O (18.01528) (University of Delaware, 2014).

Molecular weight = sum of individual residues weights - water molecular weight ´ (number of residues - 1)

   a) Homo sapiens ($C_{5337}H_{8024}N_{1410}O_{1452}S_{30}$): 116131.78
   b) Mesocricetus auratus ($C_{5357}H_{8070}N_{1434}O_{1449}S_{24}$): 116514.11
   c) Mus musculus ($C_{5338}H_{8043}N_{1431}O_{1454}S_{28}$): 116424.91
   d) Danio rerio ($C_{5178}H_{7752}N_{1380}O_{1437}S_{30}$): 113287.91
   e) Chelonia mydas ($C_{3896}H_{5817}N_{979}O_{1045}S_{23}$): 83826.69
   f) Heterocephalus glaber ($C_{3528}H_{5347}N_{919}O_{973}S_{21}$): 76876.45

3. Results

3.1 Reconstructing a Phylogeny Based on Klotho Gene Variants

The resulting base-pair alignments yielded nearly identical outputs. At first glance, *type a* organisms and *type b* organisms are each placed together within their respective groups – with four mammalian candidates on one end of the diagram(s), and the remaining non-mammalian candidates to the other [see Figure 1]. When cross-checked with morphology, the corresponding nodes reside in correct order of taxonomy on both diagrams. On one end, three of six organisms belonging to *order rodentia* coincide in proper placement according to scientific classification; while my fourth *mammalian* species (Homo sapiens) falls within a fairly close proximity; followed by the remaining candidates.

The phylogenies illustrated in this study depict gene families within paralogy regions consistent with the early evolution of vertebrates. Because similar variations of KL-derivative proteins occur in all species of *Chordata* – ranging from fish, reptiles to mammals – we may safely infer that, in fact, Klotho family proteins belong to ancient paralogy lineages. While having the same basic molecular functions, Klotho protein variants may have undergone only tiny degrees of modification throughout evolutionary history (originating at the nucleotide level). As the next section will demonstrate, these tiny variations are further observed among my candidate sequences. Alas, the current MSA models help further verify group categorization, and the extent of evolutionary relatedness among group *type a* and group *type b* candidates, including node length and probable time scales in decimal format, are detailed below.

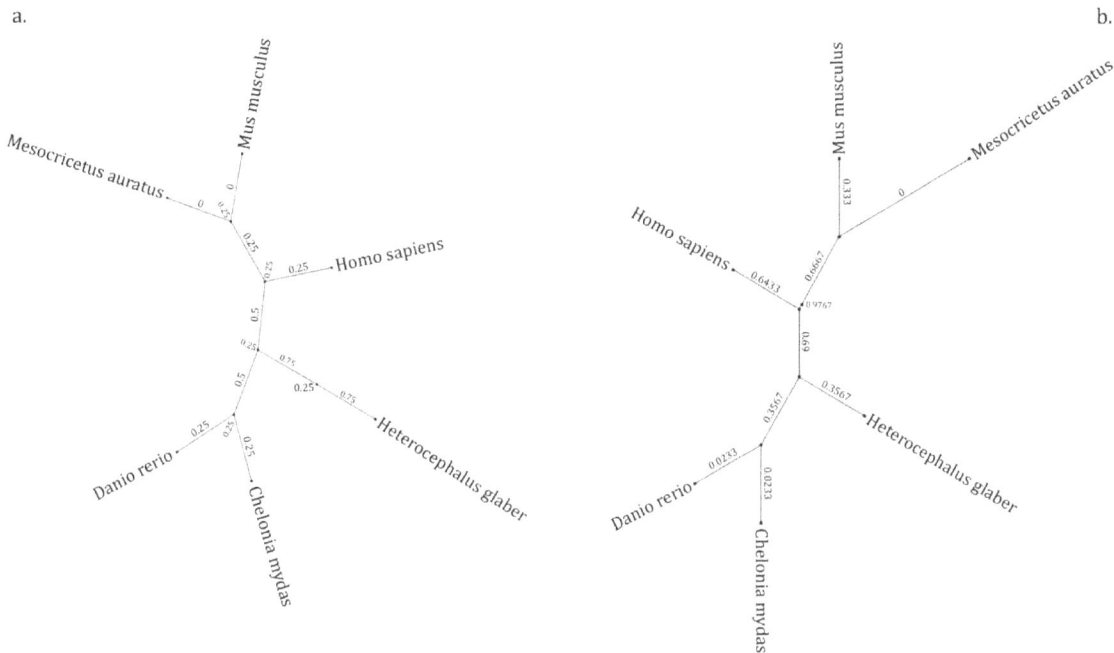

Figure 1. Radial phylogenetic trees. a) Protein-protein output (Klotho). b) Reverse protein-DNA output (Klotho)

3.2 Minimal Degrees of Variation in Evolutionary Conserved Chemical Properties of Klotho

The chemical properties of the α-amino acids of proteins determine the biological activity of the protein, and it conveys a vast array of chemical versatility (University of Arizona, 2014). Evolution itself exerts pressure to preserve α-amino acid residues that bear invaluable functional roles. According to Dokholyan et al. (2002), proteins may retain significantly critical chemical properties (or molecular signatures) that contribute, in lower or higher frequency values, to protein stability via their interaction with other α-amino acid residues. To this end, we might expect to observe a large percentage of evolutionary conserved α-amino acids across entire protein families, with only minimal degrees of variation occurring among wide ranging taxonomical groups.

Indeed, in the context of evolutionary conservation, only three of twenty α-amino acids in this study exceed disparity ratios that average more than 0.093 percent. Comparative analyses of α-amino acid compositions highlight only three strongly distinct molecular signatures indicative of each group. As Figure 2 and Figure 3 illustrate, a correlation between elevated levels of L-Lysine coupled with low quantities of L-Alanine and L-Arginine are featured in group *type a* [as compared to group *type b*]; where the average disparity ratio is ±1.8 percent (L-Lysine), ±1.6 percent (L-Alanine), and ±1.34 percent (L-Arginine). A max disparity ratio of >3.1 percent occurs in L-Lysine molecules [*type a*]. Increased levels of L-Lysine are significant in terms of optimal structure and fold, and suggest that natural selection may have acted to increase the frequency of this molecule.

It is worthwhile to note that L-Alanine and L-Arginine are manufactured internally via biosynthetic pathways; whereas L-Lysine must be ingested as lysine or lysine-containing proteins (plants synthesize L-Lysine from aspartic acid) (Beals, 1999). While the implications of dietary L-Lysine are inconsequential to a study involving techniques in comparative sequence analysis, various other studies suggest that increased dietary L-Lysine plays a role in calcium absorption; a function that is also inadvertently linked to KLOTHO (Tan et al., 2012). As noted in the sections above, "Klotho stimulates calcium reabsorption in the distal convoluted tubule by deglycosylating and stabilizing the epithelial calcium channel TRPV5 on the surface of cellular membrane (Robinson, 2012)." This, and proficiencies in tissue regeneration and other immunity-related health benefits are also linked to increased levels of dietary L-Lysine (Falco, 1995).

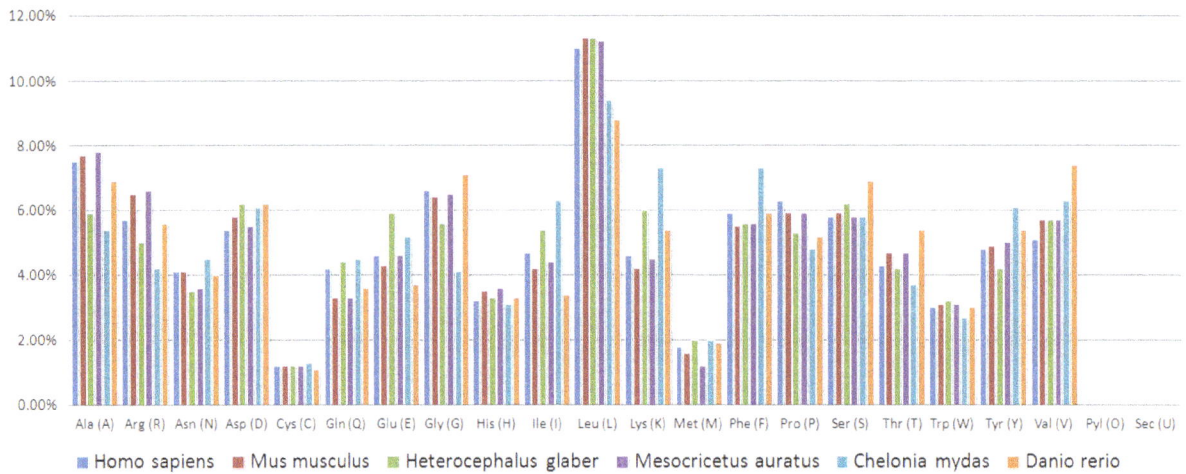

Figure 2. Amino acid composition analysis (six sequences)

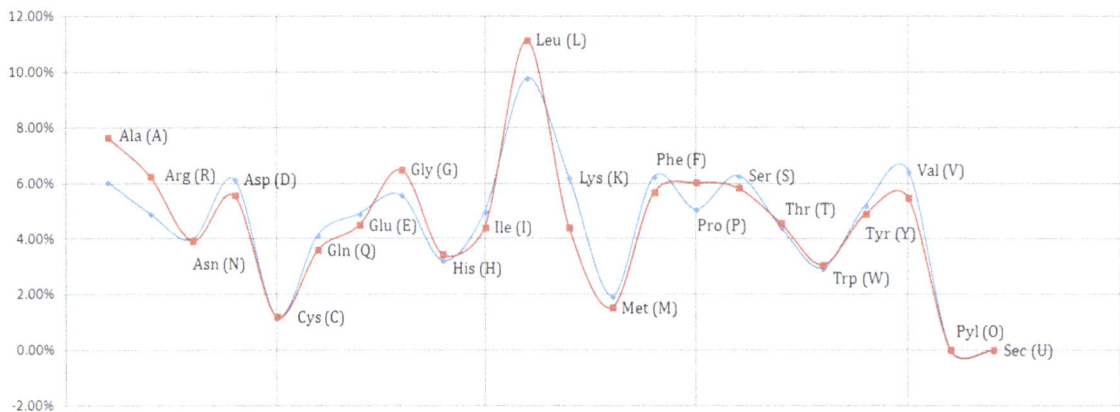

Figure 3. Amino acid composition analysis (*type a* highlighted in blue; *type b* highlighted in red)

4. Discussion

Proteins are essential building blocks of living cells; and life itself can be viewed as resulting substantially from the chemical activity of proteins (Lupas, 2014). Systematic approaches directed toward the evolution of proteins suggest that protein interactions were possibly the first forms of self-reproducing life on earth. As such, the chemical properties of each protein have been developed over billions of years. The α-amino acid composition of proteins contributes to the overall quality of a protein. If certain α-amino acids are optimal for protein structure, natural selection should have acted over evolutionary time to increase the frequency of those α-amino acids (Anthis et al., 2013).

As previously noted, the chemical properties of proteins are also conserved over much longer periods than individual codons; and frequency patterns in Klotho gene variants across wide ranging taxonomy confirm it. Yet, conflicting distribution ratios among L-Lysine, L-Alanine, and L-Arginine [in group *type a* versus group *type b*] is indicative of natural selection acting upon frequency distribution. Could this have consequences in terms of protein function that contribute toward Klotho-associated anti-aging mechanisms?

In fully examining the data presented in this study, it is tempting for me to draw premature [and definitive] conclusions based upon the preliminary appearance of a modest correlation; but correlation does not necessarily imply causation. Instead of over-speculating on functional cause-and-effect and probable implications thereof, I would simply stress the primary scope of this paper is not to presume on solutions, but rather, on associations and identification. From the perspective of comparative sequence analysis, I have outlined a set of significantly distinct biomarkers in molecular variation patterns of KLOTHO across wide ranging taxonomy; and, in the broadest sense, I demonstrated homology and relatedness based on protein variants of Klotho that is consistent with the early evolution of vertebrates. There is still considerable work to do to improve our understanding of age-related mechanisms and the protein functions that play a decisive role. And while a great deal about protein evolution remains unresolved, much can be learned about protein evolution by trying to reconstruct ancient paralogy lineages via comparative analysis of modern proteins. With this in mind, I believe that my results may help provide meaningful reference toward a fully systematic investigation.

References

United States Fish and Wildlife Service. (2005). *Green sea turtle (Chelonia mydas)*. North Florida Field Office.

American Federation for Aging Research. (2006). *Theories of aging -- Why and how do we age?*

Florida State University. (2007). Chemists Kill Cancer Cells With Light-activated Molecules. *ScienceDaily*. Retrieved from http://www.sciencedaily.com/releases/2007/08/070808132019.htm

Minnesota State University. (2012). Molecular Weight Estimations and SDS PAGE.

University of Arizona. (2014). *The Chemistry of Amino Acids*.

Agnisola, C., & Femiano, S. (2006). Studies on routine metabolism in adult zebrafish, Danio rerio. *Acta Physiol., 187*.

Aldrich, J. (1995). Correlations genuine and spurious in Pearson and Yule. *Statistical Science*, 364-376.

Andreollo, N. A., Santos, E. F. D., Araújo, M. R., & Lopes, L. R. (2012). Rat's age versus human's age: what is the relationship? ABCD. *Arquivos Brasileiros de Cirurgia Digestiva (São Paulo), 25*(1), 49-51. http://dx.doi.org/10.1590/S0102-67202012000100011

Anthis, N. J., & Clore, G. M. (2013). Sequence-specific determination of protein and peptide concentrations by asorbance at 205 nm. *Protein Science, 22*, 851-858. http://dx.doi.org/10.1002/pro.2253

Bastien, O, Aude, J. C., Roy, S., & Marechal, E. (2004). Fundamentals of massive automatic pairwise alignments of protein sequences: theoretical significance of Z-value statistics. *Bioinformatics, 20*(4), 534–537. http://dx.doi.org/10.1093/bioinformatics/btg440

Beals, M., Gross, L., & S. Harrell, S. (1999). *Amino Acid Frequency*. University of Tennessee. Retrieved from http://www.tiem.utk.edu/~gross/bioed/webmodules/α-amino acid.htm

Bouchard, C., Tremblay, A., Nadeau, A., Despres, J. P., Theriault, G., Boulay, M. R., ... & Fournier, G. (1989). Genetic effect in resting and exercise metabolic rates. *Metabolism, 38*(4), 364-370. http://dx.doi.org/10.1016/0026-0495(89)90126-1

Bryan, J. K. (1980). Aspartate family and branched-chain amino acids. In B. J. Miflin (Ed.), *The Biochemistry of Plants* (Vol. 5, pp. 403-452). New York: Academic Press.

Buffenstein, R. (2008). Negligible senescence in the longest living rodent, the naked mole-rat: insights from a successfully aging species. *Journal of Comparative Physiology B, 178*(4), 439-445. http://dx.doi.org/10.1007/s00360-007-0237-5

Chao, K. M., & Zhang, L. (2008). *Sequence comparison: theory and methods* (Vol. 7). Springer.

Chen, C., Sander, J. E., & Dale, N. M. (2003). The effect of dietary lysine deficiency on the immune response to Newcastle disease vaccination in chickens. *Avian diseases, 47*(4), 1346-1351. http://dx.doi.org/10.1637/7008

Chen, L., Bush, D. R. (1997). LHT1, a lysine- and histidine-specific amino acid transporter in Arabidopsis. *Plant Physiol., 115*, 1127-1134. http://dx.doi.org/10.1104/pp.115.3.1127

Composition/Molecular Weight Calculation tool. (ND). University of Delaware. Retrieved from http://pir.georgetown.edu/pirwww/search/comp_mw.shtml

Csiszar, A., Labinskyy, N., Orosz, Z., Xiangmin, Z., Buffenstein, R., & Ungvari, Z. (2007). Vascular aging in the longest-living rodent, the naked mole rat. *American Journal of Physiology-Heart and Circulatory Physiology, 293*(2), H919-H927. http://dx.doi.org/10.1152/ajpheart.01287.2006

DALY, T. J. M., WILLIAMS, L. A., & BUFFENSTEIN, R. (1997). Catecholaminergic innervation of interscapular brown adipose tissue in the naked mole‐rat (Heterocephalus glaber). *Journal of anatomy, 190*(3), 321-326. http://dx.doi.org/10.1046/j.1469-7580.1997.19030321.x.

Dena, B. (2014). Life Extension Factor Klotho Enhances Cognition, Cell Reports, http://dx.doi.org/10.1016/j.celrep.2014.03.076.

Dokholyan, N. V., Mirny, L. A., & Shakhnovich, E. I. (2002). Understanding conserved amino acids in proteins. *Physica A: Statistical Mechanics and its Applications, 314*(1), 600-606. http://dx.doi.org/10.1016/S0378-4371(02)01079-8

Enstipp, M. R., Ciccione, S., Gineste, B., Milbergue, M., Ballorain, K., Ropert-Coudert, Y., ... & Georges, J. Y. (2011). Energy expenditure of freely swimming adult green turtles (Chelonia mydas) and its link with body acceleration. *The Journal of experimental biology, 214*(23), 4010-4020. http://dx.doi.org/10.1242/jeb.062943

Falco, S. C., Guida, T., Locke, M., Mauvais, J., Sanders, C., Ward, R. T., & Webber, P. (1995). Transgenic canola and soybean seeds with increased lysine. *Biotechnology (N.Y.), 13*, 577-582. http://dx.doi.org/10.1038/nbt0695-577

Felsenstein, J. (1981). Evolutionary trees from DNA sequences: A maximum likelihood approach. *Journal of Molecular Evolution, 17*(6), 368–376. http://dx.doi.org/10.1007/BF01734359.

Galili, G. (1995). Regulation of lysine and threonine synthesis. *Plant Cell, 7*, 899-906. http://dx.doi.org/10.1105/tpc.7.7.899

Gibbs, R. A., Weinstock, G. M., Metzker, M. L., Muzny, D. M., Sodergren, E. J., Scherer, S., ... & Gill, R. (2004). Genome sequence of the Brown Norway rat yields insights into mammalian evolution. *Nature, 428*(6982), 493-521. http://dx.doi.org/10.1038/nature02426

Griffith, R. S., Norins, A. L., & Kagan, C. (1978). A multicentered study of lysine therapy in herpes simplex infection. *Dermatology, 156*(5), 257-267. http://dx.doi.org/10.1159/000250926

Hirschberg, D. S. (1975). A linear space algorithm for computing maximal common subsequences. *Communications of the ACM, 18*(6), 341–343. http://dx.doi.org/10.1145/360825.360861.

IUPAC, I. (1984). Nomenclature and symbolism for amino acids and peptides. *Eur J Biochem, 138*, 9-37.

Jopling, C., Sleep, E., Raya, M., Martí, M., Raya, A., & Belmonte, J. C. I. (2010). Zebrafish heart regeneration occurs by cardiomyocyte dedifferentiation and proliferation. *Nature, 464*(7288), 606-609. http://dx.doi.org/10.1038/nature08899

Jumpertz, R., Hanson, R. L., Sievers, M. L., Bennett, P. H., Nelson, R. G., & Krakoff, J. (2011). Higher energy expenditure in humans predicts natural mortality. *The Journal of Clinical Endocrinology & Metabolism, 96*(6), E972-E976. http://dx.doi.org/10.1210/jc.2010-2944

Kim, B. (2008). Thyroid hormone as a determinant of energy expenditure and the basal metabolic rate. *Thyroid, 18*(2), 141-144. http://dx.doi.org/10.1089/thy.2007.0266

Kurosu, H., Yamamoto, M., Clark, J. D., Pastor, J. V., Nandi, A., Gurnani, P., ... Kuro-o, M. (2005). Suppression of aging in mice by the hormone Klotho. *Science, 309*(5742), 1829-1833. http://dx.doi.org/10.1126/science.1112766

Lassmann, T., & Sonnhammer, E. L. (2005). Kalign—An Accurate and Fast Multiple Sequence Alignment Algorithm. *BMC Bioinformatics, 6*, 298. http://dx.doi.org/10.1186/1471-2105-6-298

Little, A. G., & Seebacher, F. (2014). Thyroid hormone regulates cardiac performance during cold acclimation in zebrafish (Danio rerio). The Journal of experimental biology, 217(5), 718-725. http://dx.doi.org/10.1242/jeb.096602

Mannakee, B., & Gutenkunst, R. (2012). *Influence of Protein Network Dynamics on Protein Evolution.* University of Arizona.

Marsh, J. A., & Teichmann, S. A. (2010). How do proteins gain new domains. *Genome Biol, 11*(7), 126. http://dx.doi.org/10.1186/gb-2010-11-7-126

Martin, A. P., & Palumbi, S. R. (1993). Body size, metabolic rate, generation time, and the molecular clock. *Proceedings of the National Academy of Sciences, 90*(9), 4087-4091. http://dx.doi.org/10.1073/pnas.90.9.4087

Mueller, P., & Diamond, J. (2001). Metabolic rate and environmental productivity: well-provisioned animals evolved to run and idle fast. *Proceedings of the National Academy of Sciences, 98*(22), 12550-12554. http://dx.doi.org/10.1073/pnas.221456698

NHLBI Communications Office. (2004). *Scientists Compare Rat Genome With Human, Mouse -- Analysis Yields New Insights into Medical Model, Evolutionary Process.* National Institutes of Health.

Pérez, V. I., Buffenstein, R., Masamsetti, V., Leonard, S., Salmon, A. B., Mele, J., ... & Chaudhuri, A. (2009). Protein stability and resistance to oxidative stress are determinants of longevity in the longest-living rodent, the naked mole-rat. *Proceedings of the National Academy of Sciences, 106*(9), 3059-3064. http://dx.doi.org/10.1073/pnas.0809620106

Robinson, M. (2012). *In response to: Which is more informative: a phylogenetic tree based on alignment of protein α-amino acid sequences or one based on the corresponding DNA sequences?* Los Alamos National Laboratory. Retrieved from Research Gate.

Rozing, M. P., Westendorp, R. G., de Craen, A. J., Frölich, M., Heijmans, B. T., Beekman, M., ... & van Heemst, D. (2009). Low serum free triiodothyronine levels mark familial longevity: the Leiden Longevity Study. *The Journals of Gerontology Series A: Biological Sciences and Medical Sciences*, glp200.

Speakman, J. R., Selman, C., McLaren, J. S., & Harper, E. J. (2002). Living fast, dying when? The link between aging and energetics. *The Journal of nutrition, 132*(6), 1583S-1597S.

Sun, P., Zhang, Y., Yu, F., Parks, E., Lyman, A., Wu, Q., ... & Hsiai, T. K. (2009). Micro-electrocardiograms to study post-ventricular amputation of zebrafish heart. *Annals of biomedical engineering, 37*(5), 890-901. http://dx.doi.org/10.1007/s10439-009-9668-3

Tan, T. C., Rahman, R., Jaber-Hijazi, F., Felix, D. A., Chen, C., Louis, E. J., & Aboobaker, A. (2012). Telomere maintenance and telomerase activity are differentially regulated in asexual and sexual worms. *Proceedings of the National Academy of Sciences, 109*(11), 4209-4214. http://dx.doi.org/10.1073/pnas.1118885109

Torres, P. U., Prie, D., Molina-Bletry, V., Beck, L., Silve, C., & Friedlander, G. (2007). Klotho: an antiaging protein involved in mineral and vitamin D metabolism. *Kidney international, 71*(8), 730-737. http://dx.doi.org/10.1038/sj.ki.5002163

Vogt, G., Etzold, T., & Argos, P. (1995). An assessment of amino acid exchange matrices in aligning protein sequences: the twilight zone revisited. *Journal of molecular biology, 249*(4), 816-831. http://dx.doi.org/10.1006/jmbi.1995.0340

Vogt, G., Etzold, T., & Argos, P. (1995). An assessment of amino acid exchange matrices in aligning protein sequences: the twilight zone revisited. *Journal of molecular biology, 249*(4), 816-831. http://dx.doi.org/10.1006/jmbi.1995.0340

Warwicker, J., Charonis, S., & Curtis, R. A. (2013). Lysine and Arginine Content of Proteins: Computational Analysis Suggests a New Tool for Solubility Design. *Molecular pharmaceutics, 11*(1), 294-303. http://dx.doi.org/10.1021/mp4004749

8

Chromosomal Q-Heterochromatin and Sex in Human Population

Ibraimov A. I.[1,2]

[1] Institute of Balneology and Physiotherapy, Bishkek, Kyrgyz Republic

[2] Kazakh National Medical University after name S. D. Asfendyarov, Almaty, Kazakhstan

Correspondence: Ibraimov A. I., Laboratory of Human Genetics, National Center of Cardiology and Internal Medicine, Bishkek, Kyrgyzstan. E-mail: ibraimov_abyt@mail.ru

Abstract

Individuals in the population differ from each other on the number, size, location and intensity of fluorescence of Q-heterochromatin regions (Q-HRs) of chromosomes. It is known that there is Y chromosome in a karyotype of males, which contains the largest block of Q-heterochromatin in human genome, and for this reason it is taken for granted that in total there is on average twice more Q-HRs in the genome of male than of female. However, the question on the existence of differences between the sexes on the quantitative content of chromosomal Q-HRs in the population still remains open. The fact is that the Y chromosome differs in its broad interindividual and interpopulation variability on the size of Q-heterochromatin material. A comparative analysis of the amount of chromosomal Q-HRs in the genome of male and female of different racial, ethnic and age groups showed that sex differences at the population level is much more complicated than heretofore assumed.

Keywords: chromosomal Q-heterochromatin, human Y chromosome, Q-heterochromatin polymorphism, Q-heterochromatin dose compensation, Q-heterochromatin selection

1. Introduction

A fundamental feature of chromosomes of higher eukaryotes, including man, is the presence of two evolutionary consolidated types of genetic material: euchromatin and heterochromatin. Euchromatin – the conservative portion of the genome – contains transcribed structural genes, while heterochromatin – the variable portion of the genome – predominantly consists of nontranscribed repeated DNA sequences (John, 1988; Verma, 1988).

To-date, two types of heterochromatin are known: C- and Q-heterochromatin. There are several significant differences between them, including the fact that C-heterochromatin is encountered in chromosomes of all higher eukaryotes, while Q-heterochromatin is only present in man, the chimpanzee and gorilla. In man C-heterochromatin is present in all his chromosomes, varying mainly in size, while Q-heterochromatin can only be detected on seven autosomes and the Y-chromosome (Paris Conference, 1971, 1975).

Individuals in a population differ from each other on the number, location, size and intensity of staining (fluorescence) of chromosomal Q-heterochromatin regions (Q-HRs). It has been shown that certain human populations differ significantly as concerns this feature (Geraedts & Pearson, 1974; Müller et al., 1975; Buckton et al., 1976; Lubs et al., 1977; Yamada & Hasegawa, 1978; Al-Nassar, 1981; Ibraimov & Mirrakhimov, 1982a, 1982b, 1982c; Ibraimov et al., 1982, 1986, 1990, 1991, 2000, 2013; Stanyon et al., 1988; Kalz et al., 2005; Decsey et al., 2006).

It is known that there is Y chromosome in a karyotype of males, which contains the largest block of Q-heterochromatin in human genome, and for this reason it is taken for granted that in total there is on average twice more Q-HRs in the genome of male than of female (Paris Conference, 1971, 1975). However, the question on the existence of differences between the sexes on the quantitative content of chromosomal Q-HRs on autosomes in the population still remains open. The fact is that there is a link between the size of Q-heterochromatin on Y chromosome and the amount of Q-HRs on autosomes (Ibraimov et al., 2000). Moreover, the Y chromosome differs in its interindividual and interpopulation variability on the size of Q-heterochromatin material (Paris Conference, 1971, 1975; Verma, 1988). A comparative analysis of the amount of chromosomal Q-HRs on autosomes in the genome of male and female showed that sex differences at the population level is much more complicated than heretofore assumed. This fact is very important for understanding the possible biological role of the broad quantitative variability of human chromosomal Q-HRs.

2. Materials and Methods

2.1 Sample Characteristics

Using identical methods we studied neonates, as well as students from Kazakh National Medical University in Almaty. The group of elderly subjects (aged 60 years and over) consisted of individuals living in the Old People's Homes of Almaty.

2.2 Cytogenetic Methods

Chromosomal preparations were made using short-term cultures of peripheral blood lymphocytes, with the exception of newborn infants where umbilical blood was used. The cell cultures were processed according to slightly modified (Ibraimov, 1983a) conventional methods (Hungerford, 1965). The dye used was quinacrine mustard. All the chromosomal preparations were analyzed by one and the same cytogeneticist (A.I.I.) to investigate chromosomal Q-HR variability. Calculation and registration of chromosomal Q-HRs was performed using the criteria and methods described in detail elsewhere (Ibraimov et al., 1982, 1990).

2.3 Quantitative Characteristics of Q-HR Variability and Methods Used for Comparisons

Chromosomal Q-HR variability in populations is usually described in the form of five main quantitative characteristics:

1) The distribution of the number of Q-HRs in a population, i.e., distribution of individuals having different numbers of Q-HRs in the karyotype regardless of their location on seven Q-polymorphic autosomes, which also reflected the range of Q-HRs variability in the population genome;

2) The derivative of this distribution, an important population characteristic, is the mean number of Q-HRs per individual;

3) The frequency of Q-HRs on seven Q-polymorphic autosomes (3, 4, 13-15, 21 and 22) in the population. Despite the fact that in human autosomes there are twelve loci in which Q-HRs can be detected (3 cen, 4 cen, 13 p11, 13 p13, 14 p11, 14 p13, 15 p11, 15 p13, 21 p11, 21 p13, 22 p11, 22 p13), individuals with 24 Q-HRs in their genome could exist, but such cases have not as yet been reported. In individuals of a population the number of Q-HRs on the autosomes usually ranges from zero to ten (Yamada & Hasegawa, 1978; Al-Nassar et al., 1981; Ibraimov & Mirrakhimov, 1985; Ibraimov, 2010).

4) The distribution of Q-HRs on autosomes according to their size and intensity of fluorescence (type of Q-variants), estimated as described by Paris Conference (1971, 1975).

5) The size of the Y chromosome, being (a) large ($Y \geq F$); (b) medium ($F > Y > G$); and (c) small ($Y \leq G$) according to (Ibraimov & Mirrakhimov, 1985).

The distribution of the numbers and mean number of Q-HR per individual in samples were compared using the Student t – test.

3. Results

It should be emphasized that in comparative population cytogenetic studies only autosomal Q-HRs are considered. Nevertheless, variability of the largest Q-heterochromatin band in the human genome, localized on the q12 segment of the Y chromosome, has been mainly considered separately from the quantitative variability of autosomal Q-HRs (Geraedts & Pearson, 1974; Müller et al., 1975; Buckton et al., 1976; Lubs et al., 1977; Yamada & Hasegawa, 1978; Al-Nassar, 1981; Ibraimov & Mirrakhimov, 1982 a, b, c; Ibraimov et al., 1982; 1986; 1990; 1991; 2000; 2013). Table 1 shows the distribution of the numbers and mean number of Q-HRs on autosomes in males and females in two age groups of Kazakh nationality.

Table 1. Distribution of the numbers and mean number of Q-HRs on autosomes in males and females in newborns and 18 - 25 years individuals of Kazakh nationality

Number of Q-HRs	Newborns Boys (n = 207)	Newborns Girls (n = 182)	Males 18 - 25 years (n = 49)	Females 18 - 25 years (n = 190)
	I	II	III	IV
0	3	1		
1	5	4	9	7
2	39	21	12	24
3	47	38	13	45
4	51	46	11	49
5	37	39	4	40
6	18	20		17
7	7	13		7
Total	770	750	136	745
Mean number	3.72 ± 0.102	4.12 ± 0.111	2.78 ± 0.176	3.92 ± 0.104
Statistics	$t_{I, II} = 2.649$; df = 387; P = 0.008*			
	$t_{II, III} = 5.775$; df = 229; P = <0.001*			
	$t_{III, IV} = 5.119$; df = 237; P = <0.001*			
	$t_{I, III} = 4.137$; df = 254; P = <0.001*			
	$t_{II, IV} = 1.313$; df = 370; P = 0.190			

* - these differences are statistically significant.

As can be seen from this Table, in every case females are characterized by the highest mean number value, and by a broad range of variability in the distribution of the numbers of chromosomal Q-HRs as compared to males. These differences are statistically significant. Such regularities as in Kazakhs was also found in Russian samples (data not shown).

And the question arises of why still comparative population cytogenetic studies did not show statistically significant differences in the number of autosomal Q-HRs between the sexes, where such attempts have been made (Geraedts & Pearson, 1974; Müller et al., 1975; Kalz et al., 2005; Decsey et al., 2006). The answer to this question could perhaps be drawn from the data we have obtained in the previous study (Tables 2 and 3) (Ibraimov et al., 2013, 2014).

Table 2. Distribution of the numbers and mean number of chromosomal Q-HRs per individual in Kazakh samples

Number of Q-HRs	Newborns (n = 389) I	18 - 25 years (n = 239) II	60 years and older (n = 33) III
0	4 (1.0%)		2 (6.0%)
1	9 (2.3%)	16 (6.7%)	5 (15.1%)
2	60 (15.4%)	36 (15.1%)	7 (21.2%)
3	85 (21.9%)	58 (24.3%)	12 (36.4%)
4	97 (24.9%)	60 (25.1%)	6 (18.2%)
5	76 (19.5%)	44 (18.4%)	1 (3.0%)
6	38 (9.8%)	17 (7.1%)	
7	20 (5.1%)	7 (2.9%)	
8		1 (0.4%)	
Total	1520 (99.9%)	881 (100.0%)	84 (99.9%)
Mean number of Q-HRs	3.91 ± 0.07	3.69 ± 0.09	2.55 ± 0.213
Statistics	$t_{I, II}$ = 1.808;	df = 626;	P = 0.071;
	$t_{I, III}$ = 5.068;	df = 420;	P = <0.001;*
	$t_{II, III}$ = 4.259;	df = 270;	P = <0.001;*

* - these differences are statistically significant.

Table 3. Distribution of the numbers and mean number of chromosomal Q-HRs per individual in Russian samples

Number of Q-HRs	Newborns (n = 83) I	18 - 25 years (n = 60) II	60 years and older (n = 80) III
0	1 (1.2%)		4 (5.0%)
1	4 (4.8%)	4 (6.6%)	6 (7.5%)
2	9 (10.8%)	7 (11.7%)	28 (35.0%)
3	14 (16.9%)	14 (23.3%)	23 (28.7%)
4	30 (36.1%)	17 (28.3%)	14 (17.5%)
5	16 (19.3%)	13 (21.6%)	5 (6.2%)
6	6 (7.2%)	4 (6.6%)	
7	3 (3.6%)	1 (1.6%)	
Total	321 (99.9%)	224 (99.9%)	212 (99.9%)
Mean number of Q-HRs	3.87 ± 0.156	3.73 ± 0.177	2.65 ± 0.133
Statistics	$t_{I, II}$ = 0.502;	df = 141;	P = 0.575
	$t_{I, III}$ = 5.895;	df = 161;	P = <0.001*
	$t_{II, III}$ = 4.980;	df = 138;	P = <0.001*

* - these differences are statistically significant.

So, judging by the data presented in Tables 2 and 3, there are no statistically significant differences between newborns and samples of young adults (18-25 year olds) that contradict existing observations (Ibraimov et al., 1986, 1990, 1991, 2000). It was found that the main reasons for the lack of differences between the two age groups were: a) the combination of individuals of both sexes in the same group, and b) the predominance of girls in samples of young adults (49 ♂ and 190 ♀ Kazakhs and 12 ♂ and 48 ♀ Russian, see Table 1).

It has long been noted that at the population level, due to the largest block of Q-heterochromatin on Y chromosome in males, amount of Q-HRs on autosomes is less than in females (Ibraimov et al., 1986, 2000). This fact prompted us return to the problem of sex differences in the number of chromosomal Q-HRs at the level of human populations to get an answer to the question: why the number of Q-HRs on autosomes in female individuals more than in males?

First we show the results of all our previous observations on the populations of Eurasia and Africa, which, in general, indicate that there is some connection between the mean number of Q-HRs and sizes of Q-heterochromatin bands on Y chromosome (Table 4).

Table 4. Frequency of the Y chromosome of various sizes and mean number of autosomal Q-HRS in certain populations of Eurasia and Africa

Populations	n	m	$Y \geq F$ Large	$F > Y > G$ Medium	$Y \leq G$ Small
Mozambique Negroes	148	4.72	20 (13.5%) m = 4.40	109 (73.6%) m = 4.69	19 (12.8%) m = 5.26
Zimbabwe Negroes	34	4.65	4 (11.8%) m = 2.75	27 (79.4%) m = 4.93	3 (8.8%) m = 4.67
Angola Negroes	132	4.64	6 (4.5%) m = 4.00	111 (84.1%) m = 4.61	15 (11.4%) m = 5.21
Northern India Indians	41	3.11	12 (29.3%) m = 1.91	29 (70.7%) m = 4.31	0 (0.0%)
Ethiopians	17	3.06	4 (23.5%) m = 2.00	13 (76.5%) m = 3.40	0 (0.0)%)
Southern Kazakhstan Kazakhs	119	2.95	16 (13.4%) m = 2.31	84 (70.6%) m = 3.02	19 (16.0%) m = 3.21
Southern Siberia Khakass	49	2.43	11 (22,4%) m = 2,18	29 (59,2%) m =2,28	9 (18,4%) m =3,22
Western Siberia Nenets	60	2.30	9 (15,0%) m =1,30	47 (78.3%) m =2,40	4 (6,7%) m =2,50
Far North Chukchi	55	2.09	20 (36.4%) m = 1.30	32 (58.2%) m = 2.59	3 (5.5%) m = 2.00
Russians	360	2.09	56 (15.6%) m = 1.53	280 (77.8%) m = 2.18	24 (6.6%) m = 2.58
Chinese	27	2.07	8 (29.6%) m = 1.38	19 (70.4%) m = 2.37	0 (0.00%)
Kyrgyz of Pamir and Tien-Shan	152	1.91	16 (10.5%) m = 1.43	125 (82.2%) m = 2.13	11 (7.3%) m = 2.18
Western Siberia Yakuts	46	1.69	1 (2/2%) m = 1.00	42 (91.3%) m = 1.71	3 (6.5%) m = 1.67
Western Siberia Selkups	66	1.56	13 (19.7%) m = 1.77	48 (72.7%) m = 1.52	5 (7.6%) m = 1.40
Western Siberia Khanty	22	1.50	2 (9.1%) m = 1.00	17 (77.3%) m = 1.50	3 (13.6%) m = 1.30

n - sample size; m - mean number of Q-HRs per individual in the population.

These data gave us reason to believe that: 1) Q-heterochromatin on the Y chromosome, being the largest in the human genome, somehow "restricts" the total content of Q- HR on the autosomes in males and 2) Q-HRs on human chromosomes appears to have a common nature, regardless of the characteristics of their localization in the karyotype. The legitimacy of such an assumption is confirmed by our new data.

Thus, Table 5 presents the distribution of the numbers and mean number of autosomal Q-HRs in males with Y chromosomes of various sizes in Kazakh newborns. If until now the existence of a close connection between the number of Q-HRs on autosomes and the size of Q-heterochromatin on the Y chromosome at the population level were showed by only adult individuals (Table 4), this time we found the same pattern in newborns. The same patterns were found in Russian samples (data not shown).

Table 5. Distribution of the numbers and mean number of autosomal Q-HRs in males with Y chromosomes of various sizes in Kazakh newborns

Number of Q-HRs	Large $Y \geq F$ (n = 53) I	Medium $F > Y > G$ (n = 102) II	Small $Y \leq G$ (n = 32) III
0	3		
1	5	1	
2	21	12	2
3	11	26	1
4	6	28	10
5	7	22	13
6		10	4
7		3	2
Total	139	406	150
Mean number of Q-HRs	2.62 ± 0.185	3.98 ± 0.129;	4.69 ± 0.203
Statistics	t I, II = 6.077;	df = 153	P = < 0.001 *
	t II, III = 2.748;	df = 132;	P = 0.007 *
	t I, III = 7.223;	df = 83;	P = < 0.001 *

* - these differences are statistically significant.

The amount of chromosomal Q-HRs in the human genome can be additionally evaluated according to their size and intensity of fluorescence on a five-point degree (Paris Conference, 1971, 1975). Later, recommendations of the Paris Conference on this issue have been further developed in a number of papers (Muller et al., 1975; Buckton et al., 1976; Lubs et al., 1976; Yamada & Hasegawa, 1978) and comparative population studies began to consider Q-bands with only 4 and 5 degrees of fluorescence intensity. In estimating the size of Q-HRs we adhered to the recommendations of Yamada and Hasegawa (1978) according to which the sizes of Q-bands were divided into five degrees, comparing them with the short arm of chromosome 18. However, mostly Q-HRs of medium (> 0,5 – 0,75 × 18p) and large (> 0,75 - 1,0 × 18p) sizes are found in a population. Abbreviation "QFQ" stands for "Q-bands by fluorescence using quinacrine" (Paris Conference, 1970, 1975). For example, "QFQ 15" indicates that the given Q-band has the medium size with very bright fluorescence. For clarity, Figure 1 shows an example of a karyotype with 3 chromosomal Q-HRs.

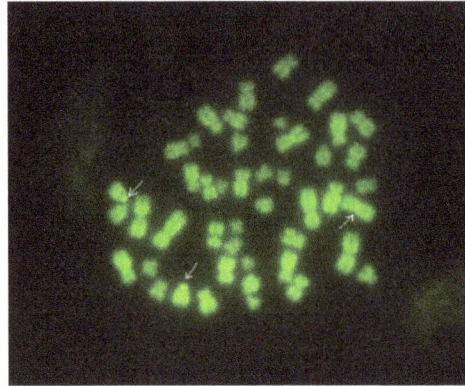

Figure 1. Q-HRs of large sizes and 5th degree intensity of fluorescence on the short arms of both chromosomes 13 and Q-HR of medium size and 5th degree of fluorescence intensity in the pericentric region of chromosome 3

Table 6 presents data on the size and intensity of fluorescence of chromosomal Q-HRs in newborns and young adults (18-25 year-olds) of the Kazakh nationality. In the examined samples of individuals QFQ points ranged from 15 to 112. Here QFQ 15 indicates that the genome of a given individual has one Q-HR of medium size with the intensity of fluorescence equal to 5 degrees. While, QFQ 98 means that the karyotype of this individual has seven Q-HRs of medium size with the fluorescence intensity of 5 degrees. For statistical analysis types of Q-HRs we divided into 7 groups: Group 1 (14, 15 points); Group 2 (28, 29 points), Group 3 (40 to 45 points); Group 4 (53 to 58 points); Group 5 (from 67 to 73 points), Group 6 (80 to 86 points); Group 7 (98 to 112 points). In other words, Group 1 included individuals the karyotype of which had only one Q-HR, Group 2 with two, and the remaining groups with three or more Q-bands with intensity of fluorescence of 5 degrees.

Table 6. Types of Q-HRs on autosomes in males and females in newborns and 18 - 25 years individuals of Kazakh nationality

Types of Q-HRs	Newborns Boys (n = 180) I	Newborns Girls (n = 159) II	Males 18- 25 years (n = 43) III	Females 18 - 25 years (n = 120) IV
0	3	2		
1	6	4	8	7
2	34	15	7	14
3	40	33	13	28
4	42	36	12	30
5	36	40	2	26
6	14	14	1	11
7	5	13		4
8		4		
Total	661	668	125	463
Mean number	3.67 ± 0.110	4.20 ± 0.668	2.91 ± 0.193	3.83 ± 0.132
Statistics	t I, II = 3.164;	df = 337;	P = 0.002*	
	t III, IV = 3.807;	df = 161;	P = < 0.001*	

* - these differences are statistically significant.

As shown in this Table, in the genome of females the amount of chromosomal Q-HRs significant more not only in number but also in size and intensity of fluorescence. Similar data were obtained in Russian samples (data not shown).

The literature shows the absence of statistically significant differences between the sexes in frequencies of chromosomal Q-HRs in Q-polymorphic loci of seven autosomes (Geraedts & Pearson, 1974; Buckton et al., 1976; Lin et al., 1976; Schwinger, 1976; Yamada & Hasegawa, 1978). However, the quantitative variability of chromosomal Q-HRs can be assessed, in addition to their frequency on seven Q-polymorphic autosomes, through their mean numbers, size and intensity of fluorescence (Paris Conference, 1971, 1975). Unfortunately, not all authors provide primary quantitative data indicating the number, size and intensity of fluorescence of chromosomal Q-HRs, to allow additional comparative statistical analysis.

It is generally known that the size of the Q-heterochromatin on the long arm of a Y chromosome of even medium size is greater than that of the Q-HRs on any of seven Q-polymorphic autosomes in the human karyotype, especially as the morphological variability of the Y chromosome (large, medium, small) is mainly determined by the size of the Q-heterochromatin segment on its long arm. Based on the data presented above, we assume, that the Q-heterochromatin band on the Y chromosome, being the largest Q-heterochromatin segment in the human genome, somehow 'restricts' the overall number of Q-HRs on autosomes in males. Apparently, for the same reason, amount of autosomal Q-HRs increases in females compared to males within each individual population (Ibraimov, 1983b). In addition, human chromosomal Q-HRs may have a basically similar role regardless of their location in the karyotype (Ibraimov, 2010).

4. Discussion

Why the detection of the existence of significant differences between the sexes in the number of Q-HRs on autosomes is not so simple? We believe that this is due to following facts: 1) so far all authors compared males and females only on frequencies of Q-HRs in twenty Q-polymorphic loci of seven autosomes, leaving aside such important quantitative characteristics as the distribution of the numbers, size and intensity of fluorescence of chromosomal Q-HRs in population; 2) it is taken for granted that in total there is on average twice more Q-HRs in the genome of males than of females, for the reason that there is Y chromosome in karyotype of males, which contains the largest block of Q- heterochromatin in human genome; 3) compared samples of male and female individuals were of different age groups; 4) extremely wide variability of sizes of Q-heterochromatin on Y chromosomes in human populations were not taken into account. In other words, the possible existence of a close connection between the quantitative content of Q-HRs on autosomes and the size of Q- heterochromatin of the Y chromosome at the population level were not expected. Specifically, no one tried to assess the possible impact of a large Q-heterochromatin material of the Y chromosome on the amount of Q-HRs on autosomes in the human genome.

So far it seems correct to us to explain the increasing number of chromosomal Q-HRs on autosomes in females at the population level by the existence of some evolutionary established mechanism that "compensates" the difference in the "dose" of Q-heterochromatin material in the female genome due to the lack of chromosomes in their karyotype, which carries largest Q-HR, as Y chromosome (Ibraimov, 1983b, 2010). Apparently, there is some mechanism that limits the "dose" of chromosomal Q-HRs in the human genome to a certain level. Indeed, the human karyotype has 25 loci (3 cen, 4 cen, 13 p11, 13 p13, 14 p11, 14 p13, 15 p11, 15 p13, 21 p11, 21 p13, 22 p11, 22, p13 and Yq12), where Q-heterochromatin can potentially be detected. However, as yet no one could found 25 chromosomal Q-HRs in the human karyotype; usually their number varies from 0 to 10 (Yamada & Hasegawa, 1978; Al-Nassar et al., 1981; Ibraimov & Mirrakhimov, 1985; Ibraimov, 2010).

The existence of significant sex differences at the population level raises the question of the revision of the published literature data on the mean number of Q-HRs per individual in population. Let's just take one example from our own experience. Earlier, after examining the children aged 13-16 years we reported that the mean number for the Kazakhs of Kazakhstan is 3.56 ± 0.168 (Ibraimov et al., 1982). We obtained similar data 30 years later, examining the students of the Kazakh nationality in Almaty (Ibraimov et al., 2013). Now, after studying the amount of chromosomal Q-HRs in newborns and elderly Kazakhs, we realized that our conclusion does not correspond to the reality (see Table 2). Therefore, while presenting and discussing the data on the amount of chromosomal Q-HRs in the genome of a given population we should show the mean number values separately for male and female samples with the mandatory indication of their age.

In conclusion, as expected, in general, the total amount of chromosomal Q-HRs in the male genome is greater than that of female. Indeed, the size of Q-HRs of Y chromosome of even medium size is so large that it is equal to two-three average autosomal Q-HRs in human karyotype. However, this does not mean that the total amount of chromosomal Q-HRs in the males' genome at the population level on average is twice greater than of females (Paris Conference, 1971, 1975). Our previous observations (Ibraimov, 1983b; Ibraimov et al., 1986, 1990, 1991, 2000, 2014) as well as the results of this study show that, apparently, there is some mechanism that compensates for the deficiency of Q-heterochromatin material in the female genome due to the lack of Y chromosome in their karyotype by increasing the amount of Q-HRs on autosomes. This pattern persists regardless of age and racial-ethnic characteristics of human populations. Therefore, a preponderance of the total amount of Q-HRs in the genome of the male on the level of human populations still persists.

Acknowledgements

We wish to express our gratitude to the students of Kazakh National Medical University and the elderly people from the Old People's Home of Almaty city for their cooperation.

References

Al-Nassar, K. E., Palmer, C. G., Connealy, P. M., & Pao-Lo, Yu. (1981). The genetic structure of the Kuwaiti population. II. The distribution of Q-band chromosomal heteromorphisms. *Hum. Genet., 57*, 423-427. http://dx.doi.org/10.1007/BF00282021

Buckton, K. E., O'Riordan, M. L., Jacobs, P. A., Robinson, J. A., Hill, R., & Evans, H. J. (1976). C- and Q-band polymorphisms in the chromosomes of three human populations. *Ann. Hum. Genet., 40*, 90-112. http://dx.doi.org/10.1111/j.1469-1809.1976.tb00168.x

Decsey, K., Bellovits, O., & Bujdoso, G. M. (2006). Human chromosomal polymorphism in a Hungarian sample. *Int. J. Hum. Genet., 6*(3), 177-183.

Geraedts, J. P. M., & Pearson, P. L. (1974). Fluorescent chromosome polymorphisms: frequencies and segregation in a Dutch population. *Clin. Genet., 6*, 247-257. http://dx.doi.org/10.1111/j.1399-0004.1974.tb02086.x

Hungerford, D. A. (1965). Leucocytes cultured from small inocula of whole blood and preparation of metaphase chromosomes by treatment with hypotonic KCl. *Stain Technol., 40*, 333-338.

Ibraimov, A. I. (1983a). Chromosome preparations of human whole lymphocytes – an improved technique. *Clin. Genet., 24*, 240-242. http://dx.doi.org/10.1111/j.1399-0004.1983.tb00077.x

Ibraimov, A. I. (1983b). Human chromosomal polymorphism. VII. The distribution of chromosomal Q-polymorphic bands in different human populations. *Hum. Genet., 63*, 384-391. http://dx.doi.org/10.1007/BF00274767

Ibraimov, A. I. (2010). Chromosomal Q-heterochromatin regions in populations and human adaptation. In M. K. Bhasin & C. Susanne (Eds.), *Anthropology Today: Trends and Scope of Human Biology* (pp. 225-250). Delhi: Kamla- Raj Enterprises.

Ibraimov, A. I., Akanov, A. A., Meymanaliev, T. S., Karakushukova, A. S., Kudrina, N. O., Sharipov, K. O., & Smailova, R. D. (2013). Chromosomal Q-heterochromatin polymorphisms in 3 ethnic groups (Kazakhs, Russians and Uygurs) of Kazakhstan. *Int. J. Genet., 5*(1), 121-124.

Ibraimov, A. I., Akanov, A. A., Meymanaliev, T. S., Smailova, R. D., & Baigazieva, G. D. (2014). Chromosomal Q-heterochromatin and age in human population. *JMBR* (in press).

Ibraimov, A. I., Axenrod, E. I., Kurmanova, G. U., & Turapov, O. A. (1991). Chromosomal Q-heterochromatin regions in the indigenous population of the northern part of West Siberia and new migrants. *Cytobios., 67*, 95-100.

Ibraimov, A. I., Karagulova, G. O., & Kim, E. Y. (2000). The relationship between the Y chromosome size and the amount of autosomal Q-heterochromatin in human populations. *Cytobios., 102*, 35-53.

Ibraimov, A. I., Kurmanova, G. U., Ginsburg, E. Kh., Aksenovich, T. I., & Axenrod, E. I. (1990). Chromosomal Q-heterochromatin regions in native highlanders of Pamir and Tien-Shan and in newcomers. *Cytobios., 63*, 71-82.

Ibraimov, A. I., & Mirrakhimov, M. M. (1982a). Human chromosomal polymorphism. III. Chromosomal Q-polymorphism in Mongoloids of northern Asia. *Hum. Genet., 62*, 252-257. http://dx.doi.org/10.1007/BF00333531

Ibraimov, A. I., & Mirrakhimov, M. M. (1982b). Human chromosomal polymorphism. IV. Chromosomal Q-polymorphism in Russians living in Kyrghyzia. *Hum. Genet., 62*, 258-260. http://dx.doi.org/10.1007/BF00333532

Ibraimov, A. I., & Mirrakhimov, M. M. (1982c). Human chromosomal polymorphism. V. Chromosomal Q-polymorphism in African populations. *Hum. Genet., 62*,261-265. http://dx.doi.org/10.1007/BF00333533

Ibraimov, A. I., & Mirrakhimov, M. M. (1985). Q-band polymorphism in the autosomes and the Y chromosome in human populations. In A. A. Sandberg & R. Alan (Eds.), *Progress and Topics in Cytogenetics. The Y chromosome. Part A. Basic Characteristics of the Y chromosome* (pp. 213-87). New York: Liss Inc.

Ibraimov, A. I., Mirrakhimov M. M., Axenrod E. I., & Kurmanova, G. U. (1986). Human chromosomal polymorphism. IX. Further data on the possible selective value of chromosomal Q-heterochromatin material. *Hum. Genet., 73*,151-156. http://dx.doi.org/10.1007/BF00291606

Ibraimov, A. I., Mirrakhimov, M. M., Nazarenko, S. A., Axenrod, E. I., & Akbanova, G. A. (1982). Human chromosomal polymorphism. I. Chromosomal Q-polymorphism in Mongoloid populations of Central Asia. *Hum. Genet., 60*, 1-7. http://dx.doi.org/10.1007/BF00281254

John, B. (1988). The biology of heterochromatin. In R. S. Verma (Ed.), *Heterochromatin: molecular and structural aspects* (pp. 1-147). Cambridge University Press.

Kalz, L., Kalz-Fuller, B., Hedge, S., & Schwanitz, G. (2005). Polymorphism of Q-band heterochromatin; qualitative and quantitative analyses of features in 3 ethnic groups (Europeans, Indians, and Turks). *Int. J. Hum. Genet., 5*(2), 153-163.

Lin, C. C., Gedeon, M. M., & Griffith, M. M. (1976). Chromosome analysis on 930 consecutive newborn children using quinacrine fluorescent banding technique. *Hum. Genet., 31*, 315-328. http://dx.doi.org/10.1007/BF00270861

Lubs, H. A., Patil, S. R., Kimberling, W. J., Brown, J., Hecht, F., Gerald, P., & Summitt, R. L. (1977). Racial differences in the frequency of Q- and C-chromosomal heteromorphism. *Nature, 268*, 631-632. http://dx.doi.org/10.1038/268631a0

Mckenzie, W. H., & Lubs, H. A. (1975). Human Q and C chromosomal variations: distribution and incidence. *Cytogenet. Cell Genet., 14*, 97-115. http://dx.doi.org/10.1159/000130330

Müller, H. J., Klinger, H. P., & Glasser, M. (1975). Chromosome polymorphism in a human newborn population. II. Potentials of polymorphic chromosome variants for characterizing the idiogram of an individual. *Cytogenet. Cell Genet, 15*, 239-255. http://dx.doi.org/10.1159/000130522

Paris Conference and Supplement. (1971, 1975). Standartization in human cytogenetics. *Birth Defects, 6*, 1-84.

Schwinger, E., & Wehner, H. (1976). Frequency of chromosomal fluorescence polymorphism in normal persons and in clinical patients with diagnosed chromosome aberrations. *Hum. Genet., 32*, 115-119. http://dx.doi.org/10.1007/BF00291493

Stanyon, R., Studer, M., Dragone, A., De Benedictis, G., & Brancati, C. (1988). Population cytogenetics of Albanians in Cosenza (Italy): frequency of Q- and C-band variants. *Int. J. Anthropol., 3*, 19-29.

Verma, R. S. (1988). Heteromorphisms of heterochromatin. In R. S. Verma (Ed.), *Heterochromatin: molecular and structural aspects* (pp. 1-147). Cambridge University Press.

Yamada, K., & Hasegawa, T. (1978). Types and frequencies of Q-variant chromosomes in a Japanese population. *Hum. Genet., 44*, 89-98. http://dx.doi.org/10.1007/BF00283578

αS1-Casein Lineage Assessed by RFLP in the Endangered Goat Breed "Retinta Extremeña"

José L. Fernández-García[1] & María P. Vivas Cedillo[1]

[1] Genetics and Animal Breeding, Veterinary Faculty, Universidad de Extremadura, 10071 Cáceres, Spain

Correspondence: José L. Fernández-García, Genetics and Animal Breeding, Veterinary Faculty, Universidad de Extremadura, 10071 Cáceres, Spain. E-mail: pepelufe@unex.es

Abstract

The Retinta Extremeña goat is a well-adapted breed to "Dehesa" environment. Traditionally their raw milk is used to make artisan cheese. However, crosses with specialized breeds are occurring since the eighties, this goat breed has been declared of special protection by the Spanish Agriculture Ministry (R.D. 1682/1997 and R.D. 229/2008). Genetic studies about casein variants have been mainly performed on Spanish goats of high milk yields because the caseins are a relevant fraction of milk. But recent studies claimed to study the caseins in all breeds, including threatened goat breeds to decide about its conservation value. This study was focused on the αS1-casein in the endangered "Retinta Extremeña" goat for the first time to enhance its conservation interest. Genomic DNA of seventy five pureblood goats was studied. A PCR-RFLP assay was designed to find a BmyI target that distinguishing A versus B2 lineages (including recombinant variant M and B1, respectively) of the αS1-casein locus. The allelic frequency of variants related to A lineage (CAG triplet) was 14.0% similar to other southwestern Spanish breeds. It is suggested that individuals or families carrying A lineage should be more studied to detect less allergen null alleles while the opposite allele pools of the B2 lineages should be tested for alleles associated to unsaturated fatty acid content. Therefore, the priorities for conservation plans of animal genetic resources as threatened goat breeds; more investigation is claimed in the aim to study for proved useful alleles of certain genes, as casein variants.

Keywords: αS1-lineage, casein variant, endangered goat breed, genetic resources

1. Introduction

The Retinta (due to show uniform red coat color) Extremeña goat breed is geographically located almost exclusively in Extremadura Autonomy at the Southwestern Spanish territory. This goat is well adapted to "Dehesa" environment and often related to the most depressed socioeconomic uses within their distribution area. Moreover, it has been historically exploited as dual purpose (meat/milk) in a context of familiar subsistence and exclusively under extensive production systems based on browsing and grazing. More recently, the high quality cheese "Queso Ibores" (a Protected Designation of Origin artisan product) used raw milk of this breed and other autochthonous breeds. Despite, its substitution and crossing with more specialized milk yield breeds has occurred since the eighties. This fact led a census reduction of pureblood below 2,000 animals, but in geographically fragmented flocks. From here, the Retinta goat was considered to be on genetic threatening situation reason why the Spanish Agriculture Ministry declared of special protection this breed (R.D. 1682/1997 and R.D. 229/2008). Breeders Association and herd book have been established (Decreto 296/2011) for which the breed was first described with morphological, reproductive and productive data available. However, there is a shallow genetic characterization of the breed in general, much less on casein variants in particular. This is important because this goat is being used to product artisan cheese.

Goat milk synthesized in the mammary gland has six different types of milk proteins (the four caseins -αS1, αS2, β and κ - being 80% of milk protein), among which αS1-casein (αS1-cn: localized to the fourth goat chromosome; Hayes, Petit, Bouniol & Popescu, 1993) has been largely studied due to its impacts and positive correlation with goat milk composition (the amount of total protein, total solids, milk fat concentration and fatty acid composition; see Valenti, Pagano & Avondo, 2012; and references there in) and cheese-making properties (Remeuf, 1993; Pirisi, Colin, Laurent, Scher, & Parmentier, 1994; Clark & Sherbon, 2000; Chilliard et al., 2006). The 18 allelic variants of αS1-cn have been subdivided into four categories according to its quantity in goat milk (Moioli,

D'Andrea, & Pilla, 2007) as follow: (1) high expressing or strong alleles (A, B1, B2, B3, B4, C, H, L and M), (2) intermediate alleles (E and I) and (3) weak alleles (D, F and G) and null alleles (01, 02 and N) with levels of 3.6 g/L, 1.1 g/L and 0.45 g/L (cero for nulls) per allele, respectively (Valenti, Pagano, Pennisi, Lanza & Avondo, 2010). So, strong variants triple the performance levels associated to the αS1-cn and to which should be added the corresponding by the correlation with other milk components already mentioned. Also it has been reported the positive relationship between variants and the physicochemical properties of milk that affects the technological characteristics related to the clotting time of curd, firmness and rennet yield (Grosclaude et al., 1987).

Therefore, genetic studies about casein in threatened goat breed may be reasoned not only to analyze for high yield genes but also to decide about its conservation value based on traits of especial interest for its sustainable use in rural areas. This last statement includes one important of several priorities for conservation of specific animal genetic resources (Boettcher et al., 2010).

The aims of this study was to show for the first time in pureblood "Retinta Extremeña" goat breed valuables casein αS1-lineages to enhance the interest for its conservation.

2. Materials and Methods

Seventy five complete blood samples were collected of the pureblood "Retinta Extremeña" goats from the Selection Centre of Animal Selection and Reproduction (CENSYRA) of the Extremadura Government (4 males and 71 females, around the 45% of the total herd at CENSYRA). Although limited pedigree information was available, only three partial generations was able to assess Mendelian inheritance of the locus. Since all sampled animals belonged to the CENSYRA herd, where this and other endangered breeds are maintained institutionally as pureblood nucleus, the data presented here can be considered to represent this pureblood breed.

Genomic DNA was isolated by non-commercial procedure as in Fernández-García et al. (2012). The identification of natural variants for αS1-cn precursor was obtained in the SWISS-PROT database (CAS1-CAPH1 locus, acc. n° P18626). At 77 (amino acid position) all phylogenetically related casein variants to the most ancestral A variant shows a CAG triplet (glutamine: Q) but other variants wear the triplet GAG (glutamic: E) (Devilacqua et al., 2001). This polymorphism can be identified at 1045-1047 triplet in a segment of the casein coding sequence expands from exon 9 to exon 11 (Leroux, Mazure & Martin,1992) (accession number: GenBank X56462) (Figure 1). Two restriction targets for the endonuclease B*my*I (GDGCH/C) was predicted using software PROPHET (The Prophet Group at BBN Systems and Technologies; http://www-prophet.bbn.com). A primer pair was designed with PRIMER3 software (Koressaar and Remm 2007) that extend partly the intron 9, the exon 10, the intron 10, the exon 11 and partly the intron 11, but flanking both B*my*I target (C-A1 Forward 5′AAGCTATGATGTGTCTGGTT and C-A1Reverse 5′AACATTCTTGCTCATTCCCT). The PCR reaction is performed in 50 µL of a mixture containing template DNA (20-50 ng), primers [10pmol] and other general PCR reagents. Thermal cycling profile was as follow: [97° 5'] / [94°C - 1'; 59° - 1'; 72° - 1'] * 33 cycles / 72° - 5' final extension; 4° C-∞. 10 µL PCR products were run in agarose gel to assess amplification. After, 10 µL of purified PCR products were digested overnight with 4 U of B*my*I according to the manufacturer (Boehringer Mannheim, Germany). For detection of genotypes the PCR-RFLP fragments were separated in 2.0 % agarose gel. CERVUS ver 3.0.3 (Marshall, Slate, Kruuk, & Pemberton, 1998) was used to verify HWE for the B*my*I polymorphic site of this locus.

3. Result.

One fragment of 318 bp was always present as expected after PCR assays based on Leroux et al. (1992) sequence (GenBank X56462). Based on knowledge obtained from databases (Figure 1), sequences from the F variant have a 3 bp insertion at intron nine between the selected primers but these should not be resolvable by agarose gels two percent (Figure 1). Digestion with B*my*I revealed the two target sites in the PCR product but only one of them was polymorphic (Figure 2). Accordingly, profiles for three possible genotypes were observed after digestion.

The homozygous goats for the GAG triplet showed profiles consisting of two fragments of 84 and 234 bp (n = 55 genotype GAG/GAG). Therefore, the 234 bp band was present in goats carrying the B2 lineage (see Bevilacqua, Ferrant, Garro, Veltri & Lagonigro, 2002) (Figure 2). In heterozygous goats (n = 19 genotypes GAG/CAG) were seen four bands with 234 bp, 122 bp, 112 bp, but these two last bands almost co-migrant and worst resolved in 2 % agarose gels but signaling the A lineage gene, and 84 bp (Figure 2). The homozygous goats for the CAG triplet have had profiles with 122, 112 and 84 bp bands (n = 1 genotype). Furthermore, the presence of another target for B*my*I was conserved in all samples and therefore it could be used as an internal control of the digestion or enzymatic activity.

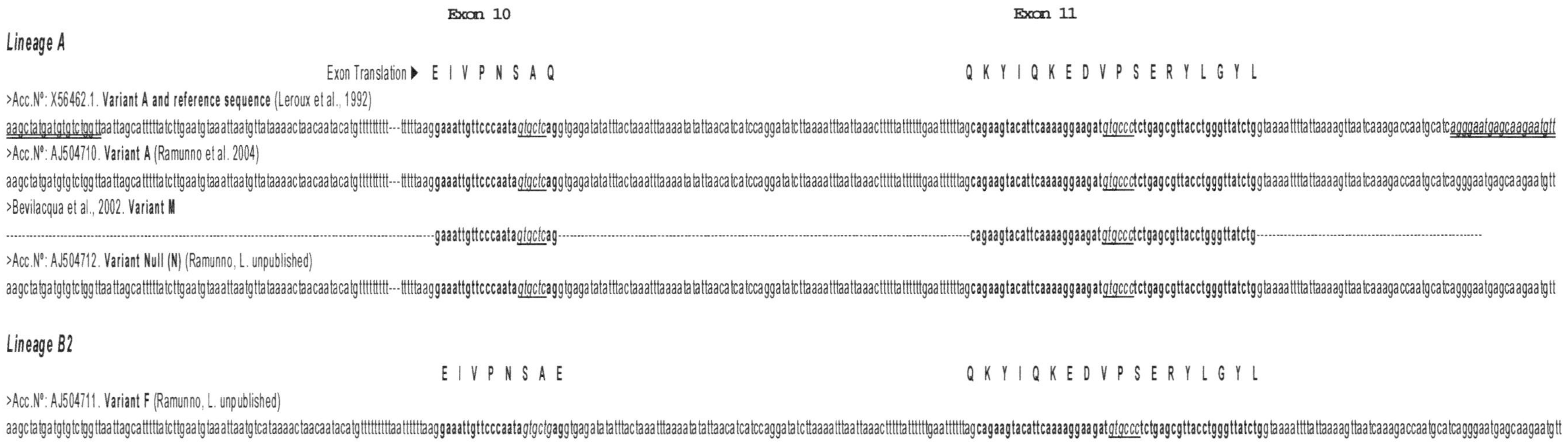

Exon 10

Exon 11

Lineage A

Exon Translation ▶ E I V P N S A Q

Q K Y I Q K E D V P S E R Y L G Y L

>Acc. N°. X56462.1. **Variant A and reference sequence** (Leroux et al. 1992)

>Acc. N°. AJ504710. **Variant A** (Ramunno et al. 2004)

>Bevilacqua et al., 2002. **Variant M**

>Acc. N°. AJ504712. **Variant Null (N)** (Ramunno, L. unpublished)

Lineage B2

E I V P N S A E

Q K Y I Q K E D V P S E R Y L G Y L

>Acc. N°. AJ504711. **Variant F** (Ramunno, L. unpublished)

Figure 1. Sequence alignment of different alleles of the αS1-casein. The sequence of the variant A with Accession Number X56462.1 in GenBank (Leroux et al., 1992) was used to design primers and size predictions. Partial sequence of the allele M was transcribed from Bevilacqua et al. (2002). Exon translation: IUPAC code for amino acids of the exon 10 and 11. Double underline for primers sequences, bold type for the nucleotide sequence of each exon, underline-italics for B*mny*I target

Co-dominant segregation of B*my*I target was observed in only three parental- descendant available families (data not showed). No discrepant segregation was observed for mendelian transmission of the polymorphic site. The allelic frequency of the CAG triplet was 14.0 % in the total sample, but 37.5 % in the four breeding males. By χ^2 test it was verified HWE equilibrium of the locus using only unrelated animals (p = 0.801). So the higher percentage of the GAG triplet in breeding males should be attributed to stochasticity.

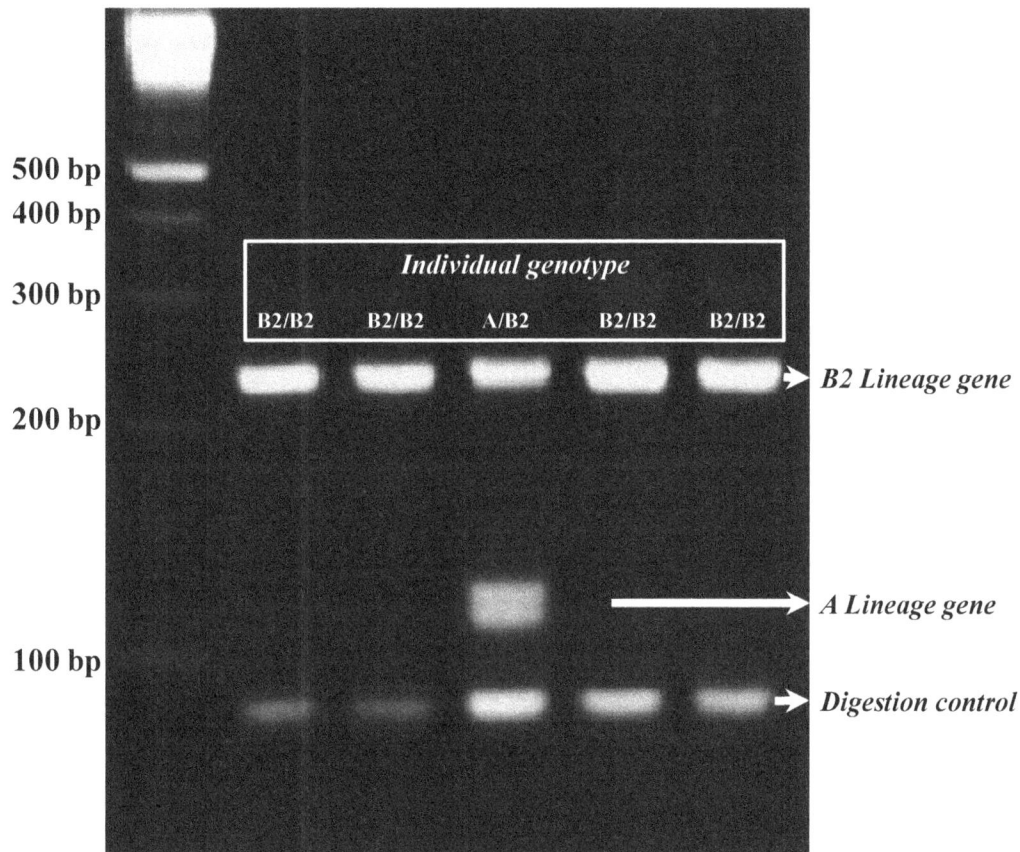

Figure 2. RFLP profile with B*my*I endonuclease from five individuals of the Retinta Extremeña goat. (Right) Restriction fragment corresponding to the gene of each lineage (B2 or A) and the 84 bp fragment digestion Control at the bottom and (left) fragment size of the molecular weight marker

4. Discussion

Structural analysis of goat casein performed both at the protein and the genomic level (Leroux, Martin, Mahè, Leveziel, & Mercier, 1990, Grosclaude & Martin 1997, Sacchi et al., 2005) have showed high variability and complex relation among polymorphic variants for the casein of the goat milk (Mahè & Grosclaude, 1989). This is especially certain for the locus αs1 and κ -casein which showed the highest number of variants (number of variant ≥ 16; Moioli et al., 2007). This high polymorphism feature provides further evidence that the allelic diversity come from multiple pathways, including recombination events between both ancestral lineages (A vs B2) as it has been hypothesized for the αs1- casein locus (Bevilacqua et al., 2002 and references there in)(Figure 3). In this study it was showed that the BmyI-RFLP can be useful to discriminate between lineages A and B2 (including recombinant variant M and B1, respectively) and thus for screening the "allele pools" associated with each one of both in the individual, familiar or population scale (Figure 3). This is important because null alleles (O1, O2 and N) have been exclusively associated to the A lineage (Grosclaude & Martin 1997; Bevilacqua et al., 2002).

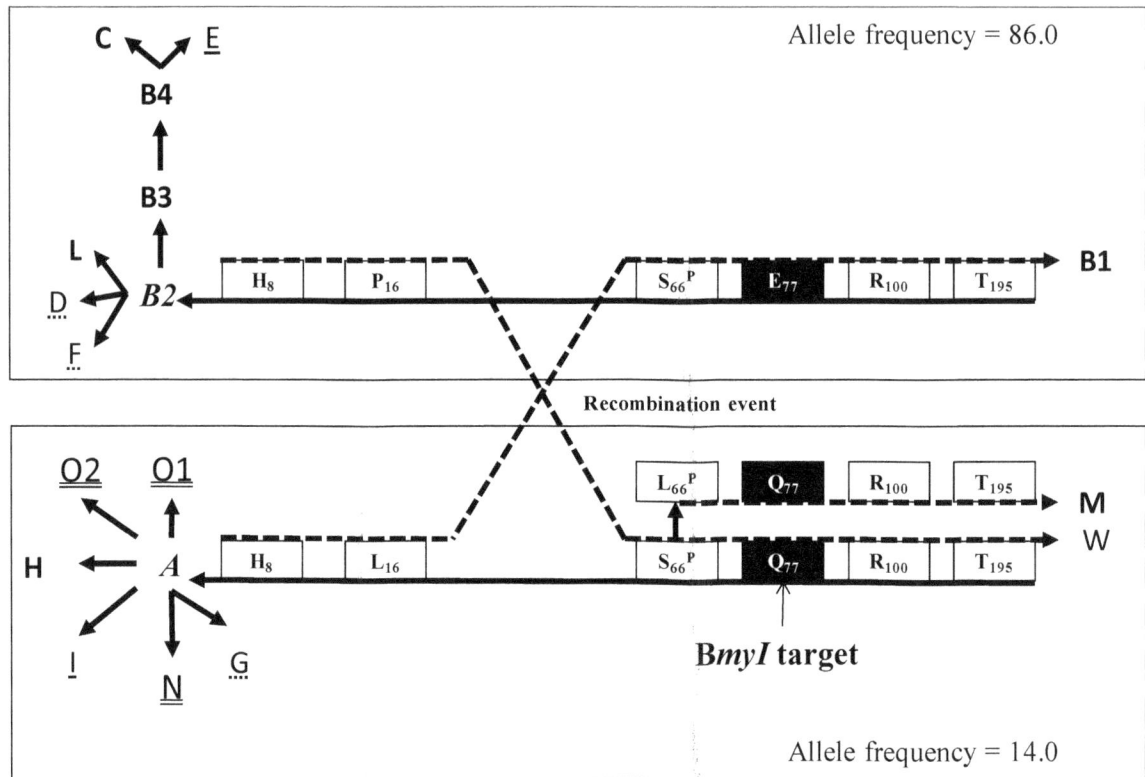

Figure 3. Schematic view of lineages of the αS1-casein locus using a simplified formula for haplotypes. The small boxes mimic exons but in some of them were annotated the polymorphic amino acid residues that occurred between lineages and also it is indicated the amino-acid residue targeted by Bmyl. Both large boxes regroup variants of casein: top for lineage B2 and bottom for lineage A. The numbers at right (down and top) are the frequency of each lineage in "Retinta Extremaña" goat. The continuous lines under the small boxes regroup the amino acid residues belonging to the respective ancestral lineages A or B2 and derived variants but dashed lines are signalling sites of inter-allelic recombination event between both as causes of the B1, W and M variants. Dark arrows indicate the possible evolutionary pathway scenario. Underline of variants: not underlined, simple underline dotted underline and double underlined for those of high yield, average yield, low yield and null (adapted from Bevilacqua et al., 2002)

Several studies revealed that the genotypes with strong or medium alleles of the B2 lineage of the αs1-casein predominated among goat breeds from the Mediterranean range (Sarda, Maltese, Moroccan, Tunisian; Moioli et al., 2007 and references there in) and, within the Iberian Peninsula, in southern breeds as the Murciano-granadina, Malageña or Payoya (Jordana et al., 1996; Caravaca et al., 2009) but allele E being at high frequency (Jordana et al., 1996). Regarding αs1-casein lineages, the genotypic frequencies observed in the Retinta Extremeña breed was similar to other Mediterranean breeds. For example, by pooling alleles belonging to the lineages B2, which carried the GAG triple that coded for the E (glutamic acid) (Bevilacqua et al., 2002), for the murciano-granadina breed (Caravaca et al., 2009) the estimated frequency was around the 88% (the complementary 12% for lineage A). As a whole, the native goat breeds from the Iberian Peninsula showed frequencies of the lineage A within a range between 10% and 20% (Jordana et al., 1996), what includes the results of the Retinta Extremeña reported here. Then, alleles of low frequency but high technological interest might be easily screened for a first glance by the Bmyl-RFLP and thus uncovering hidden αs1-casein lineages on endangered and/or bottlenecked breeds.

Contradictory results about the effect on total protein, fat and casein contents reported for this casein locus (Mahé, Manfredi, Ricordeau, Piacere, & Grosclaude, 1993; Manfredi, Ricordeau, Barbieri, Amigues, & Bibe, 1995; Grosclaude & Martin 1997; Caravaca et al., 2009; Valenti et al., 2010) suggested to be cautious in making decisions about this particular locus regarding the Retinta breed. Both, complicated interactions among genotypes, other *loci* and environmental factors as certain dependence on the particular polymorphism present in populations under consideration should be deemed (Caravaca et al., 2009, Valenti et al., 2010). Recent studies suggested that although some interaction genotype x diet should be expected; comparisons between AA vs FF genotypes affected to nine different fatty acids. But it should stress that monounsaturated fatty acids (especially C16:1 and cis-9 C18:1) were lower in milk from AA (lineage A) than FF genotypes (lineage B2) (Valenti et al., 2010). Other studies suggested that milk defective in αs1-cn (null alleles) could be less allergenic as reported by Bevilacqua et al. (2001). The goats with these particular genotypes may to be of interest because these have been proposed as substitute of cow milk for people having allergies to cow proteins (Moioli et al., 2007). Regarding to the conservation of goat breeds, all these studies are suggestive because somehow give support to research on the different casein variants within endangered breeds, such as the retina breed. In fact, actions aimed to genotype for proved useful genes should be among the priorities for conservation of specific animal genetic resources (Boettcher et al., 2010), especially for breeds well adapted to difficult territories as the Retinta Extremeña breed. As claimed Caravaca et al. (2009), more investigations in different breeds regarding the effects of casein polymorphism on goat milk yields and the features of their products should be addressed, because it may have an especial value in future comprehensive conservation plans.

Acknowledgements

We thank to CENSYRA (Badajoz, Junta de Extremadura) for providing all samples of the endangered Retinta Extremeña breed and to support this study. We also would like to thank anonymous reviewers for their valuables advices.

References

Bevilacqua, C., Ferrant, P., Garro, G., Veltri C., & Lagonigro, R. (2002). Interallelic recombination is probably responsible for the occurrence of a new αs1-casein variant found in the goat specie. *European Journal* of *Biochemistry, 269*, 1293-1303. http://dx.doi.org/10.1046/j.1432-1033.2002.02777.x

Bevilacqua, C., Martin, P., Candalh, C., Fauquant, J., Piot, M., Roucayrol, A. M., ... Heyman, M. (2001). Goats' milk of defective alpha(s1)-casein genotype decreases intestinal and systemic sensitization to beta-lactoglobulin in guinea pigs. *Journal of Dairy Research, 68*, 217-227. http://dx.doi.org/10.1017/S0022029901004861

Boettcher, P. J., Tixier-Boichard, M., Toro, M., Simianer, H., Eding, H., Gandini, G., ... the GLOBALDIV Consortium. (2010). Objectives, criteria and methods for using molecular genetic data in priority setting for conservation of animal genetic resources. *Animal Genetics, 41*(Suppl. 1), 1-14. http://dx.doi.org/10.1111/j.1365-2052.2010.02050.x

Caravaca, F., Carrizosa, J., Urrutia, B., Baena, F., Jordana, J., Amills, M., Badaoui, B., Sanchez, A., Angiolillo, A., & Serradilla, J. M. (2009). Effect of αS1-casein (*CSN1S1*) and κ-casein (*CSN3*) genotypes on milk composition in Murciano-Granadina goats. *Journal of Dairy Science*, 92, 2960-2964. http://dx.doi.org/10.3168/jds.2008-1510

Chilliard, Y., Rouel, J., & Leroux, C. (2006). Goat's alpha-s1 casein genotype influences its milk fatty acid composition and delta-9 desaturation ratios. *Animal Feed Science and Technology, 131*, 474-487. http://dx.doi.org/10.1016/j.anifeedsci.2006.05.025

Clark, S., & Sherbon, J. W. (2000). Alpha S1- casein, milk composition and coagulation properties of goat milk. *Small Ruminant Research, 38*, 123-134. http://dx.doi.org/10.1016/S0921-4488(00)00154-1

Fernández-García, J. L. (2012). The endangered *Dama dama mesopotamica*: genetic variability, allelic loss and hybridization signals. *Contributions to Zoology, 81*(4), 223-233.

Grosclaude, F., & Martin, P. (1997). Casein polymorphisms in the goat. IDF Seminar`Milk Protein Polymorphism II' (pp. 241-253). International Dairy Federation, Palmerston North, New Zealand.

Grosclaude, F., Mahé, M. F., Brigonon, G., Di Stasio, L., & Jeunet, R. (1987). A. Mendelian polymorphism underlying quantitative variation of goat αs1-CN. *Genetics Selection Evolution, 19*, 399-412. http://dx.doi.org/10.1186/1297-9686-19-4-399

Hayes, H., Petit, E., Bouniol, C., & Popescu, P. (1993). Localization of the alpha S2-casein gene (CASAS2) to the homologous cattle, sheep and goat chromosomes 4 by *in situ* hybridization. *Cytogenetics and Cell Genetics, 64*, 281-285. http://dx.doi.org/10.1159/000133593

Jordana, J., Amills, M., Díaz, E., Angulo, C., Serradilla, J. M., & Sanchez, A. (1996). Gene frequencies of caprine αs1-casein polymorphism in Spanish goat breeds. *Small Ruminant Research, 20*, 215-221. http://dx.doi.org/10.1016/0921-4488(95)00813-6

Koressaar, T., & Remm, M. (2007). Enhancements and modifications of primer design program Primer3. *Bioinformatics, 23*, 1289-1291. http://dx.doi.org/10.1093/bioinformatics/btm091

Leroux, C., Martin, P., Mahè, M. F., Leveziel, H., & Mercier, J. C. (1990) Restriction fragment length polymorphism identification of the goat αs1-casein alleles. A potential tool in selection of individuals carrying alleles associated with a high level protein synthesis. *Animal Genetics, 21*, 341-351. http://dx.doi.org/10.1111/j.1365-2052.1990.tb01979.x

Leroux, C., Mazure, N., & Martin, P. (1992). Mutations away from splice site recognition sequences might cis-modulate alternative splicing of goat αs1-casein transcripts. Structural organization of the relevant gene. *Journal of Biological Chemistry, 267*, 6147-6157.

Mahè, M. F., & Grosclaude, F. (1989). aS1-CnD, another allele associated with a decreased synthesis rate at the caprine αS1-casein locus. *Genetics Selection Evolution, 21*, 127-129. http://dx.doi.org/10.1186/1297-9686 -21-2-127

Mahè, M. F., Manfredi, E., Ricordeau, G., Piacere, A., & Grosclaude, F. (1993). Effets du polymorphisme de la caseine αs1 caprine sur les performances laitieres: Analyse intradescendance de boucs de race Alpine. *Genetics Selection Evolution, 26*, 151-157. http://dx.doi.org/10.1186/1297-9686-26-2-151

Manfredi, E., Ricordeau, G., Barbieri, M. E., Amigues, Y., & Bibe, B., (1995). Genotype caseine αs1 et selection des boucs sur descendance dans les races Alpine et Saanen. *Genetics Selection Evolution, 27*, 451-458. http://dx.doi.org/10.1186/1297-9686-27-5-451

Marshall, T. C., Slate, J., Kruuk, L. E. B., & Pemberton, J. M. (1998). Statistical confidence for likelihood-based paternity inference in natural populations. *Molecular Ecology, 7*, 639-655. http://dx.doi.org/10.1046/j.1365 -294x.1998.00374.x

Moioli, B., D'Andrea, M., & Pilla, F. (2007). Candidate genes affecting sheep and goat milk quality. *Small Ruminant Research, 68*, 179-192. http://dx.doi.org/10.1016/j.smallrumres.2006.09.008

Pirisi, A., Colin, O., Laurent, F., Scher, J., & Parmentier, M. (1994). Comparison of milk composition, cheesemaking properties and textural characteristics of the cheese from two groups of goats with high or low rate of alfa-s1-casein synthesis. *The International Dairy Journal, 4*, 329-345. http://dx.doi.org/10.1016/ 0958-6946(94)90030-2

Ramunno, L., Cosenza, G., Rando, A., Illario, R., Gallo, D., Di Berardino, D. & Masina, P. (2004). The goat alphas1-casein gene: gene structure and promoter analysis. *Gene, 334*, 105-111. http://dx.doi.org/10.1016/ j.gene.2004.03.006

Remeuf, F. (1993). Influence of genetic polymorphism of caprine αs1-casein on physicochemical and technological properties of goat's milk. *LAIT, 73*, 549-557. http://dx.doi.org/10.1051/lait:19935-652

Sacchi, P., Chessa, S., Budelli, E., Bolla, P., Ceriotti, G., Soglia, D., ... Caroli, A. (2005). Casein haplotype structure in five Italian goat breeds. *Journal of Dairy Science, 88*, 1561-1568. http://dx.doi.org/10.3168/jds. S0022-0302(05)72825-3

Valenti, B., Pagano R. I., & Avondo, M. (2012). Effect of diet at different energy levels on milk casein composition of Girgentana goats differing in CSN1S1 genotype. *Small Ruminant Research, 105*, 135-139. http://dx.doi.org/10.1016/j.smallrumres.2011.11.013

Valenti, B., Pagano, R. I., Pennisi, P., Lanza, M., & Avondo, M. (2010). Polymorphism at αs1-casein locus. Effect of genotype diet × interaction on milk fatty acid composition in Girgentana goat. *Small Ruminant Research, 94*, 210-213. http://dx.doi.org/10.1016/j.smallrumres.2010.07.023

10

In-vitro Characterization of an L-Kynurenine-Responsive Transcription Regulator of the Oxidative Tryptophan Degradation Pathway in *Burkholderia xenovorans*

Richard S Hall[1], Ricardo Martí-Arbona[1], Scott P Hennelly[1], Tuhin S Maity[1], Fangping Mu[1], John M Dunbar[1], Clifford J Unkefer[1] & Pat J Unkefer[1]

[1] Los Alamos National Laboratory, Los Alamos, NM 87545, United States

Correspondence: Pat J Unkefer, Los Alamos National Laboratory, Los Alamos, NM 87545, United States. E-mail: punkefer@lanl.gov

Abstract

To study the transcriptional regulation of oxidative tryptophan degradation in *Burkholderia*, we used comparative genomics that focused on the operon containing the genes annotated as *kynA*, *kynU* and *kynB*. In all sequenced β-proteobacteria to-date, including *Burkholderia*, *Ralstonia*, *Collimonas*, and *Cupriavidus* species, there is a conserved AsnC/Lrp family transcriptional regulator (TR) gene located upstream and in the opposite strand of the operon encoding for the oxidative tryptophan degradation genes. In *Burkholderia xenovorans* the TR is Bxe_A0736. GST-Bxe_A0736 binds L-kynurenine with greater affinity and specificity than any other amino acid or tryptophan degradation product with a dissociation constant of ~82 ± 11 μM. *DNase*I footprinting suggested that Bxe_A0736 protects a set of four degenerate, palindromic sequences within the intergenic region between Bxe_A0735 (*kynB*) and Bxe_A0736. The optimal consensus sequence obtained by analysis for these sites, ATATTCCGAATAT, closely resembles the sequence obtained with a protein binding microarray. Under our fluorescence anisotropy experimental conditions, 1 mM L-kynurenine increased the affinity of Bxe_A0736 for a portion of its promoter region. Our results are consistent with Bxe_A0736 acting as the TR that promotes the transcription of the oxidative tryptophan degradation genes in the presence of L-kynurenine while inhibiting the transcription of its own gene.

Keywords: Bxe_A0736, L-tryptophan degradation, kynurenine, fluorescence anisotropy, *Burkholderia xenovorans*, TR, transcriptional regulator, FAC-MS, frontal affinity chromatography coupled to mass spectrometry, COG, clusters of orthologous groups, PBM, protein binding microarray, footprinting, TRE, transcriptional regulator effector

1. Introduction

The kynurenine pathway oxidatively degrades L-tryptophan to provide important precursors for molecules such as quinolobactin siderophores and quorum signaling quinolones. KynA, KynB and KynU catalyze the first steps in the kynurenine pathway. The products of these first pathway steps are formic acid, anthranilic acid and L-alanine, where N-formyl-L-kynurenine and L-kynurenine are intermediates (Figure 1). L-Tryptophan oxidation activity is induced by L-kynurenine in the γ-proteobacterium *Pseudomonas fluorescens* (Palleroni & Stanier, 1964). Mutants of the L-kynurenine pathway in *P. aeruginosa* had a reduced ability to kill *Staphylococcus aureus*, possibly through a reduced production of 4-quinolones such as the *Pseudomonas* quinolone signal (PQS) 2-heptyl-3-hydroxy-4-quinolone (Farrow & Pesci, 2007). The L-kynurenine route of tryptophan degradation has been implicated as a contributor to bacterial pathogenesis through quinolobactin siderophore biosynthesis (Matthijs et al., 2004) and L-kynurenine has been implicated in contributing to a state of infectious tolerance within the host (Belladonna, Orabona, Grohmann, & Puccetti, 2009).

The β-proteobacterium *Burkholderia xenovorans* shares much in common with pathogens such as *B. mallei*, *B. pseudomallei*, *P. aeruginosa* and *P. fluorescens*. These similarities are evident in the genomic organization around the genes *kynA*, *kynB* and *kynU* (Figure 2). In *B. xenovorans*, the TR Bxe_A0736 is located 130 bases upstream of the start codon for *kynB* on the opposite strand. Bxe_A0735 is a very well conserved TR annotated as *asnC*. Proteins sharing a greater than 60% sequence identity, with similar operon organization, are present throughout β-proteobacteria, including *Burkholderia* species, *Cupriavidus taiwanensis*, *Cupriavidus metallidurans*, *Ralstonia solanacearum*, *Ralstonia pickettii*, *Ralstonia metallidurans*, *Ralstonia eutropha*, and in the γ-proteobacteria *Pseudomonas fluorescens* and *Pseudomonas aeruginosa*. The AsnC/Lrp family TR PA2082 from *P. aeruginosa* is responsible for the L-kynurenine-dependent transcription of the kynurenine pathway genes *kynA* and *kynB* and should be annotated as *kynR* (Knoten, Hudson, Coleman, Farrow & Pesci, 2011). KynR from *P. aeruginosa* and Bxe_A0736 share a 66% protein sequence identity.

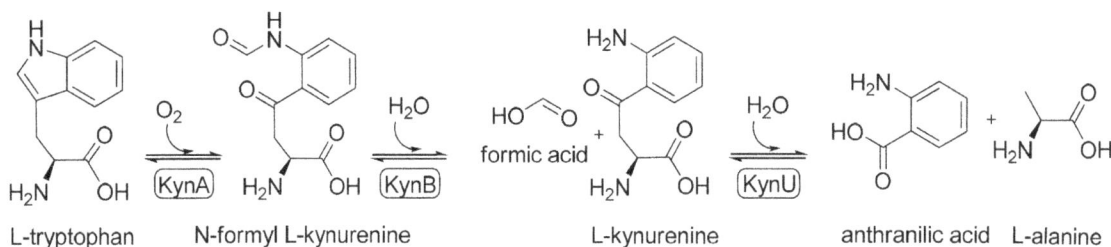

L-tryptophan   N-formyl L-kynurenine   L-kynurenine   anthranilic acid   L-alanine

Figure 1. The kynurenine pathway of oxidative degradation of L-tryptophan. The products of the pathway are anthranillic acid, alanine and formic acid

The AsnC/Lrp family of TRs is named after two of the three TRs in this family that are present in *E. coli*. AsnC binds L-asparagine (de Wind, de Jong, Meijer, & Stuitje, 1985; Kolling & Lother, 1985) to control the transcriptional repression of asparagine synthetase while acting as an autorepressor of its own transcription (Kolling & Lother, 1985). Lrp binds L-leucine, controlling a global metabolic response, particularly in amino acid synthesis and degradation in response to feast or famine (Brinkman, Ettema, de Vos, & van der Oost, 2003; Calvo & Matthews, 1994; Landgraf, Wu, & Calvo, 1996). The third family member, YbaO has no known effector. Effectors have been indentified for other members of the AsnC/Lrp family including L-leucine for Lrp (Willins, Ryan, Platko & Calvo, 1991), L-proline for PutR (Keuntje, Masepohl, & Klipp, 1995), L-glutamate for Grp (Peekhaus, Tolner, Poolman, & Kramer, 1995), L-methionine for MdeR (Inoue et al., 1997) and branched chain amino acids for BkdR (Madhusudhan, Lorenz, & Sokatch, 1993).

X-ray crystal structures have shown that members of the AsnC/Lrp family of TRs form an octameric disc shape where the helix-turn-helix DNA binding domains are displayed on the outside of the structure (Koike, Ishijima, Clowney & Suzuki, 2004; Ren et al., 2007; Thaw et al., 2006). Based on those structures, it has been proposed that DNA wraps around the perimeter of the octameric protein complex. This structure may explain why several of the TRs in this family have been shown to have rather long regions of DNA interacting with these TRs (Enoru-Eta, Gigot, Thia-Toong, Glansdorff, & Charlier, 2000; Wang & Calvo, 1993; Yokoyama, Ebihara, Kikuchi, & Suzuki, 2005). The closest homologue of Bxe_A0736 with a crystal structure deposited to the Protein Data Bank is Lrp from *Neisseria meningitides* (Ren et al., 2007). X-ray crystal structures of Lrp showed an absence of significant conformational differences between L-leucine- bound Lrp and ligand-free Lrp. It was stated that effector binding appears to stabilize the octameric assembly of the TR (Ren et al., 2007). The mechanism of effector-controlled transcriptional regulation by members of the AsnC/Lrp family is in need of further study.

Reliable functional knowledge is lacking for many proteins. This is especially true for transcriptional regulators due to the difficulty in uncovering their function and mechanism. To characterize the TR Bxe_A0736 in *B. xenovorans* we utilized a battery of techniques for transcriptional regulator prediction and characterization, effector prediction, screening and characterization as well as DNA regulatory site identification and characterization. Here we report that TR Bxe_A0736 binds L-kynurenine but not other metabolites in tryptophan degradation and that it interacts with a rather long region of DNA that contains repeated pallindromic sequences that are found upstream of Bxe_A0735. We provide evidence in support of Bxe_A0736 as involved in regulating the oxidative tryptophan degradation genes *kynA*, *kynB* and *kynU*.

2. Methods

2.1 Sequence Similarity Network Analysis

A sequence similarity network was constructed according to the method of Atkinson and coworkers (Atkinson, Morris, Ferrin, & Babbitt, 2009). Their method of independent pairwise alignments between sequences allows functional relationships to be observed over very large sets of evolutionarily related proteins, such as members of specific COGs (clusters of orthologous groups). The COG database accessible on the NCBI website contains most of the predicted proteins in 66 genomes of unicellular organisms (Tatusov et al., 2003). According to the classification in this database, COG1522 contains the diverse members of the AsnC/Lrp TR family. The method requires that increasingly stringent BLAST E value cutoffs be tested until the desired degree of clustering is observed. At a BLAST E value cutoff of 10^{-40}, some clustering is achieved; the proline-binding PutR and the leucine-binding Lrp proteins grouped together and the glutamate-binding Grp from *Zymomonas mobilis* grouped with YbaO from *E. coli*, a TR of unknown function. At a BLAST E value cutoff of 10^{-45}, the different characterized functions for COG1522 separate into the distinct groups. This E value was used. The results are visualized using the organic layout in Cytoscape (Cline et al., 2007).

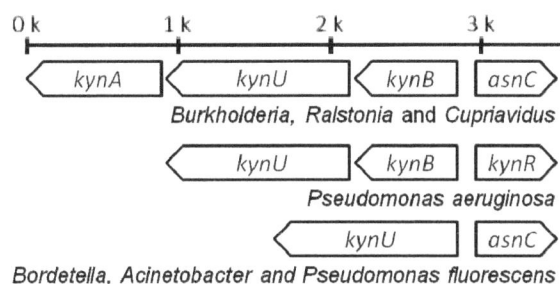

Figure 2. Genomic organization around *kynA*, *kynB* and *kynU* from several bacteria. The *Burkholderia xenovorans* TR Bxe_A0735 is annotated as *asnC*. The positions of TR-encoding genes annotated as *asnC* and *kynR* relative to other genes in the kynurenine pathway are conserved in these organisms

2.2 Cloning and Protein Production

2.2.1 Cloning and Overexpression of GST-Bxe_A0736

Bxe_A0736 was synthesized and sub-cloned into pET42a(+) by Genscript with codon usage optimized for protein expression in *E. coli* (the optimized DNA sequence for the protein can be found in the supplemental material, Figure S1). The *PshA*I and *EcoR*I cloning sites were utilized to incorporate an N-terminal GST protein fusion tag which is cleavable by Factor Xa digestion. The plasmid DNA was then transformed into ArcticExpress (DE3) competent cells from Agilent Technologies and plated onto an LB/Agar plate containing50 μg/mL kanamycin. A single colony was used to inoculate LB media supplemented with 50 μg/mL kanamycin and 20 μg/mL gentamycin for overnight growth with shaking at 37°C. 10 mL of the overnight culture was used to inoculate each liter of sterile LB broth in a 2 liter shaker flask and supplemented with 50 μg/mL kanamycin. The cells were incubated at 37°C with shaking until they yielded an O.D. of 0.5 at 600 nm. The temperature of the growth chamber was then reduced to 15°C and isopropyl-β-d-thiogalactopyranoside (IPTG) was added to a final concentration of 0.5 mM.

2.2.2 Cloning and Overexpression of Bxe_A0736-His$_8$

The Bxe_A0736 gene was amplified by conventional PCR methods (manufacturer's protocol for Platinum *Pfx* DNA polymerase- Invitrogen) from the pET42a(+) construct synthetized by Genscript (section 2.1.1). The Bxe_A0736 gene was amplified using the primers: forward 5'-CATATGAATGCGATCTCCCTGGAC-3' and reverse 5'-CTCGAGGCGATCATTATGGGCAGC-3', containing *Nde*I and *Xho*I sites, respectively. The resulting PCR product was digested with the respective endonucleases and ligated into the pET42a(+) vector at the *Nde*I and *Xho*I sites. The C-terminal His$_8$ tagged Bxe_A0736 (Bxe_A0736-His$_8$) encoding recombinant plasmid was transformed into ArcticExpress (DE3) competent cells from Agilent Technologies and plated onto an LB/Agar plate containing 50 μg/mL kanamycin. A single colony was used to inoculate LB media supplemented with 50 μg/mL kanamycin and 20 μg/mL gentamycin for overnight growth with shaking at 37°C. 10 mL of the overnight culture was used to inoculate each liter of sterile LB broth in a 2 liter shaker flask and supplemented with 50 μg/mL kanamycin. The cells were incubated at 37°C with shaking until they yielded an O.D. of 0.5 at 600 nm. The temperature of the growth chamber was then reduced to 15°C and isopropyl-β-d-thiogalactopyranoside (IPTG) was added to a final concentration of 0.5 mM.

2.2.3 Protein Purification

After incubation at 15°C for 24 hours, the cells (expressing either GST-Bxe_A0736 or Bxe_A0736-His$_8$) were harvested with centrifugation at 3,000 × g for 25 min. The cells were then re-suspended in 10 mM Na$_2$HPO$_4$, 1.8 mM KH$_2$PO$_4$, 140 mM NaCl, 2.7 mM KCl, 1 mM DTT, 1 mM EDTA, pH 7.3 (Buffer A) supplemented with 100 µg/mL of the protease inhibitor phenylmethanesulfonyl fluoride and 1 unit/mL of DNase I. The cells were then lysed with sonication while being stirred on ice and then the cell debris was removed from solution by centrifugation at 14,000 × g for 30 min. The remaining solution was passed through a 0.45 um syringe filter and loaded onto an AKTA Purifier HPLC from GE Healthcare with two stacked 5 mL GSTrap FF HiTrap affinity columns for the GST-Bxe_A0736 or two stacked 5 mL HisTrap HP (Ni) affinity columns for the Bxe_A0736-His$_8$. The bound protein was washed with 10 column volumes of binding buffer and then eluted with a linear gradient (0-50%) of 20 mM Tris at pH 8.0 containing 1 mM DTT and 700 mM glutathione for GST-Bxe_A0736 or with a 0-55% gradient of Buffer A containing 700 mM imidazole for Bxe_A0736-His$_8$. In each separate case, the protein containing fractions were pooled and concentrated to about 10 mL and purified in Buffer A with a HiLoad 26/60 Superdex 200 prep grade gel filtration column, this extra purification step removes the traces of elution buffer (glutathione or imidazole) as well as further purify the proteins. The purification steps were monitored with SDS PAGE analysis, and the final protein solution yielded a single visible band. The GST-Bxe_A0736 was utilized for the FAC-MS and protein binding microarray experiments and the Bxe_A0736-His$_8$ was used for the fluorescence anisotropy experiments described below. A representative gel for the expression and purification steps for the GST-Bxe_A0736 fusion protein can be found in the supplemental material, Figure S2.

2.2.4 FAC-MS Column Preparation

The GST-Bxe_A0736 protein was diluted to a concentration of 1 mg/mL and loaded onto a 1 mL GSTrap FF affinity column. After loading the GST-Bxe_A0736 protein, the FAC-MS columns were washed with 10 column volumes of buffer A. The preparation of the FAC-MS column was concluded by exchanging the binding buffer for a MS friendly buffer such as 20 mM ammonium formate at pH 7.2.

2.2.5 Preparation of WT Bxe_A0736

Wild type (non-tagged) Bxe_A0736 was prepared by cleaving the GST fusion tag from GST-Bxe_A0736 fusion protein by factor Xa from Thermo. The purified fusion protein was loaded onto a 1 mL GSTrap FF column and equilibrated with factor Xa cleavage buffer composed of 50 mM HEPES, 100 mM NaCl, 5 mM CaCl$_2$, 0.5 mM DTT, pH 7.5. Twenty units of Factor Xa were then dissolved in 1 mL of the cleavage buffer and loaded onto the column. The column was incubated at room temperature overnight after which the cleaved Bxe_A0736 and Factor Xa were eluted from the column. Factor Xa was removed with Novagen Xarrest agarose beads according to the manufacturer's instructions. The WT Bxe_A0736 was utilized for the *DNase*I footprinting experiments described below. A gel showing the purity and approximate size of the WT Bxe_A0736 protein can be found in the supplemental material, Figure S3.

2.3 Quaternary Structure Determination with Size Exclusion Chromatography

A Tricorn 5/150 column from GE Healthcare was filled with Superdex 200 prep-grade resin and equilibrated with buffer containing 50 mM Tris, 100 mM KCl, pH 7.5. The column was calibrated with a 12-200 kDa molecular weights kit for gel filtration chromatography, purchased from Sigma-Aldrich. To determine the elution volume for each protein in the kit and for WT Bxe_A0736, the protein standards were dissolved in equilibration buffer and WT Bxe_A0736 was concentrated to 1 mg/mL. 100 µL of each protein was injected onto the column. The size of purified WT Bxe_A0736 was then estimated with the use of a calibration curve through comparison of its elution volume to those of the protein size standards.

2.4 Effector Library Design and Screening with FAC-MS

Possible effectors were identified based on the frequency of occurrence of those compounds in the KEGG database (http://www.genome.jp/kegg/) as substrates and products of the five genes upstream and downstream of Bxe_A0736 and its twenty closest homologues. Commercially available compounds sharing similar structures to those compounds were included in the library. The compounds tested as effectors included all standard amino acids, 3,4-dihydroxy benzoic acid, 3-hydroxy anthranilic acid, 4-hydroxy benzoic acid, L-kynurenine, 3-hydroxy L-kynurenine as well as the deoxy nucleotide triphosphates, diphosphates and monophosphates.

For effector screening, we utilized our frontal affinity chromatography with ESI-MS detection (FAC-MS) that is designed specifically for screening for TR effector molecules (Marti-Arbona et al., 2012). The affinity column used was GST-Bxe_A0736 fusion protein immobilized on a 1 mL GSTrap FF column purchased from GE. The protein bound column was washed with 20 mM ammonium formate, pH 7.4 for about ten column volumes until the absorbance at 280 nm stabilized. High-resolution mass detection was performed using a Thermo Exactive ESI-MS equipped with a high-flow steel needle. A typical screening run was performed by infusing mixtures of eight compounds into the column at a rate of 85 μL/minute. Each compound was present at 50 μM in the ammonium formate buffer. To achieve sufficient positive mode ionization, formic acid-spiked methanol was mixed in equal ratios with the post-column eluent prior to electrospray ionization. The elution times were then noted to rank the affinity of ligands and to identify non-binders that could serve as possible non-ligand void volume markers. The small molecule metabolites have no affinity for the column matrix without immobilized Bxe_A0736.

2.5 DNA Operator Sequence Identification

2.5.1 Protein Binding Microarray Experiments

A microarray containing all possible 10mer nucleotide combinations was described by Maity and coworkers (Maity et al., 2012). This microarray (available from Agilent) can be used to find a set of sequences for which a DNA binding protein has affinity. These sequences are then analyzed to define a consensus nucleotide sequence bound by the target protein. The GST-Bxe_A0736 fusion protein was bound to the microarray prepared according to the manufacturer's instructions. Protein bound to the microarray was treated with an Alexa488-labeled GST antibody and imaged with the Genepix professional 4200A scanner at resolution of 5 μm and an excitation of 488 nm.

2.5.2 DNaseI Footprinting Experiments

To prepare DNA material for the *DNase*I footprinting experiments, the intergenic region between Bxe_A0736 and Bxe_A0735 plus 60 bp of each flanking gene was PCR amplified from genomic DNA using Alexa488-5'-TGTCGCCTGGCCACACGG-3' primer and the non-tagged reverse primer 5'-TCTTGCTGAAGCACCGTCAAGATACG-3'. This resulted in a PCR product that is a 253 bp piece of Alexa488-labeled, double stranded DNA. The fluorescently labeled DNA was then sequenced using the Thermo Sequenase Cycle Sequencing Kit, according to manufacturer's instructions. The sequencing reactions were analyzed on an ABI 310 genetic analyzer. For the DNase I footprinting reactions, 200 ng of DNA were digested on ice for 5 minutes with 0.004 units of DNase I, with and without various amounts of WT Bxe_A0736 and L-kynurenine. All reactions were performed in Factor Xa cleavage buffer supplemented with 5 mM MgCl2 and 5 mM DTT. The digested DNA was isolated by phenol/chloroform extraction and ethanol precipitation. For capillary electrophoresis analysis, the DNA samples were de-salted using Bio-Rad Micro Bio-Spin 6 columns, dried and re-suspended in formamide containing 1 mM EDTA and were heated at 95°C for 2 min.

2.6 Fluorescence Anisotropy

Fluorescence anisotropy was performed with a BioTek Synergy H4 Hybrid Reader, using the standard filter wheel setup for fluorescein. This uses an excitation wavelength of 485 nm with a 485/20 filter and an emission wavelength of 528 nm with a 528/20 filter. The reactions took place in a 96 well plate at 30°C in 200 μL of 25 mM HEPES, 50 mM KCl, 25 mM NaCl, 2 mM $MgCl_2$, 0.1 mg/mL BSA, pH 7.5, using Bxe_A0736-His_8 and 1 nM of either of the two 39 bp 5'-fluorescein labeled double-stranded DNA oligos. Each set of oligos contained two sites, representing each half of the protected intergenic region. Listed below in the 5' to 3' direction are the forward sequences of the two oligos. The reverse-complement sequences were label free. All oligos were ordered from Invitrogen.

Oligo1: Fluorescein-5'-TGATAGGTGGCTTGACGCGAAATGTGATTGCGAAAAAAT-3'

Oligo2: Fluorescein-5'-ATAATTTGAGATAAATGGGAAAGGGAGACCGAATATGAA-3'

2.7 Data Analysis

Sequence alignments were performed using Clustal W. Consensus sequence analysis was performed using the regulatory sequence analysis tool (Thomas-Chollier et al., 2008). Binding constants were determined by regression analysis using Sigmaplot 12.0. Equation 1 describes the binding of FAC-MS determination of the dissociation constant of L-kynurenine binding to GST-Bxe_A0736 where V is the elution volume of the ligand, V_0 is the elution volume of a non-ligand void marker and B_t is the amount of immobilized protein with an active binding site (Slon-Usakiewicz, Ng, Dai, Pasternak, & Redden, 2005).

$$V - V_0 = B_t / ([A_0] + K_d) \qquad (1)$$

Re-arrangement to form a linear expression yields equation 2 where plotting $[A_0]$ versus $1/(V-V_0)$ yields a straight line where the slope is B_t and the negative y intercept is K_d.

$$[A]_0 = B_t [1/(V - V_0)] - K_d \qquad (2)$$

A nonlinear regression analysis of protein concentration versus fluorescence anisotropy was used with equation 3 to determine the kinetic constants for the binding of protein to fluorescein-labeled double stranded DNA. For this equation, ΔA is the change in anisotropy at a given protein concentration, ΔA_T is the total change in anisotropy, E is the protein concentration, H is the Hill coefficient for cooperative binding, and K_d is the dissociation constant for protein-DNA binding (LiCata & Wowor, 2008).

$$\Delta A = [(\Delta A_T (E^H / K_d^H)) / (1 + (E^H / K_d^H))] \qquad (3)$$

3. Results and Discussion

3.1 Sequence Similarity Network Analysis

The annotation of Bxe_A0736 as *asnC*, a TR that regulates asparagine synthetase in other bacteria (Kolling & Lother, 1985; Yokoyama et al., 2005), was not consistent with its conserved close proximity to the genes of the kynurenine pathway (Figure 2). To examine this apparent discrepancy, we first carried out a sequence similarity network analysis (Atkinson et al., 2009) of COG 1522, which contains the diverse members of the AsnC/Lrp TR family in 66 genomes of unicellular organisms (Figure 3). Bxe_A0736 and the L-kynurenine responsive TR KynR from *P. aeruginosa* are together in group 24 while AsnC is in group 17. Other TRs of known function, including Lrp, PurR, MdeR and NirH were in separate groups. YbaO and many Grp anotated proteins were in yet another group. Most of the groups contain only a few proteins not associated with any known function. This analysis combined with its genomic proximity with the kynurenine pathway genes encouraged us to experimentally characterize Bxe_A0736 as a potential kynurenine-responsive TR.

3.2 Native Molecular Weight and Quaternary Structure Examination

To examine the native oligomeric state of Bxe_A0736, we analyzed the native protein by size exclusion chromatography. According to its elution volume through a calibrated size-exclusion column, the protein eluted as a 183 ± 2 kDa protein regardless of the presence or absence of L-kynurenine. Given the calculated mass of 19.4 kDa for a single subunit, the estimated molecular weight of 183 kDa corresponds to 9.4 subunits for a globular shaped protein. Thus Bxe-A0736 is an oligomeric protein; an octameric quaternary structure could be expected (Yokoyama et al., 2005). It also appears that the method may have over estimated the native molecular weight. AsnC/Lrp family of TRs is known to be frequently disc shaped, not the much more familiar spherical shape. Size exclusion chromatography provides good estimates of the native molecular weight for spherical (globular) shaped proteins. However, it over-esimates the molecular weights for disc-shaped proteins because these proteins tumble and therefore move through the column as a function of the diameter of their disc. As a result, they chromatograph like larger spherical proteins. Our data are consistent with Bxe-A0736 being an octameric protein that is disc shaped, like all structurally characterized members of the AsnC/Lrp family of transcriptional regulators.

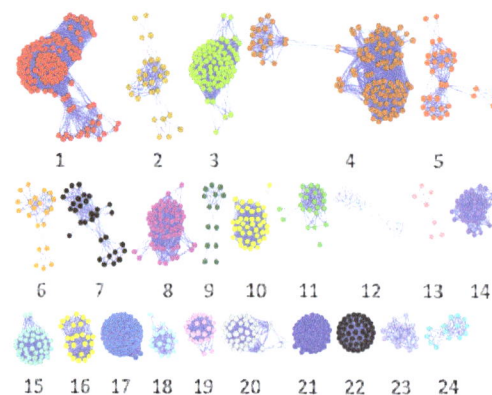

Figure 3. Cytoscape's organic layout sequence similarity network representation for COG1522 which contains AsnC/Lrp family of TRs. The kynurenine-responsive *knyR* and Bxe_A0736 are together in group 24 while *asnC* (the current annotation for Bxe_A0736) is in group 17. Each node represents a distinct sequence and each line (edge) is a pairwise connection between two sequences at an *E* value better than the cutoff

3.3 Effector Identification

We generated a list of possible effectors from the substrates and products listed in the KEGG database for the 5 genes upstream and downstream of the transcriptional regulator Bxe_A0736 and its 40 closest homologues in other bacteria. This list is highly enriched in the tryptophan oxidative metabolic products. The most common metabolite identified is L-formylkynurenine which has 73 occurrences. Anthranilate is the second most common metabolite with 43 occurrences. L-kynurenine, 3-hydroxy-L-kynurenine, formylanthranilate, 3-hydroxyanthranilate and L-alanine all have 41 occurrences. Structural analogues and all amino acids and nucleotides were included in the screen to test for possible effectors.

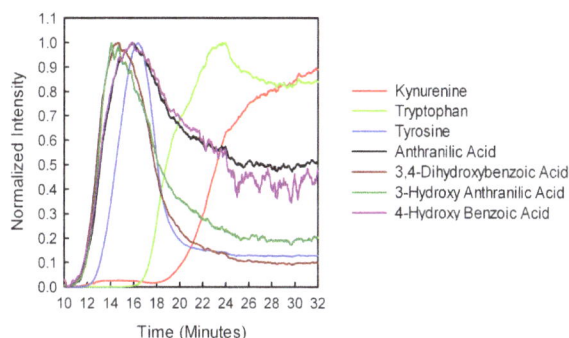

Figure 4. Screening for effector(s) of Bxe_A0736. The elution of the library of potential effectors (50 uM) after flowing past immobilized GST-Bxe-A0736 is shown. The affinity of Bxe_A0736 for these compounds is L-kynurenine (red) > 4-hydroxybenzoic acid (dark red) = anthranilic acid (black) > L-tryptophan (green) > L-tyrosine (pink) > 3,4-dihydroxybenzoic acid (brown) = 3-hydroxyanthranilic acid (dark green)

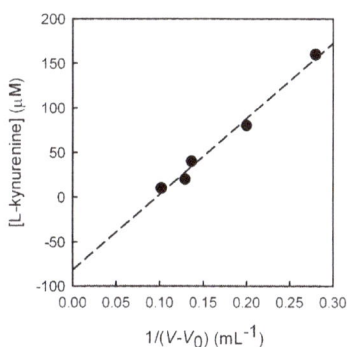

Figure 5. Estimation of the dissociation constant of kynurenine for Bxe_A0736. The elution volumes for kynurenine (10, 20, 40, 80 and 160 μM) when flowed past immobilized GST-Bxe_A0736 were used. The K_d for L-kynurenine was estimated to be 82 ± 11 μM. The column contained 850 ± 60 nanomoles of binding sites (B_t). K_d and B_t were estimated from fits to equation 2

FAC-MS screening identified L-kynurenine, L-phenylalanine, L-tryptophan, L-arginine, and L-lysine as binders of Bxe_A0736. Competition experiments showed that L-kynurenine binds to the immobilized GST-Bxe_A0736 fusion protein with high affinity as shown in Figure 4. The elution volume of L-kynurenine was consistent when present in different mixtures of possible effectors indicating specific binding. L-Phenylalanine, L-tryptophan, L-lysine and L-arginine also showed large elution volumes but were decreased when present in different mixtures, indicating non-specific binding. L-kynurenine was consistently the strongest binder and always had a sigmoidal shaped elution profile. We estimated the dissociation constant for GST-Bxe_A0736 with L-kynurenine to be 82 ± 11 μM (Figure 5). AMP was used as an indicator of the column void volume (V_0) due to its lack of interaction with the column; it was included in all titrations to estimate the breakthrough volume. We found all of the phosphorylated nucleotides were non-binders; they always eluted immediately and with consistent elution volumes.

3.4 Identification of Regulatory Sequence

A protein-binding microarray containing all possible 10mer nucleotides was used to find the set of sequences for which the Bxe_A0736 protein had affinity. By analyzing the 36 sequences for which Bxe_A0736 had affinity, we found a slightly degenerate A/T rich 6 bp sequence. The consensus sequence for the forward and reverse sequences of the top 36 binding sequences yields the consensus operator sequence (C/G)ATAT(A/G) (Figure 6, panel A).

DNase I footprinting applied to the intergenic region upstream of Bxe_A0735 revealed a rather large protected region composed of ~105 nucleotides base pairs. This region begins at -35 bp above the ATG start site for *kynU*. Within this protected region were four areas that showed enhanced protection. Large (up to 200 bps) regions of DNA involved in binding the TR and containing repeats of a consensus binding site have been observed in other members of this same TR family. These regions for Bxe-A0735 are illustrated in Figure 7.

Analysis of the four protected regions in the intergenic sequence did not yield an obvious consensus sequence, so we evaluated a sequence alignment of homologous intergenic regions from multiple *Burkholderia* species. The conserved regions from the eight closest homologues shown in Figure 7 correspond to the areas of enhanced protection. Utilizing these conserved and protected regions, we found the degenerate palindromic consensus sequence shown in the panel B of Figure 6. The degenerate sequence is: DBHDDNSNHHDVH where D = G, A or T; B = G, C or T; H = A, C or T; N = any nucleotide; S = C or G; and V = G, A or C. Based on the most common base found at each position, the optimal sequence is: ATATTCCGAATAT. This sequence is similar to the smaller half-site consensus sequence found using the protein-binding microarray. The fact that the sequences of these protein-binding regions are not highly conserved may be an indication that the TR recognizes common features of the DNA structure that may form from various DNA sequences.

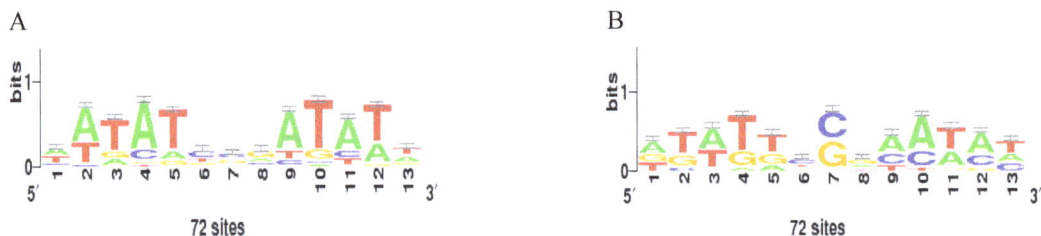

Figure 6. Consensus sequences for the regulatory sites derived for Bxe_A0736. Panel A shows the consensus sequence as determined from the protein binding microarray.et al.,Panel B shows the consensus sequence of the four binding sites protected during the phylogenetic footprinting using the intergenic sequence upstream from Bxe_A0736 and the eight homologues from different *Burkholderia* species

3.5 Fluorescence Anisotropy

Two oligos were constructed and used to test if the presence of kynurenine affected the binding by Bxe_0736 to portions of the much larger DNA region identified to interact with this TR. Because fluorescence anisotropy cannot be used with large DNA sequences, we divided the DNA interaction region into two sequences (oligos1 and 2); each contained two of the proposed binding sites. For each oligo with Bxe_A0736-His$_8$, the presence of L-kynurenene increased the affinity (ie decreased the dissociation constant) for the oligo. The titration of 1 nM fluorescein labeled oligo 1 with Bxe_A0736- His$_8$, yielded an apparent dissociation constant of 145 nM for Bxe_A0736-His$_8$. With 1 mM L-kynurenine present, the dissociation constant was lowered to 97 nM. The Hill coefficient was also lowered from 5.2 to 4.2 with L-kynurenine. Examination of Bxe_A0736-His$_8$ with 1 nM oligo2 yielded a dissociation constant of 154 nM and a Hill coefficient of 5.3 In the presence of 1 mM L-kynurenine, the dissociation constant was reduced to 128 nM and the Hill coefficient was lowered to 4.8. Even when higher concentrations of NaCl, KCl or MgCl$_2$ were used in the assay, the presence of L-kynurenine lowered the dissociation constant and Hill coefficient. The values obtained with these oligos cannot be expected to quantitatively reflect the affinity of Bxe_A0736 for the entire interaction region that contained four, not two binding sites as did the oligos. These values can be expected to be under estimations. In addition, the conditions of the experiment may not accurately reproduce the intracellular conditions. The data do support L-kynurenene increasing the affinity of Bxe_A0736 for the identified binding sites.

```
GENE NAME                                                    Kynurenine Formamidase (kynB)
BPHY_0509          ---CCCGGCCAGACGGGCGTCGCCGTGTCCACGGCAGGCGTGATATCCCAAAGTGTGTCC
Bxe_A0736          --GCCTGGCCACACGGGCGTGGCGGTATCGACGGCGGGTGTGATGTCCCAAAGTGTTCGC
BPHYT_3226         --GCCGGGCCAGACGGGCGTGGCGGTATCGACGGCAGGGGTGATGTCCCAAAGTGTTGGC
BC1002_2483        --GCCCGGCCAGACGGGCGTGGCGGGATCGACGGCGGGGGTGATGTCCCAGAGTGCTCGC
BH160DRAFT_4294    --CCCCGGCCAGACGGGCGTGGCGGTATCGACGGCGGGGGTAATGTCCCAAAGTGCTTGC
BOKLC_010100004123 TCGCCCGGCCAGACGGGCGTTGCGGGGTCGATTGACGGGGAGATGTCCCAGATCGTGTCC
BUBOB_010100009881 --GCCCGGCCACACGGGGGTGGCGGGGTCGACCGGCGGCGAGATGTCCCAGAGGGTGTCC
BAMMEX5DRAFT_5259  --GCCCGGCCACACGGGGGTGGCGGGGCTGACGGGCGGCGAGATGTCCCAGAGGGTGTCC
BCAL2789           -CGCCCGGCCACACGGGGGTGGCGGGGCTGACGGCGCGGTGAGATGTCCCAGAGTGTGTCC


BPHY_0509          ATGT-GTCCCCGGTGCATCTGGCAGCTAT-GGAACGATGATAGGCCGCATCAGACGAAAT
Bxe_A0736          ATAGTCGCATCGGCAGTTTGTG--TCTAT-CCGGGAATGATAGGTGGCTTGACGCGAAAT
BPHYT_3226         ATAGTCGCATCGGCGGTTTGTG--TCTAT-CCGGGAATGATAGGTGGCTTGCCGCGAAAT
BC1002_2483        ATGACGGTTTCGACGGAAGGTG--ACTAT-CTCTAAATGATAGAGGTCAAGCCGTGAAAT
BH160DRAFT_4294    ATGACGGTTTCGGCGAAAGGTG--ACTCT-CTTTGAATGATAGAGGTCAAGCCGCGAAAT
BOKLC_010100004123 ATGC-GATATCGAAGAACGTTG----GATGCTTCGGATGATAGGCAAACCTGCGAAGCAG
BUBOB_010100009881 ATGG-CATGGCGAAG--CAGCGAAATGATGCTTCGCATGATAGGCGGACGCTCGCAAAAA
BAMMEX5DRAFT_5259  ATGG-CATGACGAAT--CGGTGAATCGATCCTTCGAATGATAGGTGGACACTCGCAGAAA
BCAL2789           ATGG-CATGACGAAG---CGTGAGTCGATCCTTCGAATGATAGGTGGACGCTCGCAGAAA


BPHY_0509          GTGCTTGCTAAAAAAACTGTTTATTTCGCCCCCTTTGGAACATAATTTGAGGCAAACGGG
Bxe_A0736          GTGATTGCGAAAAAATCTCCTTCCAGGCCGTGTCTTGGAACATAATTTGAGATAAATGGG
BPHYT_3226         GTGATTGCGAAAAAATCTCCTTCTGCGCCGTGTCTTGGAACATAATTTGATTCAAACGGG
BC1002_2483        GTGATTGCGAAAAAATCTCCTAACCAGCCGTGTCTTGGAACATAATTTGAGCCAAACCGG
BH160DRAFT_4294    GTGATTGCGAAAAAATCTCCTTCCTAGCCGTGTCTTGGAACATAATTTGAGTCAAACAGG
BOKLC_010100004123 GTGCTTGCGAAATAAGCATGGACAATCCACCAATATGGCAGATAATTCGAAATAGACCTA
BUBOB_010100009881 GTGCTTGCGAAATAACCATGGACAAACCGCCAGATTGGCAGATAATTCGAAGCAGAGAGG
BAMMEX5DRAFT_5259  GTGCTTGCGAAATAAACATGGACAATCCGCCGGATTGGCAGATAATTCGAATGAGAGGGA
BCAL2789           GTGCTTGCGAAATAAACATGGACAATCCGCCGGATTGGCAGATAATTCGACTCAGAGGGA


BPHY_0509          AAAGGGAGACCGAAAATGAACGCGATCTCACTAGACGCCACCGATTGCCGTATCTTGACG
Bxe_A0736          AAAGGGAGACCGAATATGAACGCGATCTCGCTCGACGCCACCGATTGCCGTATCTTGACG
BPHYT_3226         AAAGGGAGACCGAATATGAACGCGATCTCGCTCGACGCCACCGATTGCCGTATCTTGACG
BC1002_2483        AAATGGAGACCGAATATGAACGCGATCTCGCTCGACGCCACCGATTGCCGTATCTTGACG
BH160DRAFT_4294    AAAGGGAGACCGAATATGAACGCGATCTCGCTCGACGCCACCGATTGCCGTATCTTGACG
BOKLC_010100004123 AAACGGAGACCAAAAATGCATGCGATCACGCTTGACGCAACCGATTGCCGTATTCTCGCG
BUBOB_010100009881 AAACGGAGACCAAAAATGCACGCGATCACGCTCGACGCCACCGACTGCCGGATTCTCGCG
BAMMEX5DRAFT_5259  AAACGGGGACCAAAAATGCACGCGATCACGCTCGACGCCACCGACTGCAGGATTCTGGCG
BCAL2789           AAACGGGGACCAAAAATGCACGCGATCACGCTCGATGCCACCGACTGCCGGATTCTGGCG
                                  L-kynurenine TR (kynR)
```

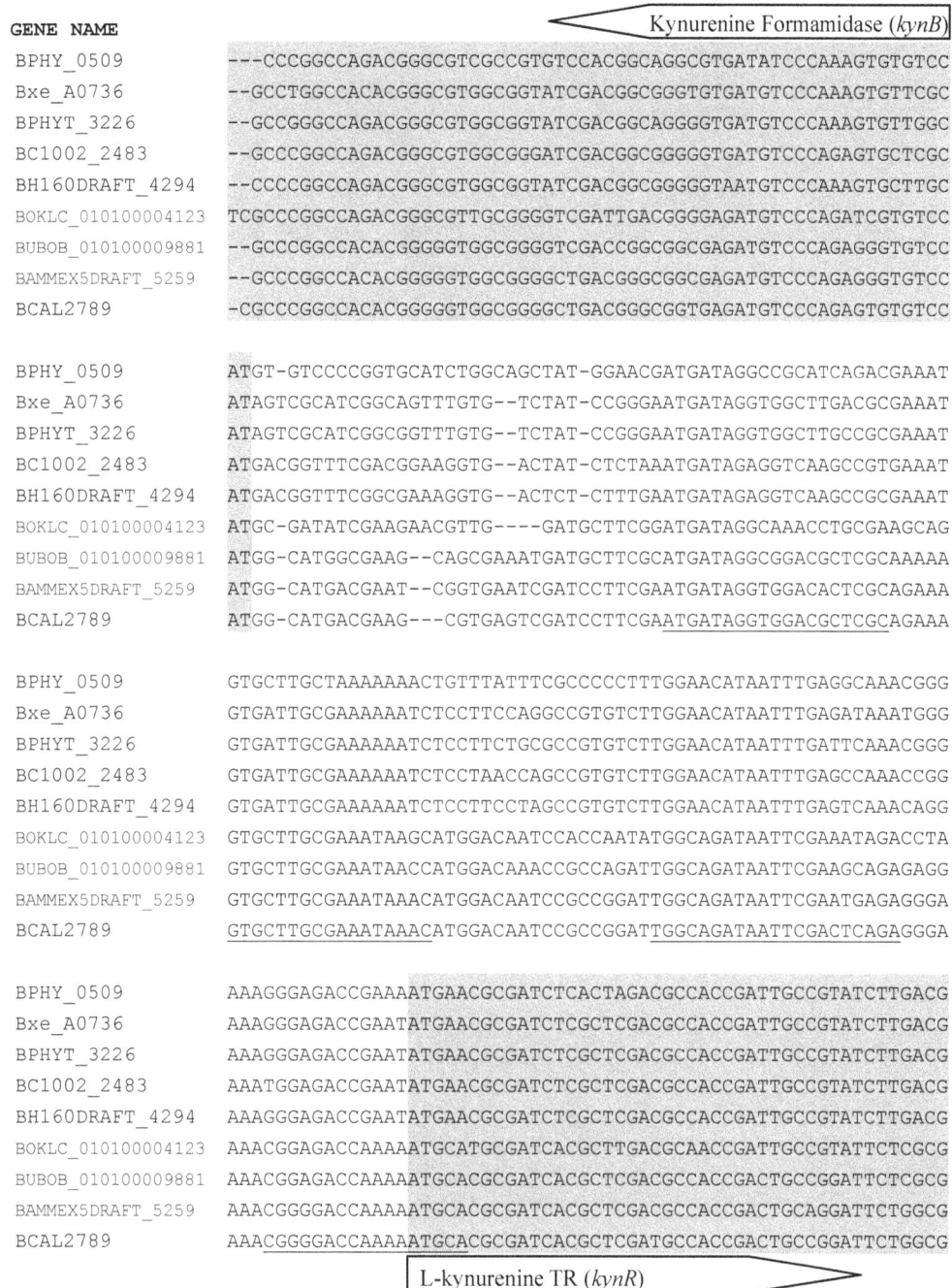

Figure 7. Multiple sequence alignment of the intergenic regions between kynurenine formamidases and the L-kynurenine responsive AsnC/Lrp like transcriptional regulators. The protein coding regions are highlighted in gray. The four degenerate palindromic sequences identified by *DNase*I footprinting are underlined

4. Conclusions

Our results support that Bxe_A0736 is a TR of the kynurenine pathway of L-tryptophan degradation in *B. xenovorans* and that it should be designated as KynR. Bxe_A0736 (KynR) interacts with a rather long region of DNA that contains four small repeated pallindromic sequences for which a consensus sequence could be defined. The presence of kynurenine was accompanied by a moderate increase in affinity of Bxe_A0736 for a portion of the much longer region with which it interacts. One of the small pallindromic regulatory sites for Bxe_A0736 (KynR) lies over the ATG start codon for *kynR*, suggesting that Bxe_A0736 (KynR) binding may block its own transcription.

Acknowledgements

We thank Dr. Virginia A. Unkefer for editorial improvements to this manuscript. This work was conducted in part under the auspices of the US Department of Energy and supported by the LDRD program at the Los Alamos National Laboratory (Grant No. 20090107DR).

References

Atkinson, H. J., Morris, J. H., Ferrin, T. E., & Babbitt, P. C. (2009). Using sequence similarity networks for visualization of relationships across diverse protein superfamilies, *PLoS One, 4*(2), e4345. http://dx.doi.org/10.1371/journal.pone.0004345

Belladonna, M. L., Orabona, C., Grohmann, U., & Puccetti, P. (2009). TGF-beta and kynurenines as the key to infectious tolerance, *Trends Mol Med, 15*(2), 41-49. http://dx.doi.org/S1471-4914(09)00013-6

Brinkman, A. B., Ettema, T. J., de Vos, W. M., & van der Oost, J. (2003). The Lrp family of transcriptional regulators. *Mol Microbiol, 48*(2), 287-294. http://dx.doi.org/3442

Calvo, J. M., & Matthews, R. G. (1994). The leucine-responsive regulatory protein, a global regulator of metabolism in *Escherichia coli, Microbiol Rev, 58*(3), 466-490.

Cline, M. S., Smoot, M., Cerami, E., Kuchinsky, A., Landys, N., Workman, C., … Bader, G. D. (2007). Integration of biological networks and gene expression data using Cytoscape. *Nat Protoc, 2*(10), 2366-2382. http://dx.doi.org/nprot.2007.324

De Wind, N., de Jong, M., Meijer, M., & Stuitje, A. R. (1985). Site-directed mutagenesis of the *Escherichia coli* chromosome near oriC: identification and characterization of asnC, a regulatory element in *E. coli* asparagine metabolism. *Nucleic Acids Res, 13*(24), 8797-8811.

Enoru-Eta, J., Gigot, D., Thia-Toong, T. L., Glansdorff, N., & Charlier, D. (2000). Purification and characterization of Sa-lrp, a DNA-binding protein from the extreme thermoacidophilic archaeon Sulfolobus acidocaldarius homologous to the bacterial global transcriptional regulator Lrp. *J Bacteriol, 182*(13), 3661-3672.

Farrow, J. M., & Pesci, E. C. (2007). Two distinct pathways supply anthranilate as a precursor of the *Pseudomonas* quinolone signal. *J Bacteriol, 189*(9), 3425-3433. http://dx.doi.org/JB.00209-07

Inoue, H., Inagaki, K., Eriguchi, S. I., Tamura, T., Esaki, N., Soda, K., & Tanaka, H. (1997). Molecular characterization of the mde operon involved in L-methionine catabolism of *Pseudomonas putida*. *J Bacteriol, 179*(12), 3956-3962.

Keuntje, B., Masepohl, B., & Klipp, W. (1995). Expression of the putA gene encoding proline dehydrogenase from Rhodobacter capsulatus is independent of NtrC regulation but requires an Lrp-like activator protein. *J Bacteriol, 177*(22), 6432-6439.

Knoten, C. A., Hudson, L. L., Coleman, J. P., Farrow, J. M., & Pesci, E. C. (2011). KynR, a Lrp/AsnC-Type Transcriptional Regulator, Directly Controls the Kynurenine Pathway in Pseudomonas aeruginosa. *J Bacteriol, 193*(23), 6567-6575. http://dx.doi.org/JB.05803-11

Koike, H., Ishijima, S. A., Clowney, L., & Suzuki, M. (2004). The archaeal feast/famine regulatory protein: potential roles of its assembly forms for regulating transcription. *Proc Natl Acad Sci U S A, 101*(9), 2840-2845. http://dx.doi.org/10.1073/pnas.0400109101

Kolling, R., & Lother, H. (1985). AsnC: an autogenously regulated activator of asparagine synthetase A transcription in Escherichia coli. *J Bacteriol, 164*(1), 310-315.

Kolling, R., & Lother, H. (1985). AsnC: an autogenously regulated activator of asparagine synthetase A transcription in *Escherichia coli*. *J Bacteriol, 164*(1), 310-315.

Landgraf, J. R., Wu, J., & Calvo, J. M. (1996). Effects of nutrition and growth rate on Lrp levels in *Escherichia coli*. *J Bacteriol, 178*(23), 6930-6936.

LiCata, V. J., & Wowor, A. J. (2008). Applications of fluorescence anisotropy to the study of protein-DNA interactions. *Methods Cell Biol, 84*, 243-262. http://dx.doi.org/S0091-679X(07)84009-X

Madhusudhan, K. T., Lorenz, D., & Sokatch, J. R. (1993). The bkdR gene of *Pseudomonas putida* is required for expression of the bkd operon and encodes a protein related to Lrp of *Escherichia coli*. *J Bacteriol, 175*(13), 3934-3940.

Maity, T. S., Close, D. W., Valdez, Y. E., Nowak-Lovato, K., Marti-Arbona, R., Nguyen, T. T., ... Dunbar, J. (2012). Discovery of DNA operators for TetR and MarR family transcription factors from Burkholderia xenovorans. *Microbiology, 158*(Pt 2), 571-582. http://dx.doi.org/10.1099/mic.0.055129-0

Marti-Arbona, R., Teshima, M., Anderson, P. S., Nowak-Lovato, K. L., Hong-Geller, E., Unkefer, C. J., & Unkefer, P. J. (2012). Identification of new ligands for the methionine biosynthesis transcriptional regulator (MetJ) by FAC-MS. *J Mol Microbiol Biotechnol, 22*(4), 205-214. http://dx.doi.org/10.1159/000339717

Matthijs, S., Baysse, C., Koedam, N., Tehrani, K. A., Verheyden, L., Budzikiewicz, H., ... Cornelis, P. (2004). The *Pseudomonas* siderophore quinolobactin is synthesized from xanthurenic acid, an intermediate of the kynurenine pathway. *Mol Microbiol, 52*(2), 371-384. http://dx.doi.org/10.1111/j.1365-2958.2004.03999.x

Palleroni, N. J., & Stanier, R. Y. (1964). Regulatory mechanisms governing synthesis of the enzymes for tryptophan oxidation by *Pseudomonas fuorescens. J Gen Microbiol, 35*, 319-334.

Peekhaus, N., Tolner, B., Poolman, B., & Kramer, R. (1995). The glutamate uptake regulatory protein (Grp) of *Zymomonas mobilis* and its relation to the global regulator Lrp of *Escherichia coli. J Bacteriol, 177*(17), 5140-5147.

Ren, J., Sainsbury, S., Combs, S. E., Capper, R. G., Jordan, P. W., Berrow, N. S., ... Owens, R. J. (2007). The structure and transcriptional analysis of a global regulator from *Neisseria meningitidis. J Biol Chem, 282*(19), 14655-14664. http://dx.doi.org/M701082200

Slon-Usakiewicz, J. J., Ng, W., Dai, J. R., Pasternak, A., & Redden, P. R. (2005). Frontal affinity chromatography with MS detection (FAC-MS) in drug discovery. *Drug Discov Today, 10*(6), 409-416. http://dx.doi.org/S1359644604033604

Tatusov, R. L., Fedorova, N. D., Jackson, J. D., Jacobs, A. R., Kiryutin, B., Koonin, E. V., ... Natale, D. A. (2003). The COG database: an updated version includes eukaryotes. *BMC Bioinformatics, 4*(41). http://dx.doi.org/10.1186/1471-2105-4-41

Thaw, P., Sedelnikova, S. E., Muranova, T., Wiese, S., Ayora, S., Alonso, J. C., ... Rafferty, J. B. (2006). Structural insight into gene transcriptional regulation and effector binding by the Lrp/AsnC family. *Nucleic Acids Res, 34*(5), 1439-1449. http://dx.doi.org/34/5/1439

Thomas-Chollier, M., Sand, O., Turatsinze, J. V., Janky, R., Defrance, M., Vervisch, E., ... van Helden, J. (2008). RSAT: regulatory sequence analysis tools. *Nucleic Acids Res, 36*(Web Server issue), W119-127. http://dx.doi.org/gkn304

Wang, Q., & Calvo, J. M. (1993). Lrp, a global regulatory protein of Escherichia coli, binds co-operatively to multiple sites and activates transcription of ilvIH. *J Mol Biol, 229*(2), 306-318. http://dx.doi.org/S0022-2836(83)71036-3

Willins, D. A., Ryan, C. W., Platko, J. V., & Calvo, J. M. (1991). Characterization of Lrp, an *Escherichia coli* regulatory protein that mediates a global response to leucine. *J Biol Chem, 266*(17), 10768-10774.

Yokoyama, K., Ebihara, S., Kikuchi, T., & Suzuki, M. (2005). Binding of the feast/famine regulatory protein (FFRP) FL11 (pot0434017) to DNA in the "promoter to coding" region of gene fl11. *Proceedings of the Japan Academy Series B-Physical and Biological Sciences, 81*(2), 64-75.

S. Supplementary Information

S.1 Bxe_A0736:GST Fusion Purification

The *E. coli* optimized nucleotide sequence for Bxe_A0736 is shown in Figure S1.

```
ATGAATGCGATCTCCCTGGACGCGACGGACTGCCGTATCCTGACGGTTCTGCAGCAAGAAGGCCGTATCTCGAA
TCTGGACCTGGCGGAACGTATTAGCCTGTCTCCGAGTGCATGCCTGCGTCGCCTGCGTCTGCTGGAAGAACAGG
GCGTCATCGAACATTATCGTGCATGTCTGAACCGCGAAGTGCTGGGTTTTGAACTGGAAGCTTTCGTTCAGGTC
TCGATGCGTAACGATCAAGAAAATTGGCATGAACGTTTTGCAGATGCAGTGCGTGACTGGCCGGAAGTGGTTGG
TGCGTTCGTCGTGACCGGTGAAACGCACTACCTGCTGCGCGTTCTGGCCCATAACCTGAAACACTATTCTGACT
TTGTCCTGCAGCGTCTGTACAAAGCCCCGGGCGTGATGGATATTCGCTCCAATATCGTTCTGCAAACCCTGAAA
GAAGACTCAGGTGTTCCGGTTTCACTGGTCAAAAAAGCATCGGGTCACGGCGCTGCCCATAATGATCGCTAA
```

Figure S1. Coding DNA sequence for Bxe_A0736 optimized for *E. coli* expression

S.2 Expression and Purification of the GST-Bxe_A0736

As described in the manuscript, the purification of the GST-Bxe_A0736 fusion protein involved two steps. The first step was affinity chromatography using a GSTrap FF HiTrap affinity column. Protein that bound to the affinity resin was eluted with a glutathione gradient and fractions containing protein were pooled, concentrated and chromatographed using a HiLoad 26/60 Superdex 200 prep grade gel filtration column. This extra purification step removes the traces of the glutathione elution buffer as well as further purifies the protein. The purification was monitored using SDS PAGE analysis (Figure S2). The final protein solution yielded predominately a single visible band, with electrophoretic mobility consistent calculated mass of 50.7 kDa for the GST-Bxe_A0736 fusion protein.

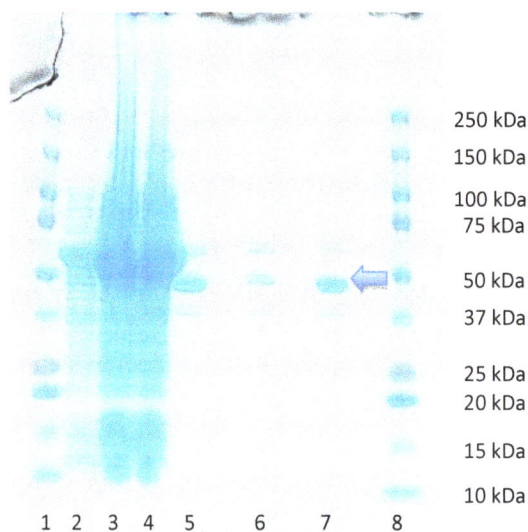

Figure S2. SDS PAGE analysis of the purification of the GST-Bxe_A0736 fusion protein.   Lanes 1 and 8 show the positions of molecular mass standards labeled on the right.   Lane 2 shows the protein profile of *E. coli* induced to express the GST-Bxe_A0736. Lane 3 shows the protein profile of the same *E. coli* cells lysed for protein purification.   Lane 4 shows the protein profile of the solution that flows through the GST affinity column.   Lane 5 shows the protein profile of the pooled protein-containing fractions eluted from the affinity resin with reduced glutathione.   Two protein-containing fractions that eluted from the gel filtration column were pooled independently and their protein profiles are represented in lanes 6 and 7. Lane 7 represents the purified solution of the GST-Bxe_A0736 fusion protein that was used for our experiments. The blue arrow points to the band corresponding to the GST-Bxe_A0736 fusion protein

S3: Preparation of WT Bxe_A0736

As described in the manuscript, the GST-Bxe_A0736 was bound to a GSTrap FF HiTrap affinity column and incubated at room temperature with twenty units of Factor Xa. After overnight incubation the cleaved Bxe_A0736 and Factor Xa were eluted from the column. Factor Xa was removed with Novagen Xarrest agarose beads according to the manufacturer's instructions. The cleavage and purification were monitored with SDS PAGE analysis shown in Figure 3S. The cleaved protein solution yielded a single visible band whose electrophoretic mobility was consistent with the calculated mass of 19.4 kDa for the Bxe_A0736.

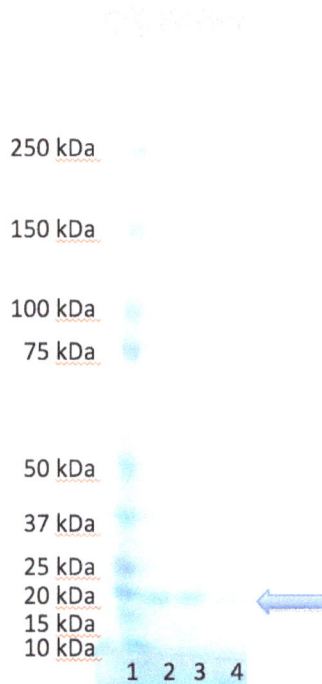

Figure S3. SDS PAGE analysis of the cleavage of Bxe_A7036 from the Bxe_A7036:GST fusion protein. Lanes 1 shows the electrophoretic mobility of molecular mass standards labeled on the left. Lane 2, 3 and 4 show the protein profile of the cleaved product after removal of Factor Xa. The blue arrow points to the band corresponding to the GST-Bxe_A0736 fusion protein

11

Atherosclerotic Events: The Role of Air Particulate Matter

Obinaju, Blessing Ebele[1]

[1] Centre for Biophotonics, Lancaster University, Lancaster, United Kingdom

Correspondence: Obinaju, B. E., Centre for Biophotonics, Lancaster University, Lancaster, United Kingdom. E-mail: bel.obinaju@yahoo.co.uk

Abstract

Epidemiological studies associate the increase of respiratory and cardiovascular mortality and morbidity with high levels of air pollution particulate matter (PM). However, the underlying mechanisms of actions by which PM induce adverse health effects remain to be clearly elucidated. Evidence from experimental studies suggests that particle composition can play an important role in PM-toxicity; however, little is known about the specific participation of components (individually or acting in groups) present in such a complex mixture that accounts for toxicity. Correlations between exposure to PM with an aerodynamic diameter 2.5 or 10 μm ($PM_{2.5}$ and PM_{10}, respectively) with cardiovascular effects have been demonstrated. Mechanisms of PM at cellular level involve free radical production (by transition metals and organic compounds), oxidative stress, cytokine release, inflammation, endotoxin-mediated damage, stimulation of capsaicin receptors, autonomic nervous system activity, covalent modification of key cellular molecules and increased pro-coagulant activity. The various interaction between particulate matter (e.g carcinogenic polyaromatic hydrocarbon components) and biological molecules trigger cascade events which initiate or aid the progression of disease conditions through cellular responses which could give rise to oxidized and/or mutagenic lesions such as are found within the atherosclerotic plaque and cancers with the most important mechanisms possibly being reactive oxygen species (ROS) generation, oxidative stress and inflammation.

Keywords: air pollution, particulate matter, atherosclerosis, carcinogenesis, PM-toxicity, reactive oxygen species

1. Introduction

Industrialization in the various regions of the world has been greatly associated with the emission of various substances which constitute air pollutants and increase air pollution. These include substances such as metal fragments, wood chippings, dust particles, and much more. Air particulate matter (a-PM) otherwise known as aerosols is a major atmospheric pollutant [considering the vast sources of emission (Table 1)] with a composition mixture of particles (solid, liquid or both) suspended in air. (Seaton et al., 1995). Depending on emission sources (natural or anthropogenic), a-PM contains complex mixtures of chemical and/or biological components (Alfaro-Moreno et al., 2002; Soukup & Becker, 2001). This composition represents a complex mixture of organic, inorganic and biological components including viable or non-viable microorganisms and fragments of microorganisms which could include toxic components such as endotoxin and mycotoxins (Gangamma, 2012) varying in size, composition and origin with properties summarized based on their aerodynamic diameter (Table 1). Particulate matter (PM) is majorly made up of sulphate, nitrates, ammonia, sodium chloride, carbon, mineral dust and water. These particles are classified as primary or secondary depending on the mechanism of formation. Although natural processes emit primary particles into the atmosphere, anthropogenic processes such as combustion from car engines; solid fuel; combustion in households and industrial activities constitute the greater source of primary particles emitted into the atmosphere (Hammond et al., 2008; Watson & Chow, 2001).

Table 1. Summary of PM size, constituents and possible sources

Particulate matter size	Aerodynamic diameter (μm)	Constituents	Sources
coarse fraction ($PM_{2.5-10}$)	2.5 - 10	Dust, Endotoxin, pollen, fungi debris, ground materials, metals.	Agriculture, Soil, road dust, sea spray, suspension in air from grinding and erosion
fine particles $PM_{2.5}$	< 2.5	Organic/elemental carbon, organic compounds, hydrocarbons	Primary from all combustion sources including coal, oil, wood and gas
Ultra-fine particles $PM_{0.1}$	< 0.1	Primary combustion-hydrocarbons, metals, organic carbon	Fresh automobile and combustion emissions. Volatile and Semi- volatile organic carbons Secondary photochemical formation from gases

Source: Brook, 2008.

2. Health Implications

Besides the effect of particulate matter/ aerosols on climate change, the ability of atmospheric aerosols to exhibit chemical heterogenicity, spatial and seasonal variability have raised concerns regarding a variety of health impacts. These include respiratory diseases, cardiovascular (CV) diseases, eye irritation and lots more (Bell & Holloway, 2007). Significant associations have been shown to exist between excess cardiopulmonary/ CV morbidity and mortality following exposure to particulate air pollution especially ambient air particles with a mass median diameter of less than 10 μM (PM_{10}) (Bascom et al., 1996; Dockery et al., 1993; Bates, 1992). PM fractions of air pollution contain constituents that could increase reactive oxygen species (ROS) generation via reactions such as transition metal catalyses, metabolism, redox-cycling of quinones and inflammation (Knaapen et al., 2004).

Although the biological mechanism is yet to be completely understood, postulations are; that the inhalation of fine particles provokes a low-grade inflammatory response in the lung that aggravates lung disease and a change in blood coagulability thus increasing pulmonary and CV deaths (Seaton et al., 1995); alveolar macrophages (AM) are the most likely link between inflammatory processes in the lung and the systemic response due to their responsibility towards the ingestion and elimination of inhaled particles (Lohmann-Matthes et al., 1994); the phagocytic activity, oxidant production and release of inflammatory mediators such as tumor necrosis factor-alpha (TNF-α) by AMs is increased by their interaction with atmospheric particles (Imrich et al., 1998; Becker et al., 1996); PM induces the activation of the c-jun-n-terminal protein kinase (JNK) which possibly enhances DNA methyltransferase - 1 (DMNT1) transcription and p16 promoter methylation (Soberanes et al., 2012; Soberanes et al., 2009; Soberanes et al., 2006); PM alters the expression of tumor protein p53, cyclin-dependent kinase inhibitor1A gene (p21) and cyclin D1 gene (CCND1) which subsequently affects cell proliferation and apoptosis (Rosas Pérez et al., 2007; Bayram et al., 2006; Dagher et al., 2006; Soberanes et al., 2006).

The putative biological mechanism which links air pollution to heart disease involves the direct effects of pollutants on the CV system, blood/ lung receptors and/or the indirect effects mediated through pulmonary oxidative stress and inflammatory responses (Brook et al., 2004). The direct effects may likely occur via a variety of agents that readily cross the pulmonary epithelium into the systemic circulation. Within the systemic circulation, these direct effects represent a plausible explanation for the occurrence of rapid CV responses such as increased myocardial infarctions (Peters et al., 2001). The less acute and chronic indirect effects likely occur via pulmonary oxidative stress and/or inflammation induced by inhaled pollutants and results in health effects such as systemic inflammatory states capable of activating haemostatic pathways, impairing vascular function and accelerating atherosclerosis (Mutlu et al., 2007; Nemmar et al., 2003).

Dating back to the 18th century and earlier, atherosclerosis was considered a disorder due to fatty acid/lipid metabolism (Steinberg, 2005). The vascular disease is majorly characterized by endothelial dysfunction, vascular inflammation and the build-up of lipid, cholesterol, calcium and cellular debris within the intima of the vessel wall. However, the critical cellular elements of the atherosclerotic lesion are leukocytes, smooth muscle cells, endothelial cells and platelets (Falk, 2006). These components of the atherosclerotic lesions indicate the possibility of an immunologic response to tissue damage. It is therefore acceptable that atherosclerosis is no longer considered a disorder due to lipid metabolism but a chronic immuno-inflammatory, fibro-proliferative disease of large and medium-sized arteries fuelled by lipids (Hansson, 2005; Glass & Witztum, 2001). This review attempts to understand the possible role of PM in the progression of atherosclerotic events. It explores the possible mechanisms by which exposure to PM encourages atherosclerotic events and perhaps other inflammatory disease conditions.

3. The Atherosclerotic Pathway

Low density lipoprotein (LDL) oxidation is predicted as an early event in atherosclerosis. This suggests that oxidized LDL play a major role in atherogenesis (Asmis et al., 2005; Stocker & Keaney Jr, 2004; Heinecke, 2001). LDL are the major cholesterol transporters consisting of a hydrophobic core containing cholesteryl ester molecules, triacylglycerols and a surface monolayer of polar lipids (mainly phospholipids) and Apolipoprotein-B (Catapano et al., 2000). The efflux of LDL from the aterial lumen into the aterial wall and oxidation (mediated by reactive oxygen species (ROS), sphingomyelinase, secretory phospholipase-2, other lipases and myeloperoxidase) of plasma LDL in the extracellular matrix results in the production of oxidized LDL (OxLDL) believed to be the ultimate atherogenic forms of LDL (Perrin-Cocon et al., 2001).

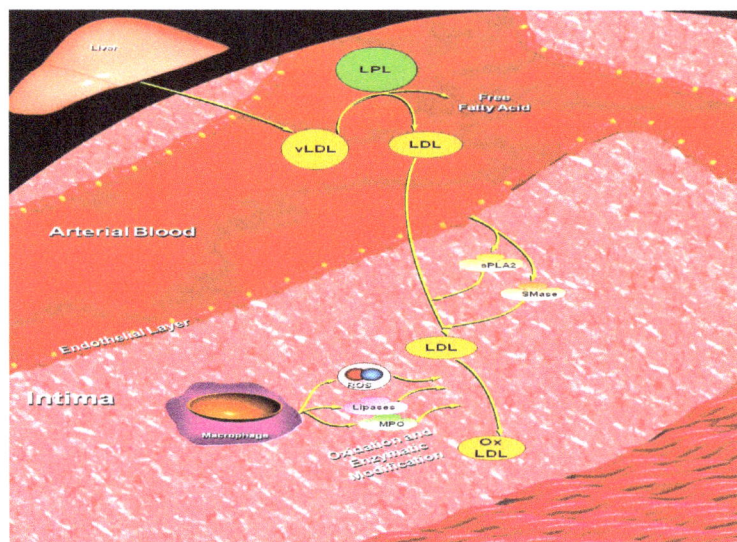

Figure 1. Oxidation of Low density lipoprotein mediated by Reactive Oxygen Species (ROS), the enzymes Sphingomyelinase (SMase), Secretory Phospholipase-2 (sPLA2), and Myeloperoxidase (MPO). Adapted from SABiosciences - Pathway Central

OxLDL induces inflammatory molecules and stimulates inflammatory signalling by endothelial cells. The release of chemotactic proteins and growth factors help the recruitment of monocytes into the arterial wall (Catapano et al., 2000). The OxLDL promotes differentiation of monocytes into macrophages (Figure 2) which engulf the OxLDL and converts them into foam cells. The necrosis of foam cells constitute part of the atherogenic plaque in fatty streak lesions (Meydani, 2001).

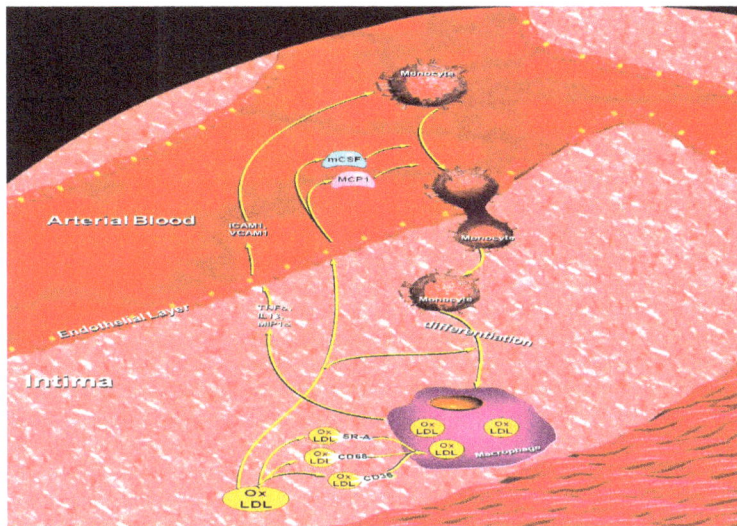

Figure 2. Inflammatory signalling via the release of Monocyte Chemotactic Protein-1 (MCP1) and release of Monocyte Colony Stimulating Factor (mCSF). The recruitment of monocytes into arterial wall, differentiation of monocytes into macrophages and phagocytic action.Adapted from SABiosciences-Pathway Central

OxLDL inhibits the production of NO (Nitric Oxide), an important mediator of vasodilation and expression of endothelial leukocyte adhesion molecules. The OxLDL particles are recognized by Macrophage Scavenger Receptors (Figure 2) Scavenger Receptor-A (SR-A), CD36 Antigen (CD36) and Macrophage Antigen CD68 (CD68). The Macrophages take up the OxLDLs, become enlarged and lipid- filled. These cells accumulate in tissue and are transformed into lipid-laden Foam cells (Figure. 3), dying and forming part of the Atherosclerotic Plaque in the fatty streak lesions (Meydani, 2001). Activation of macrophages result in the expression of cytokines such as Tumor Necrosis Factor-Alpha (TNF-Alpha), Interleukin-1Beta (IL-1Beta), Macrophage Inflammatory Protein-1Alpha (MIP1Alpha) which stimulate the expression of adhesion proteins like Vascular-Cell-Adhesion Molecule-1 (VCAM1) and Intracellular-AdhesionMolecule-1 (ICAM1) by endothelial cells (Figure 2).

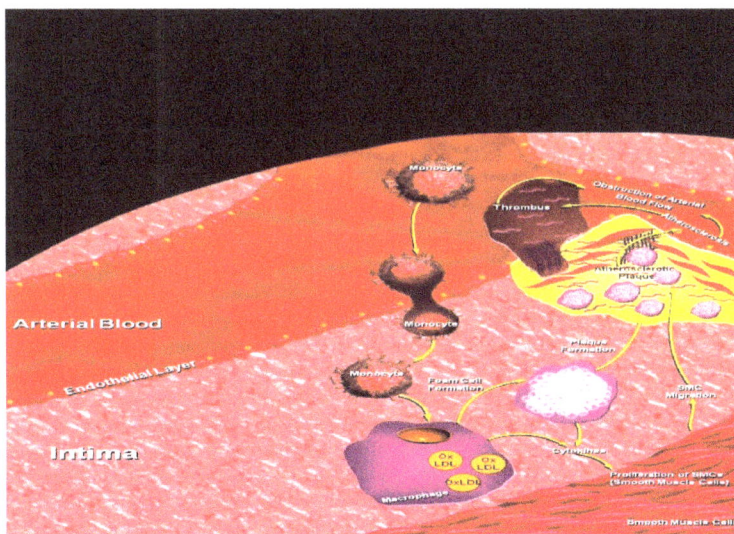

Figure 3. Foam cell formation by lipid-filled macrophages, foam cell necrosis, smooth muscle cell migration and formation of atherosclerotic plaque

Endothelial cells expressing adhesion proteins stimulate the binding of additional blood monocytes to the endothelium and recruitment into the intima. The cytokines released from the Macrophages and Foam cells also stimulate the migration of smooth muscle cells (SMC) into the Intima, then proliferate and secrete Collagen, Elastin and Proteoglycans to form a fibrous matrix resulting in the formation of Plaques with fibrous caps (Barter et al., 2004). The mature atherosclerotic plaque then consists of a fibrous cap (comprising variable numbers of SMCs, foamy macrophages, lymphocytes, extracellular matrix and a variety of inflammatory mediators) encapsulating an acellular, lipid-rich necrotic core drived partly from dead foam cells. These mature plaques protrude into the arterial lumen causing obstruction of arterial blood flow. Formation of advanced lesions and thrombi in response to rupture or erosion of the plaque results in impeded blood flow and acute occlusion with symptoms such as myocardial infarctions and stroke (Asmis et al., 2005; Catapano et al., 2000).

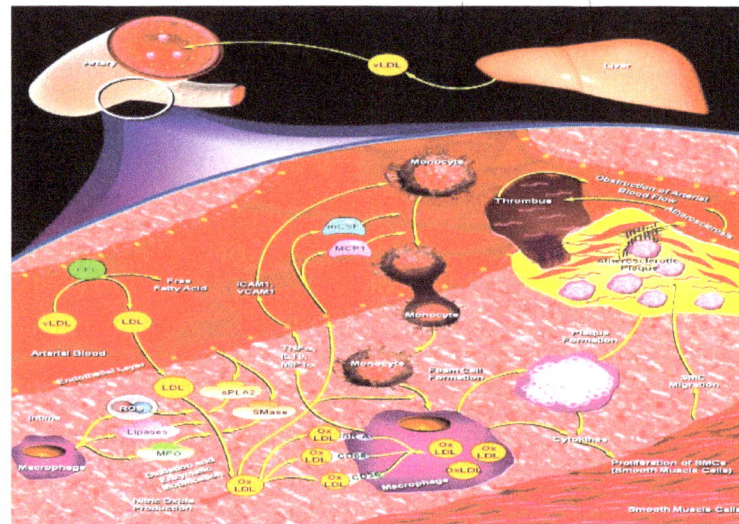

Figure 4. The complete cascade of events from atherogenesis to atherosclerosis.Adapted from SABiosciences-Pathway Central

4. Mechanisms of PM Action

The lung represents an important target tissue in the genotoxicity of pro-oxidant compounds particularly because the bronchial epithelium acts as a physicochemical barrier, playing a crucial role in initiating and augmenting defence mechanisms as well as signalling systemic responses (Vineis et al., 2004; Mills et al., 1999; Ollikainen et al., 1998). Mechanisms of PM at cellular level involve free radical production (by transition metals and organic compounds), oxidative stress, cytokine release, inflammation, endotoxin-mediated damage, stimulation of capsaicin receptors, autonomic nervous system activity, covalent modification of key cellular molecules and increased pro-coagulant activity (Araujo & Nel, 2009; Brook, 2008; Mills et al., 2008; Bhatnagar, 2006; Nel et al., 1998).

The effect of PM on organisms could depend on its chemical composition: a higher content of carcinogenic polyaromatic hydrocarbon (c-PAH) increases the genotoxicity of PM, resulting in the preferential formation of PAH-DNA adducts (Sevastyanova et al., 2008). The presence of other compounds, including o-quinones, or transition metals may lead to ROS formation and the subsequent induction of oxidative stress. Chemical composition however, may not necessarily be informative about the resulting effect of PM on the organism, because it does not take into account the interactions between various components that may cause synergistic, antagonistic, or additive effects (Donnelly et al., 1990). That said the types and sizes of a-PM inhaled may determine their toxicity and relative importance to the various mechanistic pathways. Larger fine or coarse PM cannot be transported into the circulation and would require secondary neural or pro-inflammatory response to mediate extra pulmonary actions while ultra-fine PM (or soluble constituents of larger particles) might directly enter the blood stream as a result of their ability to filter through the various biological barriers (Brook, 2008). Soukup and Becker (2001) report the induction of pro-inflamatory cytokines (IL-6 and TNF-α) in AMs by insoluble $PM_{2.5}$ and PM_{10} with higher induction levels observed in cells exposed to insoluble PM_{10}. It is possible that coarse PM particularly its insoluble components possess the potential to mediate AM functional modulation. As air pollution increases, inadvertently, the amount of particulate matter content within the atmosphere increases too. This relationship and increase in PM content has been shown to increase the incidence of CV deaths. Various studies document that CV deaths increase by approximately 1% for every $10\mu g/m^3$ short term daily increase in $PM_{2.5}$ (Pope III et al., 2006; Tonne et al., 2007; Analitis et al., 2006; von Klot et al., 2005; Zanobetti & Schwartz, 2005; Peters et al., 2004; D'Ippoliti et al., 2003; Dominici et al., 2003; Zanobetti et al., 2002; Katsouyanni et al., 2001).

Ultrafine particles (< 100nm diameter) are known for marked toxicity and may be held responsible for some of the $PM_{2.5\text{-}10}$ adverse effects. MacNee and Donaldson (2003) demonstrated that ultrafine carbon black (ufCB) does not have its effect via transition metal-mediated mechanism. Rather, ufCB and other ultrafine particles generate free radicals at their surface and are able to induce oxidative stress to cells. This ability to induce oxidative stress is likely implicated in the induction of inflammation. The hypothesis that the deposition of ultrafine particles in the lung provokes alveolar inflammation resulting in acute changes in blood coagubility and leads to morbidity and mortality in CV diseases (Seaton et al., 1995) has been supported by studies showing that exposure to ambient PM_{10} promotes inflammation in the lung and is associates with a systemic inflammatory response (Goto et al., 2004; van Eeden et al., 2001; Tan et al., 2000; Terashima et al., 1997; Seaton et al., 1995).

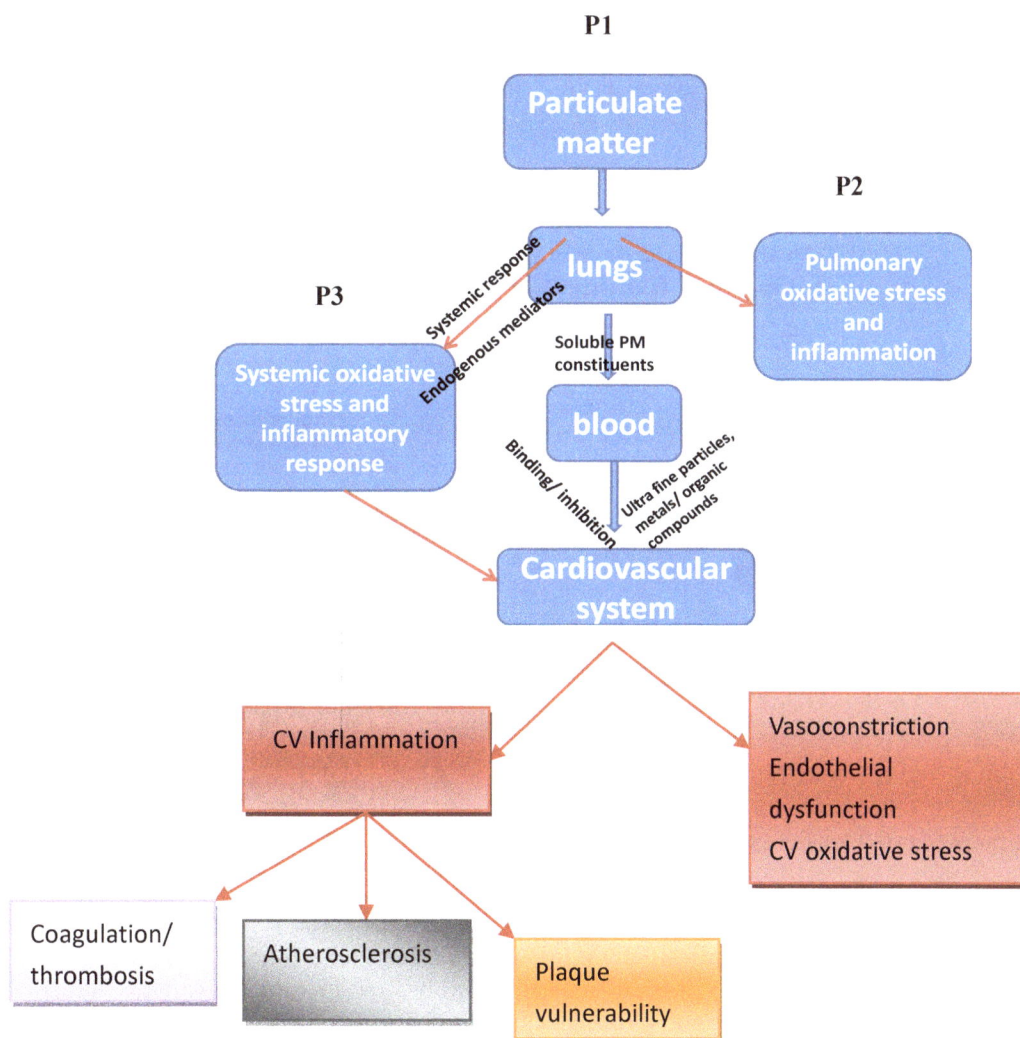

Figure 5. The possible pathways through which particulate matter exerts toxicity leading to atherosclerosis initiation and progression (**P1**: pathway 1, **P2**: pathway 2, **P3**: pathway 3)

Absorption from the lungs is usually rapid and efficient due to the surface area, excellent blood supply and barrier between the air in the alveolus and the blood stream. These properties of the lung make exposure to toxic compounds via the pulmonary vasculature toxicologically important and highly significant (Timbrell, 2000). Soluble compounds and/or Nano meter-sized PM may rapidly enter the pulmonary vasculature and subsequently be transported throughout the systemic circulation. Following inhalation, the translocated particles could directly interact with the CV system possibly via receptor – binding/ inhibition. (Figure 5. P1.). Pulmonary oxidative stress maybe responsible for instigating CV pro-oxidative (Bräuner et al., 2007; Rhoden et al., 2005; Sørensen et al., 2003; Gurgueira et al., 2002; Sharman et al., 2002) and pro-inflammatory (Behndig et al., 2006) chain reaction observed after PM exposure. Cardiac tissue oxidative stress is shown to increase within hours of $PM_{2.5}$ inhalation (Gurgueira et al., 2002). Elevated free radical generation have been found in remote non-pulmonary animal vessels hours to days following exposure to PMs (Gong et al., 2007; Nurkiewicz et al., 2006; Sun et al., 2005). Pro inflammatory mediators (cytokines and activated immune cells) released from the pulmonary into the system vasculature may then secondarily trigger a variety of adverse CV reactions (Figure 6). However, some studies have reported no signal of a systemic inflammatory response (Diez Roux et al., 2006; Pope III & Dockery, 2006) possibly because specific pollution constituents, co-pollutant levels, the duration of exposure and patient susceptibility play highly important roles in determining the subsequent responses or lack thereof. Due to PM ability to absorb thousands of chemical compounds, it is quite challenging to identify the exact chemical constituents responsible for observed genotoxic effects. As shown in Oh et al. (2011) document that crude extracts of $PM_{2.5}$ fractionated by an acid–base–neutral and silica gel fractionation procedure and divided into chemical classes of increasing polarity showed nonpolar and slightly polar extracts significantly inducing micronuclei formation and DNA breakage at a non-cytotoxic dose. Organic extracts (fractions possibly containing aliphatic chlorinated hydrocarbons, PAHs, nitro-PAHs, ketone and quinones) of $PM_{2.5}$ was observed to have induced significant increase of oxidative DNA damage including oxidized purine and pyrimidine molecules (Oh et al., 2011).

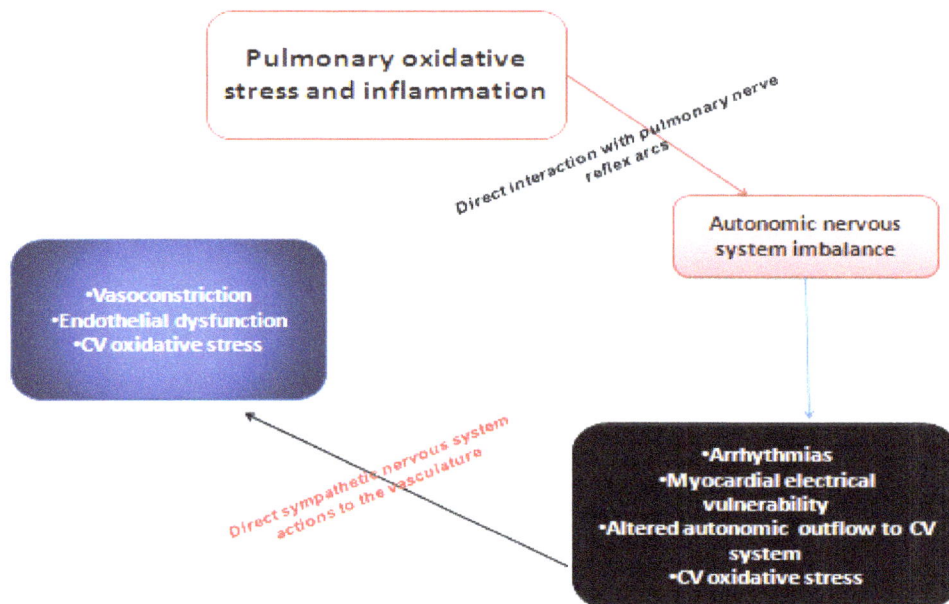

Figure 6. The possible pathways through which particulate matter enhances atherosclerosis progression via the pulmonary oxidative stress pathway

Transition metals may determine the toxic effects of PM through oxidative stress. This could result in injury via increase in airspace epithelial permeability, and inflammation via the activation of transcription factors for pro-inflammatory genes in macrophages and epithelial cells. Seaton et al. (1995) hypothesized that the deposition of ultrafine particles in the lung provokes alveolar inflammation resulting in acute changes in blood coagubility and leads to morbidity and mortality of CV diseases. This hypothesis is supported by studies showing that exposure to ambient PM_{10} promotes inflammation in the lung and is associated with a systemic inflammatory response (Goto et al., 2004; van Eeden et al., 2001; Tan et al., 2000; Terashima et al., 1997). Significantly high amounts of cytokines and chemokines including granulocyte-macrophage colony stimulating factor (GM-CSF), interleukin (IL)-6, IL-8 and chemoattractant protein (MCP)-1 were found in alveolar macrophages and lung epithelial cells incubated with PM_{10} (Fujii et al., 2002; Fujii et al., 2001; van Eeden et al., 2001). Higher levels of circulating cytokines have been observed in cells exposed to PM_{10} particles (Tan et al.,

2000). Studies showing an increase in C-reactive protein (CRP) levels (Pope, 2004; Sandhu et al., 2005) support the concept that exposure to PM_{10} is associated with a systemic inflammatory response. Chronic exposure to PM_{10} is also shown to cause downstream vascular effects resulting in the progression of atherosclerosis (Künzli et al., 2005; Suwa et al., 2002). Soberanes et al. (2012) show that concentrated ambient $PM_{2.5}$ induced oxidative stress within lungs, increased transcription of DNMT1 as well as hypermethylation of the p16 promoter in the lungs of exposed mice.

5. Conclusion

Particulate matter rarely exists by itself within ambient air pollution. However, the particles are constantly changing and in continuous interaction with gaseous, semi-volatile and volatile compounds. A wide variety of these vapour-phase compounds attach to the surface of PM and form secondary aerosol particles. The various interaction between particulate matter and biological molecules trigger cascade events which initiate or aid the progression of disease conditions through cellular responses which could give rise to oxidized and mutagenic lesions such as are found within the atherosclerotic plaque and cancers with the most important mechanisms possibly being ROS generation, oxidative stress and inflammation. Various techniques have been able to detect the effect of particulate matter *in vitro* however; the major research consideration should be the development of protocols and sensing techniques to track the activity of PM *in vivo*. The potential of nanotechnology (Quantum dot) presents an opportunity to achieve enhanced *in vivo* sensing via labelling and conjugates. However, nanoparticles seem to possess a difficult toxicity profile to overcome (Riding et al., 2012) and might limit their *in vivo* applications.

With increased technological advancement, it may be safe to anticipate the emergence of novel a-PM constituents of toxicological concern. Several studies show the need for improved air quality particularly within the urban environment (Laing et al., 2010; Oh et al., 2011) as well as further investigation towards the proper elucidation of PM mechanisms of eliciting cellular damage and possibly initiating/promoting disease conditions. Of interest would be the activity of PM and PM fractions on regulatory proteins/genes and the possible activation or suppression of cell or immune responses.

References

Alfaro-Moreno, E., Martínez, L., García-Cuellar, C., Bonner, J. C., Murray, J. C., Rosas, I., … Osornio-Vargas, Á. R. (2002). Biologic effects induced in vitro by PM10 from three different zones of Mexico City. *Environmental Health Perspectives, 110*(7), 715.

Analitis, A., Katsouyanni, K., Dimakopoulou, K., Samoli, E., Nikoloulopoulos, A., Petasakis, Y., … Cambra, K. (2006). Short-term effects of ambient particles on cardiovascular and respiratory mortality. *Epidemiology, 17*(2), 230. http://dx.doi.org/10.1097/01.ede.0000199439.57655.6b

Araujo, J. A., & Nel, A. E. (2009). Particulate matter and atherosclerosis: role of particle size, composition and oxidative stress. *Particle and Fibre Toxicology, 6*(1), 24-42. http://dx.doi.org/10.1186/1743-8977-6-24

Asmis, R., Begley, J., Jelk, J., & Everson, W. (2005). Lipoprotein aggregation protects human monocyte-derived macrophages from OxLDL-induced cytotoxicity. *Journal of Lipid Research, 46*(6), 1124. http://dx.doi.org/10.1194/jlr.M400485-JLR200

Barter, P., Nicholls, S., Rye, K., Anantharamaiah, G., Navab, M., & Fogelman, A. (2004). Antiinflammatory properties of HDL. *Circulation Research, 95*(8), 764. http://dx.doi.org/10.1161/01.RES.0000146094.59640.13

Bascom, R., Bromberg, P., Costa, D., Devlin, R., Dockery, D., & Frampton, M. (1996). A committee of environmental and occupational health assembly of the American Thoracic Society. Health effects of outdoor air pollution. *Am. J. Respir. Crit. Care Med., 153*, 3-50.

Bates, D. (1992). Health indices of the adverse effects of air pollution: The question of coherence. *Environmental Research, 59*(2), 336-349.

Bayram, H., Ito, K., Issa, R., Ito, M., Sukkar, M., & Chung, K. F. (2006). Regulation of human lung epithelial cell numbers by diesel exhaust particles. *European Respiratory Journal, 27*(4), 705-713. http://dx.doi.org/10.1183/09031936.06.00012805

Becker, S., Soukup, J., Gilmour, M., & Devlin, R. (1996). Stimulation of human and rat alveolar macrophages by urban air particulates: effects on oxidant radical generation and cytokine production. *Toxicology and Applied Pharmacology, 141*(2), 637-648. http://dx.doi.org/10.1006/taap.1996.0330

Behndig, A., Mudway, I., Brown, J., Stenfors, N., Helleday, R., Duggan, S., ... Frew, A. (2006). Airway antioxidant and inflammatory responses to diesel exhaust exposure in healthy humans. *European Respiratory Journal, 27*(2), 359. http://dx.doi.org/10.1183/09031936.06.00136904

Bell, M., & Holloway, T. (2007). Global impacts of particulate matter air pollution. *Environmental Research Letters, 2,* 045026. http://dx.doi.org/10.1088/1748-9326/2/4/045026

Bhatnagar, A. (2006). Environmental cardiology: studying mechanistic links between pollution and heart disease. *Circulation Research, 99*(7), 692. http://dx.doi.org/10.1161/01.RES.0000243586.99701.cf

Bräuner, E., Forchhammer, L., Møller, P., Simonsen, J., Glasius, M., Wåhlin, P., ... Loft, S. (2007). Exposure to ultrafine particles from ambient air and oxidative stress–induced DNA damage. *Environmental Health Perspectives, 115*(8), 1177. http://dx.doi.org/10.1289/ehp.9984

Brook, R. (2008). Cardiovascular effects of air pollution. *Clinical Science, 115,* 175-187. http://dx.doi.org/10.1042/CS20070444

Brook, R., Franklin, B., Cascio, W., Hong, Y., Howard, G., Lipsett, M., ... Smith Jr, S. (2004). Air pollution and cardiovascular disease: a statement for healthcare professionals from the Expert Panel on Population and Prevention Science of the American Heart Association. *Circulation, 109*(21), 2655. http://dx.doi.org/10.1161/01.CIR.0000128587.30041.C8

Catapano, A., Maggi, F., & Tragni, E. (2000). Low density lipoprotein oxidation, antioxidants, and atherosclerosis. *Current Opinion in Cardiology, 15*(5), 355.

Dagher, Z., Garçon, G., Billet, S., Gosset, P., Ledoux, F., Courcot, D., ... Shirali, P. (2006). Activation of different pathways of apoptosis by air pollution particulate matter (PM2. 5) in human epithelial lung cells (L132) in culture. *Toxicology, 225*(1), 12-24. http://dx.doi.org/10.1016/j.tox.2006.04.038

Diez Roux, A., Auchincloss, A., Astor, B., Barr, R., Cushman, M., Dvonch, T., ... Samson, P. (2006). Recent exposure to particulate matter and C-reactive protein concentration in the multi-ethnic study of atherosclerosis. *American Journal of Epidemiology, 164*(5), 437. http://dx.doi.org/10.1093/aje/kwj186

D'Ippoliti, D., Forastiere, F., Ancona, C., Agabiti, N., Fusco, D., Michelozzi, P., & Perucci, C. (2003). Air pollution and myocardial infarction in Rome: a case-crossover analysis. *Epidemiology*, 528-535. http://dx.doi.org/ 10.1097/01.ede.0000082046.22919.72

Dockery, D., Pope, C., Xu, X., Spengler, J., Ware, J., Fay, M., ... Speizer, F. (1993). An association between air pollution and mortality in six US cities. *The New England Journal of Medicine, 329*(24), 1753. http://dx.doi.org/10.1056/NEJM199312093292401

Dominici, F., McDermott, A., Daniels, M., Zeger, S., & Samet, J. (2003). Mortality among residents of 90 cities. *Revised analyses of time-series studies of air pollution and health. Special report. Boston, MA: Health Effects Institute*, 9-24.

Donnelly, K., Brown, K., Anderson, C., Barbee, G., Safe, S., & Mortlemans, K. (1990). Metabolism and bacterial mutagenicity of binary mixtures of benzo (a) pyrene and polychlorinated aromatic hydrocarbons. *Environmental and Molecular Mutagenesis, 16*(4), 238-245. http://dx.doi.org/10.1002/em.2850160404

Falk, E. (2006). Pathogenesis of atherosclerosis. *Journal of the American College of Cardiology, 47*(8S), 7-12. http://dx.doi.org/10.1016/j.jacc.2005.09.068

Fujii, T., Hayashi, S., Hogg, J. C., Mukae, H., Suwa, T., Goto, Y., ... van Eeden, S. F. (2002). Interaction of alveolar macrophages and airway epithelial cells following exposure to particulate matter produces mediators that stimulate the bone marrow. *American Journal of Respiratory Cell and Molecular Biology, 27*(1), 34-41.

Fujii, T., Hayashi, S., Hogg, J. C., Vincent, R., & Van Eeden, S. F. (2001). Particulate matter induces cytokine expression in human bronchial epithelial cells. *American Journal of Respiratory Cell and Molecular Biology, 25*(3), 265.

Gangamma, S. (2012). Airborne Particulate Matter and Innate Immunity Activation. *Environmental Science & Technology, 46*(20), 10879-10880. http://dx.doi.org/10.1021/es303491j

Glass, C., & Witztum, J. (2001). Atherosclerosis the road ahead. *Cell, 104*(4), 503-516.

Gong, K., Zhao, W., Li, N., Barajas, B., Kleinman, M., Sioutas, C., ... Araujo, J. (2007). Air-pollutant chemicals and oxidized lipids exhibit genome-wide synergistic effects on endothelial cells. *Genome Biology, 8*(7), R149. http://dx.doi.org/10.1186/gb-2007-8-7-r149

Goto, Y., Ishii, H., Hogg, J., Shih, C., Yatera, K., Vincent, R., & van Eeden, S. (2004). Particulate matter air pollution stimulates monocyte release from the bone marrow. *American Journal of Respiratory and Critical Care Medicine, 170*(8), 891. http://dx.doi.org/10.1164/rccm.200402-235OC

Gurgueira, S., Lawrence, J., Coull, B., Murthy, G., & González-Flecha, B. (2002). Rapid increases in the steady-state concentration of reactive oxygen species in the lungs and heart after particulate air pollution inhalation. *Environmental Health Perspectives, 110*(8), 749.

Hammond, D. M., Dvonch, J. T., Keeler, G. J., Parker, E. A., Kamal, A. S., Barres, J. A., ... Brakefield-Caldwell, W. (2008). Sources of ambient fine particulate matter at two community sites in Detroit, Michigan. *Atmospheric Environment, 42*(4), 720-732. http://dx.doi.org/10.1016/j.atmosenv.2007.09.065

Hansson, G. (2005). Inflammation, atherosclerosis, and coronary artery disease. *The New England journal of medicine, 352*(16), 1685. http://dx.doi.org/ 10.1056/NEJMra043430

Heinecke, J. (2001). Is the emperor wearing clothes? Clinical trials of vitamin E and the LDL oxidation hypothesis. *Arteriosclerosis, Thrombosis, and Vascular Biology, 21*(8), 1261. http://dx.doi.org/10.1161/hq0801.095084

Imrich, A., Danaee, H., & Ning, Y. (1998). Analysis of air pollution particulate-mediated oxidant stress in alveolar macrophages. *Journal of Toxicology and Environmental Health Part A, 54*(7), 529-545. http://dx.doi.org/10.1080/009841098158683

Katsouyanni, K., Touloumi, G., Samoli, E., Gryparis, A., Le Tertre, A., Monopolis, Y., ... Boumghar, A. (2001). Confounding and effect modification in the short-term effects of ambient particles on total mortality: results from 29 European cities within the APHEA2 project. *Epidemiology, 12*(5), 521-531.

Knaapen, A. M., Borm, P. J. A., Albrecht, C., & Schins, R. P. F. (2004). Inhaled particles and lung cancer. Part A: Mechanisms. *International Journal of Cancer, 109*(6), 799-809. http://dx.doi.org/10.1002/ijc.11708

Künzli, N., Jerrett, M., Mack, W., Beckerman, B., LaBree, L., & Gilliland, F. (2005). Ambient air pollution and atherosclerosis in Los Angeles. *Environmental Health Perspectives, 113*, 201-206. http://dx.doi.org/10.1289%2Fehp.7523

Laing, S., Wang, G., Briazova, T., Zhang, C., Wang, A., Zheng, Z., ... Chen, L. C. (2010). Airborne particulate matter selectively activates endoplasmic reticulum stress response in the lung and liver tissues. *American Journal of Physiology-Cell Physiology, 299*(4), C736-C749. http://dx.doi.org/10.1152/ajpcell. 00529. 2009

Lohmann-Matthes, M., Steinmuller, C., & Franke-Ullmann, G. (1994). Pulmonary macrophages. *European Respiratory Journal, 7*(9), 1678.

Meydani, M. (2001). Vitamin E and atherosclerosis: beyond prevention of LDL oxidation. *The Journal of Nutrition, 131*(2), 366S.

Mills, N., Donaldson, K., Hadoke, P., Boon, N., MacNee, W., Cassee, F., ... Newby, D. (2008). Adverse cardiovascular effects of air pollution. *Nature Clinical Practice Cardiovascular Medicine, 6*(1), 36-44. http://dx.doi.org/10.1038/ncpcardio1399

Mills, P. R., Davies, R. J., & Devalia, J. L. (1999). Airway epithelial cells, cytokines, and pollutants. *American Journal of Respiratory and Critical Care Medicine, 160*(Supplement 1), S38-S43.

Mutlu, G. M., Green, D., Bellmeyer, A., Baker, C. M., Burgess, Z., Rajamannan, N., ... Ghio, A. J. (2007). Ambient particulate matter accelerates coagulation via an IL-6-dependent pathway. *Journal of Clinical Investigation, 117*(10), 2952. http://dx.doi.org/ 10.1172/JCI30639

Nel, A., Diaz-Sanchez, D., Ng, D., Hiura, T., & Saxon, A. (1998). Enhancement of allergic inflammation by the interaction between diesel exhaust particles and the immune system. *Journal of Allergy and Clinical Immunology, 102*(4), 539-554. http://dx.doi.org/10.1016/S0091-6749(98)70269-6

Nemmar, A., Nemery, B., Hoet, P. H. M., Vermylen, J., & Hoylaerts, M. F. (2003). Pulmonary Inflammation and Thrombogenicity Caused by Diesel Particles in Hamsters Role of Histamine. *American Journal of Respiratory and Critical Care Medicine, 168*(11), 1366-1372. http://dx.doi.org/10.1164/rccm.200306-801OC

Nurkiewicz, T., Porter, D., Barger, M., Millecchia, L., Rao, K., Marvar, P., ... Boegehold, M. (2006). Systemic microvascular dysfunction and inflammation after pulmonary particulate matter exposure. *Environmental Health Perspectives, 114*(3), 412. http://dx.doi.org/10.1289/ehp.8413

Oh, S. M., Kim, H. R., Park, Y. J., Lee, S. Y., & Chung, K. H. (2011). Organic extracts of urban air pollution particulate matter (PM2.5)-induced genotoxicity and oxidative stress in human lung bronchial epithelial cells (BEAS-2B cells). *Mutation Research/Genetic Toxicology and Environmental Mutagenesis, 723*(2), 142-151. http://dx.doi.org/10.1016/j.mrgentox.2011.04.003

Ollikainen, T. R. J., Linnainmaa, K. I., Raivio, K. O., & Kinnula, V. L. (1998). DNA single strand breaks and adenine nucleotide depletion as indices of oxidant effects on human lung cells. *Free Radical Biology and Medicine, 24*(7), 1088-1096. http://dx.doi.org/10.1016/S0891-5849(97)00394-8

Perrin-Cocon, L., Coutant, F., Agaugué, S., Deforges, S., André, P., & Lotteau, V. (2001). Oxidized low-density lipoprotein promotes mature dendritic cell transition from differentiating monocyte. *Journal of Immunology 167*(7), 3785.

Peters, A., Dockery, D., Muller, J., & Mittleman, M. (2001). Increased particulate air pollution and the triggering of myocardial infarction. *Circulation, 103*(23), 2810. http://dx.doi.org/10.1161/01.CIR.103.23.2810

Peters, A., von Klot, S., Heier, M., Trentinaglia, I., Hormann, A., Wichmann, H., & Lowel, H. (2004). Exposure to traffic and the onset of myocardial infarction. *The New England Journal of Medicine, 351*(17), 1721. http://dx.doi.org/10.1056/NEJMoa040203

Pope, C. A. (2004). Ambient particulate air pollution, heart rate variability, and blood markers of inflammation in a panel of elderly subjects. *Environmental Health Perspectives, 112*(3), 339.

Pope III, C., & Dockery, D. (2006). Health effects of fine particulate air pollution: lines that connect. *Journal of the Air & Waste Management Association, 56*(6), 709-742. http://dx.doi.org/10.1080/10473289.2006.10464485

Pope III, C., Muhlestein, J., May, H., Renlund, D., Anderson, J., & Horne, B. (2006). Ischemic heart disease events triggered by short-term exposure to fine particulate air pollution. *Circulation, 114*(23), 2443. http://dx.doi.org/10.1161/CIRCULATIONAHA.106.636977

Rhoden, C., Wellenius, G., Ghelfi, E., Lawrence, J., & González-Flecha, B. (2005). PM-induced cardiac oxidative stress and dysfunction are mediated by autonomic stimulation. *Biochimica et Biophysica Acta (BBA)-General Subjects, 1725*(3), 305-313. http://dx.doi.org/10.1016/j.bbagen.2005.05.025

Riding, M. J., Martin, F. L., Trevisan, J., Llabjani, V., Patel, I. I., Jones, K. C., & Semple, K. T. (2012). Concentration-dependent effects of carbon nanoparticles in gram-negative bacteria determined by infrared spectroscopy with multivariateanalysis. *Environmental Pollution, 163*, 226-234. http://dx.doi.org/10.1016/j.envpol.2011.12.027

Rosas Pérez, I., Serrano, J., Alfaro-Moreno, E., Baumgardner, D., García-Cuellar, C., Martín del Campo, J. M., ... Osornio Vargas, A. R. (2007). Relations between PM< sub> 10</sub> composition and cell toxicity: A multivariate and graphical approach. *Chemosphere, 67*(6), 1218-1228. http://dx.doi.org/10.1016/j.chemosphere.2006.10.078

Sandhu, R. S., Petroni, D. H., & George, W. J. (2005). Ambient particulate matter, C-reactive protein, and coronary artery disease. *Inhalation Toxicology, 17*(7-8), 409-413. http://dx.doi.org/10.1080/08958370590929538

Seaton, A., Godden, D., MacNee, W., & Donaldson, K. (1995). Particulate air pollution and acute health effects. *The Lancet, 345*(8943), 176-178. http://dx.doi.org/10.1016/S0140-6736(95)90173-6

Sevastyanova, O., Novakova, Z., Hanzalova, K., Binkova, B., Sram, R., & Topinka, J. (2008). Temporal variation in the genotoxic potential of urban air particulate matter. *Mutation Research, 649*(1), 179. http://dx.doi.org/10.1016/j.mrgentox.2007.09.010

Sharman, J., Coombes, J., Geraghty, D., & Fraser, D. (2002). Exposure to Automotive Pollution Increases Plasma Susceptibility to Oxidation. *Archives of Environmental Health, 57*(6), 536-540.

Soberanes, S., Gonzalez, A., Urich, D., Chiarella, S. E., Radigan, K. A., Osornio-Vargas, A., ... Budinger, G. R. S. (2012). Particulate matter Air Pollution induces hypermethylation of the p16 promoter Via a mitochondrial ROS-JNK-DNMT1 pathway. [10.1038/srep00275]. *Sci. Rep., 2*.

Soberanes, S., Panduri, V., Mutlu, G. M., Ghio, A., Bundinger, G. R. S., & Kamp, D. W. (2006). p53 Mediates Particulate Matter–induced Alveolar Epithelial Cell Mitochondria-regulated Apoptosis. *American Journal of Respiratory and Critical Care Medicine, 174*(11), 1229-1238. http://dx.doi.org/10.1164/rccm.200602-203OC

Soberanes, S., Urich, D., Baker, C. M., Burgess, Z., Chiarella, S. E., Bell, E. L., ... Budinger, G. R. S. (2009). Mitochondrial Complex III-generated Oxidants Activate ASK1 and JNK to Induce Alveolar Epithelial Cell Death following Exposure to Particulate Matter Air Pollution. *Journal of Biological Chemistry, 284*(4), 2176-2186. http://dx.doi.org/10.1074/jbc.M808844200

Sørensen, M., Daneshvar, B., Hansen, M., Dragsted, L., Hertel, O., Knudsen, L., & Loft, S. (2003). Personal PM2. 5 exposure and markers of oxidative stress in blood. *Environmental Health Perspectives, 111*(2), 161.

Soukup, J. M., & Becker, S. (2001). Human alveolar macrophage responses to air pollution particulates are associated with insoluble components of coarse material, including particulate endotoxin. *Toxicology and Applied Pharmacology, 171*(1), 20-26. http://dx.doi.org/10.1006/taap.2000.9096

Steinberg, D. (2005). Thematic review series: the pathogenesis of atherosclerosis. An interpretive history of the cholesterol controversy: part II: the early evidence linking hypercholesterolemia to coronary disease in humans. *The Journal of Lipid Research, 46*(2), 179. http://dx.doi.org/10.1194/jlr.R400012-JLR200

Stocker, R., & Keaney Jr, J. (2004). Role of oxidative modifications in atherosclerosis. *Physiological Reviews, 84*(4), 1381.

Sun, Q., Wang, A., Jin, X., Natanzon, A., Duquaine, D., Brook, R., … Lippmann, M. (2005). Long-term air pollution exposure and acceleration of atherosclerosis and vascular inflammation in an animal model. *Jama, 294*(23), 3003. http://dx.doi.org/10.1001/jama.294.23.3003

Suwa, T., Hogg, J., Quinlan, K., Ohgami, A., Vincent, R., & van Eeden, S. (2002). Particulate air pollution induces progression of atherosclerosis. *Journal of the American College of Cardiology, 39*(6), 935-942. http://dx.doi.org/10.1016/S0735-1097(02)01715-1

Tan, W., Qiu, D., Liam, B., Ng, T., Lee, S., van EEDEN, S., … Hogg, J. (2000). The human bone marrow response to acute air pollution caused by forest fires. *American Journal of Respiratory and Critical Care Medicine, 161*(4), 1213.

Terashima, T., Wiggs, B., English, D., Hogg, J., & Van Eeden, S. (1997). Phagocytosis of small carbon particles (PM10) by alveolar macrophages stimulates the release of polymorphonuclear leukocytes from bone marrow. *American Journal of Respiratory and Critical Care Medicine, 155*(4), 1441.

Timbrell, J. (2000). *Principles of Biochemical Toxicology* (3rd ed.). Philadelphia: Taylor & Francis Inc.

Tonne, C., Melly, S., Mittleman, M., Coull, B., Goldberg, R., & Schwartz, J. (2007). A case-control analysis of exposure to traffic and acute myocardial infarction. *Environmental health perspectives, 115*(1), 53. http://dx.doi.org/10.1289%2Fehp.9587

van Eeden, S., Tan, W., Suwa, T., Mukae, H., Terashima, T., Fujii, T., … (2001). Cytokines involved in the systemic inflammatory response induced by exposure to particulate matter air pollutants (PM10). *American Journal of Respiratory and Critical Care Medicine, 164*(5), 826.

Vineis, P., Forastiere, F., Hoek, G., & Lipsett, M. (2004). Outdoor air pollution and lung cancer: recent epidemiologic evidence. *International Journal of Cancer, 111*(5), 647-652. http://dx.doi.org/10.1002/ijc.20292

von Klot, S., Peters, A., Aalto, P., Bellander, T., Berglind, N., D'Ippoliti, D., …Lanki, T. (2005). Ambient air pollution is associated with increased risk of hospital cardiac readmissions of myocardial infarction survivors in five European cities. *Circulation, 112*(20), 3073. http://dx.doi.org/10.1161/CIRCULATIONAHA.105.548743

Watson, J. G., & Chow, J. C. (2001). Source characterization of major emission sources in the imperial and Mexicali Valleys along the US/Mexico border. *The Science of the Total Environment, 276*(1-3), 33-47. http://dx.doi.org/ 10.1016/S0048-9697(01)00770-7

Zanobetti, A., & Schwartz, J. (2005). The effect of particulate air pollution on emergency admissions for myocardial infarction: a multicity case-crossover analysis. *Environmental Health Perspectives, 113*(8), 978. http://dx.doi.org/10.1289%2Fehp.7550

Zanobetti, A., Schwartz, J., Samoli, E., Gryparis, A., Touloumi, G., Atkinson, R., … Goren, A. (2002). The temporal pattern of mortality responses to air pollution: a multicity assessment of mortality displacement. *Epidemiology, 13*(1), 87-93.

12

Studies on Genetic Diversity of Selected Population of Hybrid Scallop *Chlamys farreri*（♀）× *Patinopecten yessoensis*（♂）by Microsatellites Markers

Biao Wu[1], Aiguo Yang[1], Ningning Cheng [1,2], Xiujun Sun[1], Zhihong Liu[1] & Liqing Zhou[1]

[1] Key Laboratory of Sustainable Development of Marine Fisheries, Ministry of Agriculture, Yellow Sea Fisheries Research Institute, Chinese Academy of Fishery Sciences, Qingdao 266071, PR China

[2] College of Fisheries and Life Science, Shanghai Ocean University, Shanghai, 201306, PR China

Correspondence: Aiguo Yang, Key Laboratory of Sustainable Development of Marine Fisheries, Ministry of Agriculture, Yellow Sea Fisheries Research Institute, Chinese Academy of Fishery Sciences, Qingdao 266071, PR China. E-mail: yangag@ysfri.ac.cn

Abstract

The growth superiority of hybrid scallop *Chlamys farreri* （♀）× *Patinopecten yessoensis* (♂), as the following successive generation selection have been reported. However, the data about the genetic diversity in those population remains unexplored. In this study, the genetic structure analysis of F_1, F_2 and F_3 were conducted by PCR with 10 Simple Sequence Repeats (SSR) primers. It showed that a total of 68 alleles were detected, and the number of alleles per locus ranged from 3 to 11, Polymorphism Information Content (*PIC*) per locus ranged from 0.4729 to 0.8429. And, the average observed heterozygosity (H_o) of the three populations were 0.6100, 0.6975 and 0.7750, while the average expected heterozygosity (H_e) were 0.7607, 0.7751 and 0.7379 respectively. F_{st} values among the three populations were also low ($F_{st}<0.05$) which suggested low genetic differentiation between each two populations. In all, those data indicated the genetic structure challenge caused by hybridization and selection, supplying a new angle to understand artificial selective breeding.

Keywords: scallop, hybrid breeding, genetic diversity, microsatellite DNA markers

1. Introduction

The scallop *Chlamys farreri*, a native bivalve in China, is one of the most important marine farming species in Northern China. However, massive mortality has caused catastrophic losses to its aquaculture since 1996, which resulted in a sharp production decline. The germplasm quality degeneration have been considered as one major reason for the massive scallop mortality. Therefore, to breed new scallop species with high resistance is an effective method to change the current status quo.

The research on distant hybridization breeding of scallop *C. farreri*(♀) × *Patinopecten yessoensis*(♂) have been carried out (Zhou, Yang, & Liu, 2003; Lv, Yang, Wang, Liu, & Zhou, 2006), and surprisingly, the first hybrid generation owned prominent heterosis performance which has been cultivated in large-scale area. The second and third generation individuals with stronger resistance, faster growth than scallop *C. farreir* have also been selected for farming (Yang, Wang, Liu, & Zhou, 2003). However, during the process of selective breeding, many uncertain factors, such as increasing risk of inbreeding, decreased number of effective groups, may lead to lower genetic diversity, even the genetic effects. Therefore, it is necessary to detect the genetic variation, to understand the changes of genetic structure for developing appropriate scientific measures so that we can smoothly control the progress of selective breeding.

Microsatellite marker (Simple Sequence Repeats, SSR), due to its simple, fast, good stability, higher polymorphism, informative genetic variation and followed Mendelian codominant genetic, has been widely used in various fields as a molecular marker (Li, Park, Endo, & Kijima, 2004; Sakamoto, Danzmann, Okamoto, Ferguson, & Ihssen, 1999; Liu et al., 2004). In terms of hybrid scallop, isoenzyme (He, Yang, Wang, Liu, & Zhou, 2006), RAPD and other labeling techniques were particularly used, however, SSR analysis of different hybrid scallop populations generated from *C. farreri* （♀）× *P. yessoensis* (♂) have not been reported. In this study, SSR was employed to analyze the genetic variation of three selective breeding population, to explore the impact of selection process on its genetic structure, which could provide a theoretical basis for molecular marker-assisted breeding.

2. Materials and Methods

2.1 Sample Collection

Mature female *C.farreri* and male *P. yessoensis* were collected as parents from Changdao, Shandong Province. The F_1 hybrid were reproduced from mother *C. farreri* and father *P. yessoensis* while F_2 generation from F_1 by self-fertilized, F_3 from F_2, respectively. 30, 40, 40 individuals were selected randomly from F_1, F_2, F_3 population, and stored at -80 °C for DNA extracted, respectively.

2.2 Preparation of Template DNA

The genome DNA was extracted from adductor muscles of hybrid individual using Phenol-chloroform method. In detail, about 100 mg tissue was sampled into a 1.5 ml centrifuge tube, and then 500 μl homogenization buffer (10 mM Tris-Cl, pH 8.0; 100 mM EDTA, pH 8.0), 50 μl 10% SDS and proteinase K with final concentration of 50 μg/ml were also added. After being mixed adequately, the sample were digested at 55 °C for 3 h, and then the proteins were extracted using phenol, phenol: chloroform (1:1), chloroform, isoamyl:alcohol (24:1), successively. And following, the nucleic acid was precipitated with ethanol, and then dissolved in ddH$_2$O. The concentration and quality of extracted DNA were detected by RNA/DNA quantitative analysis using Nanodrop 2000 and agarose gel electrophoresis, respectively. The concentration of genome DNA was diluted to 50 ng/μl and then stored at -20 °C.

2.3 PCR Amplification

10 pairs of primers (Table 1) were selected from reported SSR primers of *C.farreri* and *P. yessoensis*, to ensure its availability for amplification in hybrid offspring in this study. PCR reactions were carried out in a 10 μl reaction volume on a PCR amplification instrument, including 1 μl 10 × buffer, 0.6 μl Mg^{2+} (25 mM), 1 μl dNTP (each 2 mM), 1 μl forward/reverse primer (10 μM) each, 50 ng template DNA, 0.5 U Taq DNA polymerization enzyme (Promega), and PCR-grade water was replenished to 10 μl. PCR reaction program consisted of 95 °C for 5 min, followed by 35 cycles of 95 °C for 30 s, 72 °C for 30 s, and finally 72 °C for5 min. The PCR products were detected by 6 % denaturing polyacrylamide gel electrophoresis, silver staining for detecting polymorphism. The amplified brands were counted by manual mothed for analyzing by software.

Table 1. Microsatellite primers used in this experiment

Loci	Primer sequence	Annealing temperature	Repeat	GenBank Accession
CFMSM009	F:GTAGTCACATGATGACATAGAG R:CACAACTCCGTCAATCATTCTC	56	$(AG)_4G(AG)_5$... $(AG)_5$	DQ104704
CFMSM014	F:CATCTGATATGGCAGCTGATAC R: GAACTAACGAGGAGACAACTG	60	$(AG)_{10}$ ...$(AC)_4$	DQ104705
CFMSM020	F: CAAAGGCATTTGTAGGAAGGC R: ACGGCACTTCGTTGATTAAC	62	$(CAC)_{11}$	DQ104709
CFMSP011	F: CAAAACCAACTCCTTCACAAC R: GGCGATATTCCACCTGACC	62	$(ACAAA)_5$	AY682110
CFAD213	F: ATTAGTTGTGAAGCAGTCCT R:CTTCTCTCAATCATTTCACTATC	56	$(GA)_{14}$	EF148875
CFFD143	F: CGCCAACCTTGCAGTATCTG R: TTCTTTCCCTCTTCTGTCCC	58	$(GA)3CA(GA)_3$ $GGAA(GA)_{11}$	EF148943
CFBD075	F: TTACTATCCCTACCCCAGAG R:CACTAACCCATTACAAACACAAG	60	$(TGTC)_9(TC)_5TG(TC)_{23}$ ... $(TC)_6$ ... $(CT)_7$	EF148893
CFAD053	F:CATTGACACAGTTACAGTTCAC R: GCAACAGGATTAGGCACAAG	56	$(CT)_{20}$	EF148859
S7259	F: CGTCCTTAAATGACCTTA R: GAAATTCCAGTGTTCGTA	60	$(AACC)_6$	AY164679
S9090	F: GAGGAAGAAACATAGTAA R: CTACATCAGCTACATCTC	58	$(TTAA)_5$	AY164680

2.4 Data Analysis

According to the genotype, the value of polymorphic information content (*PIC*), heterozygosity (*Ho*), heterozygosity (*He*), allele number (a), effective number of alleles (a_e), similar coefficient and genetic distances among, and F-statistics of the three groups were calculated using Popgen 32 (Version 1.31). F_{st} value range from 0 to 0.05 was considered as low population genetic differentiation, while 0.05-0.15 as middle level, 0.15-0.25 as high level and above 0.25 as significantly high level.

3. Results

3.1 Genetic Diversity of Loci

10 pairs of primers with high polymorphism were used to perform PCR amplification with 110 individuals from F_1, F_2 and F_3 populations. In all, 68 alleles were obtained, and the numbers of allele for each locus ranged from 3 to 11. *PIC* value was from 0.4729 to 0.8429. The average observed heterozygosity was $F_2>F_1>F_3$, with the values of 0.7025, 0.6893, 0.6900, respectively, while the average expected heterozygosity were 0.8315, 0.7751, 0.7379, respectively. According to *P* values of genotype, the value of the Hardy-Weinberg equilibrium deviated significantly in three generations (as shown in Table 2), such as CFMSM009 in the F_1 and F_2 populations, CFMSP011 in F_3 population and CFAD213 in F_2 and F_3. Some representative amplified brands were exampled in Figure 1.

Table 2. Genetic variability at 10 microsatellite loci in three populations

loci	populations	a	a_e	*PIC*	*Ho*	*He*	*P*-values
CFMSM009	F1	3	2.3407	0.5089	0.3333	0.5825	0.0068[**]
	F2	3	2.0901	0.4613	0.2750	0.5282	0.0146[*]
	F3	3	2.0566	0.4380	0.1500	0.5203	0.0000[**]
CFMSM014	F1	7	4.6512	0.7613	0.6667	0.7983	0.2517
	F2	7	5.9590	0.8110	0.7000	0.8427	0.0942[*]
	F3	6	4.1995	0.7244	0.7000	0.7715	0.3040
CFMSM020	F1	11	7.2581	0.8468	0.7333	0.8768	0.2247
	F2	10	6.8085	0.8365	0.9000	0.8963	0.4101
	F3	10	6.4000	0.8258	0.9750	0.8544	0.0823[*]
CFMSP011	F1	7	3.7657	0.6956	1.0000	0.7469	0.0110[*]
	F2	7	4.3096	0.7327	0.9750	0.7775	0.0162[*]
	F3	6	3.2196	0.6387	0.9737	0.6986	0.0001[**]
CFAD213	F1	6	3.1304	0.6363	0.4667	0.6921	0.0315[*]
	F2	6	3.7915	0.6964	0.3750	0.7456	0.0001[**]
	F3	6	2.6801	0.5663	0.3750	0.6348	0.0005[**]
CFFD143	F1	8	4.6036	0.7509	0.8667	0.7960	0.2007
	F2	7	4.1290	0.7227	0.8500	0.7674	0.4974
	F3	6	3.8508	0.7009	0.8000	0.7497	0.7390
CFBD075	F1	6	4.2857	0.7328	0.6333	0.7797	0.0106[*]
	F2	6	3.6036	0.6885	0.6750	0.7316	0.2184
	F3	6	3.1809	0.6466	0.5500	0.6943	0.0290[*]
CFAD053	F1	8	6.7924	0.8213	0.7667	0.7469	1.0000
	F2	7	6.7368	0.8331	0.7000	0.8623	0.0158[*]
	F3	8	5.9530	0.8118	0.7692	0.8423	0.4489
S7259	F1	6	4.9451	0.7670	0.4333	0.8113	0.3268
	F2	6	4.9231	0.7664	0.5750	0.8070	0.0392[*]
	F3	6	4.3716	0.7390	0.6000	0.7810	0.0837
S9090	F1	6	4.2254	0.7279	1.0000	0.7763	0.0280[*]
	F2	6	5.3872	0.7873	1.0000	0.8247	0.1639
	F3	6	5.5817	0.7957	1.0000	0.8315	0.1643
Mean	F1	6.8000	4.5998	0.7241	0.6900	0.8315	
	F2	6.5000	4.7736	0.7336	0.7025	0.7751	
	F3	6.2000	4.1494	0.6930	0.6893	0.7379	

Note: * indicates significant deviation (*P* <0.05); ** indicates highly significant deviation (*P* <0.01).

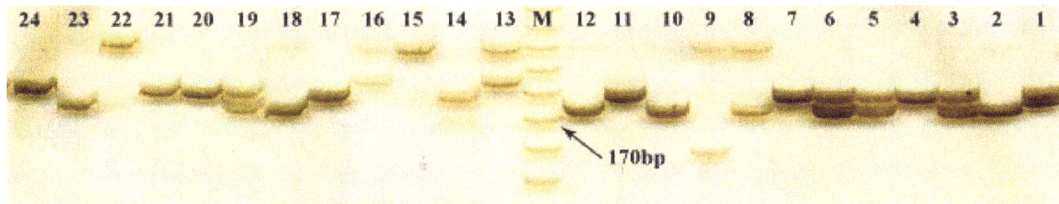

Figure 1. Demonstration of microsatellite locus amplified by CFMSM020 primer pairs in three populations
M: Marker; 1-8: F_1 population; 9-16: F_2 population; 17-24: F_3 population

3.2 Genetic Similarity Index, Genetic Distance and Cluster Analysis

The genetic distance and genetic similarity in three population was shown in Table 3. It showed that the genetic distance between F_1 and F_3 generation was the largest, while the value between F_2 and F_3 populations was the smallest. According to the genetic distance among groups, UPGMA were used to analyze relationship among three generations populations (Figure 2). It indicated that the F_2 and F_3 clustered firstly and then cluster with F_1.

Table 3. Genetic identity and genetic distance in three populations

Population	F_1	F_2	F_3
F_1	—	0.9465	0.9176
F_2	0.055	—	0.9716
F_3	0.086	0.0288	—

Notes: data below diagonal are genetic distance; data above diagonal are genetic identity.

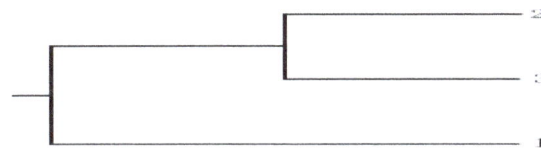

Figure 2. UPGMA dendrogram among three populations
1: F_1 population; 2: F_2 population; 3: F_3 population

3.3 Population Variation

The F_{st} values between F_1 and F_2, F_1 and F_3, F_2 and F_3 were 0.0079, 0.0196, 0.0028, respectively, which indicated that the genetic differentiation among the three generations was weak. The overall genetic differentiation coefficient value was 0.0169, which revealed that only 1.69 % genetic variation was from groups, while 98.31% variation was from individuals.

Table 4. F-statistica for three populations of hybrid scallop at ten microsatellite

loci	F_{is}			F_{st}
	F1	F2	F3	
CFMSM009	0.4180	0.4727	0.7080	0.0082
CFMSM014	0.1507	0.1588	0.0812	0.0148
CFMSM020	0.1495	-0.0549	-0.1556	0.0075
CFMSP011	-0.3616	-0.2698	-0.4124	0.0135
CFAD213	0.3143	0.4907	0.4018	0.0376
CFFD143	-0.1072	-0.1216	-0.0806	0.0235
CFBD075	-0.1739	0.0657	0.1978	0.0090
CFAD053	-0.0439	0.1780	0.0755	0.0355
S7259	0.4568	0.2784	0.2220	0.0042
S9090	-0.3100	-0.2279	-0.2180	0.0132
mean	0.0775	0.0822	0.0539	0.0169

Table 5. F_{st} values of pairwise comparison among different population at 10 microsatellite

Groups	F_1	F_2	F_3
F_1	—	0.0079	0.0196
F_2		—	0.0028
F_3			—

4. Discussion

In this study, the data revealed that the genetic diversity among three populations was not significantly different, although the proportion of polymorphic loci and genetic diversity decreased from F_1 to F_3. This situation was similar to some other reports (Hedgecock, Chow, & Waples, 1992; Mgaya, Gosling, Mercer, & Donlon, 1995; V. Sbordoni, De Matthaeis, M. C. Sbordoni, La Rosa, & Mattoccia, 1986). For example, the genetic diversity in several generations of Chinese shrimp were reported by Zhang et al. (2005) using SSR technology, which showed that the average observed heterozygosity dropped down from the first generation to the sixth. And it was same as the studied in American oysters (Yu and Guo, 2004) and Japanese flounder (Liu et al. 2005). Selective breeding is a complex process, the external environment and artificial selection pressure may cause fluctuations of population genetic variation. And on the other hand, during the artificial breeding process, due to the small effective population the inbreeding rate might increase which caused inbreeding depression and bottleneck effect. And also, high-intensity artificial directional selection might lead to the genetic deterioration and introgression, that these two factors could cause the loss of some particularly alleles, especially some rare gene allele in the breeding population. Therefore, genetic variation of breeding populations should be detected timely in the process of selective breeding.

Coefficient of genetic differentiation is an important parameter reflecting the degree of genetic differentiation among populations. Under controlled conditions, artificial selection, mutagenesis, hybridization could damage balance of genetic, which caused changes of genes and genotype, so the genetic characteristics within a population will also change. In this study, from the first generation to the third generation of breeding populations, the genetic structure within populations changed. Although the genetic structure had a certain differentiation, the degree of differentiation was not significant. The genetic variation analysis showed that genetic differentiation among generations of artificial breeding populations was smaller and differentiated was mainly from individuals. As studies in Chinese shrimp reported by Li et al. (2006), the genetic differentiation coefficient between adjacent groups showed a decreasing trend, and genetic similarity degrees of individuals within a population showed an upward trend with the increase in generation, which indicated the breeding populations tend stability after years of breeding.

Heterozygosity is an important parameter to measure the genetic diversity of populations. In this study, the mean observed heterozygosity were from 0.6893 to 0.7025 among the three generations. High significant deviation phenomena of Hardy-Weinberg equilibrium were observed, and the F_{is} values indicated that seven loci in three populations showed a certain degree loss of heterozygosity, which indicated dumb allele might exist, which was similar to the report of Chinese shrimp (Zhang et al., 2005)). This study enriches the data of hybrid breeding of scallop, opening a new angle to understand genetic diversity change of different generation during artificial selective breeding.

References

He, B., Yang, A. G., Wang, Q. Y., Liu, Z. H., & Zhou, L. Q. (2006) . A comparative study on isozymes in the hybrids of *Chlamys farreri* (♀) × *Patinopecten yessoensis* (♂) and their parent stocks. *Marine Fisheries Research, 27*(5), 23-27.

Hedgecock, D., Chow, V., & Waples, R. S. (1992). Effective population numbers of shellfish broodstocks estimated from temporal variance in allelic frequencies. *Aquaculture, 108*(3), 215-232. http://dx.doi.org/10. 1016/0044-8486(92)90108-W

Li, Q., Park, C., Endo, T., & Kijima, A. (2004). Loss of genetic variation at microsatellite loci in hatchery strains of the Pacific abalone (*Haliotis discus hannai*). *Aquaculture, 235*(1), 207-222. http://dx.doi.org/10.1016/ j.aquaculture.2003.12.018

Li, Z., Li, J., Wang, Q., He, Y., & Liu, P. (2006). The effects of selective breeding on the genetic structure of shrimp *Fenneropenaeus chinensis* populations. *Aquaculture, 258*(1), 278-282.

Liu, P., Meng, X. H., He, Y. Y., Kong, J., Li, J., & Wang, Q. Y. (2004). Genetic diversity in three wild populations of shrimp *Fenneropenaeus chinensis* in Yellow and Bohai Seas as revealed by microsatellite DNA. *Oceanologia et limnologia sinica, 35*(3), 257-262.

Liu, Y., Chen, S., & Li, B. (2005). Assessing the genetic structure of three Japanese flounder (*Paralichthys olivaceus*) stocks by microsatellite markers. *Aquaculture, 243*(1), 103-111. http://dx.doi.org/10.1016/j. aquaculture.2004.10.024

Lv, Z. M., Yang, A. G., Wang, Q. Y., Liu, Z. H., & Zhou, L. Q. (2006). Preliminary cytological identification and immunological traits of hybrid scallop from *Chlamys farreri* (♀) × *Patinopecten yessoensis* (♂) [J]. *Journal of Fishery Science of China, 13*(4), 597-602.

Mgaya, Y. D., Gosling, E. M., Mercer, J. P., & Donlon, J. (1995). Genetic variation at three polymorphic loci in wild and hatchery stocks of the abalone, Haliotis tuberculata Linnaeus. *Aquaculture, 136*(1), 71-80. http://dx.doi.org/10.1016/0044-8486(95)01037-8

Sakamoto, T., Danzmann, R. G., Okamoto, N., Ferguson, M. M., & Ihssen, P. E. (1999). Linkage analysis of quantitative trait loci associated with spawning time in rainbow trout (Oncorhynchus mykiss). *Aquaculture, 173*(1), 33-43. http://dx.doi.org/10.1016/S0044-8486(98)00463-3

Sbordoni, V., De Matthaeis, E., Sbordoni, M. C., La Rosa, G., & Mattoccia, M. (1986). Bottleneck effects and the depression of genetic variability in hatchery stocks of Penaeus japonicus (Crustacea, Decapoda). *Aquaculture, 57*(1), 239-251. http://dx.doi.org/10.1016/0044-8486(86)90202-4

Yang, A. G., Wang, Q. Y., Liu, Z. H., & Zhou, L. Q. (2003). The hybrid between the scallops *Chlamys farreri* and *Patinopecten yessoensis* and the inheritance characteristics of its first filial generation. *Marine Fisheries Research, 25*(5), 1-5.

Yu, Z., & Guo, X. (2004). Genetic analysis of selected strains of eastern oyster (*Crassostrea virginica Gmelin*) using AFLP and microsatellite markers. *Marine biotechnology, 6*(6), 575-586. http://dx.doi.org/10.1007/s10126-004-3600-5

Zhang, T. S., Wang, Q. Y., Liu, P., Li, J., & Kong, J. (2005). Genetic diversity analysis on selected populations of shrimp *Fenneropenaeus Chinensis* by microsatellites. *Oceanologia Et Limnologia Sinica, 29*(1), 6-12.

Zhou, L. Q., Yang, A. G., & Liu, Z. H. (2003). Cytological Observations on Cross Fertilization between *Chlamys farreri* (♀) and *Patinopecten yessoensis* (♂) Scallops. *Chinese Journal of Zoology Peking, 38*(4), 20-23.

Zhou, L. Q., Yang, A. G., Liu, Z. H., Du, F. Y., & Wang, Q. Y. (2003). Electron microscope observation on *Patinopecten yessoensis* sperm penetration into Chlamys farreri egg. *Journal of Fishery Sciences of China, 3*, 002.

13

DNA Underreplication in the Majority of Nuclei in the Drosophila Melanogaster Thorax: Evidence from Suur and Flow Cytometry

J. Spencer Johnston[1], Molly Schoener[2] & Dino P. McMahon[3]

[1] Department Entomology, Texas A & M University, College Station, TX, USA

[2] Department Biochemistry & Biophysics, Texas A & M University, College Station, TX, USA

[3] Department of Zoology, University of Oxford, Oxford, UK

Correspondence: J. Spencer Johnston, Department Entomology, Texas A & M University, College Station, TX 77843, USA. E-mail: spencerj@tamu.edu

Abstract

The discovery of endoreduplication in the majority of cells of the thorax of Drosophila has implications for genomics, transcriptome levels, chromatin structure and life history of these model insects. The ratio of 2C/4C DNA amounts is 2.00 for nuclei from the head, yet is 1.75 and 1.83 for nuclei from the thorax of wild type and suppressor of underreplication (SuUR) strains, respectively. The latter ratios reflect underreplication in the majority of nuclei from the thorax, which is only partially suppressed in the SuUR strain. The effect is age-dependent. Thoracic 4C DNA is significantly more underreplicated in the nuclei of 10 day old than newly emerged flies. The consequences of underreplication for the majority of thoracic cell nuclei, likely mimic those in highly endoreduplicated polytene salivary and nurse cell nuclei, which would affect expression levels in genomic, transcriptomic, methalomic and other studies based wholly or in part on Drosophila thoracic nuclei.

Keywords: underreplication, endoreduplication, thoracic nuclei, Drosophila

1. Introduction

Genomic underreplication is a phenomenon commonly associated with the polytene chromosomes of the salivary glands (Rudkin, 1969) and nurse cells of Drosophila and other Diptera (Painter & Reindorp, 1939). Underreplication of intercalated heterochromatin is apparent in polytenic salivary chromosomes as weak point regions or "fragile" sites (Laird, 1989) as well as for regions of euchromatic sequence localized in constrictions and ectopic fibers (Lamb & Laird, 1987; Lönn, 1982), whereas underreplicated pericentric heterochromatin is usually only visible as an indistinct "chromocenter" holding the various lengths of euchromatin together (Ashburner, 1970). Polyteny is also found in larval tissues where it is suggested to have evolved as "a cellular adaptation to the function of the larval period; that is, the extreme growth required by continuous, rapid food ingestion and the subsequent accumulation in the fat body of those reserves which will be later deployed in the complex morphogenetic expense of metamorphosis" (Pearson, 1974). While most larval polytene cells undergo programmed cell death (autophagy) (von Gaudecker & Schmale, 1974), a subset persists into the adult in Malphighian tubes, a select few brain cells and pericardial cells (Makino, 1938). Of these, only a few polytene brain cells have been positively shown to exhibit underreplication (Zacharias, 1993). Bosco et al. (2007) scored underreplication in 16C ovarian polytene cells in 38 species of Drosophila using flow cytometry and reported, "a strong correlation between genome size and amount of satellite replication".

Until recently, little was known about the control of underreplication. One of the first genes shown to control underreplication (Ur) was described as a naturally occurring polymorphism affecting underreplication of ribosomal genes in *Drosophila mercatorum* (DeSalle & Templeton, 1986; Malik & Eickbush, 1999). The Ur locus was shown to interact with R1 and R2 transposon inserts in the 28S rDNA of *D. mercatorum* and change life history parameters in field populations (Templeton et al., 1995). Subsequently a mutation called suppressor of underreplication (SuUR) was shown to affect underreplication of heterochromatin in polytene salivary gland chromosomes of *Drosophila melanogaster* (Belyaeva et al., 1998). The SuUR locus becomes active early in embryogenesis and maintains its activity throughout the cell cycle (Pindyurin et al., 2008). Expressed as a semi-dominant in the heterozygote, SuUR produces a maternal affect that is associated with activity during early embryogenesis, position effect variegation (PEV), and chromatin breaks during all stages of development (Belyaeva et al., 1998). The wild type allele, SuUR$^+$ acts in a dose-dependent fashion; extra doses of SuUR$^+$ enhance PEV, whereas zero or one dose suppresses variegation (Belyaeva et al., 2003). Ectopically expressed SuUR affects the overall amount of amplification of DNA in polytene chromosomes and of chorion protein genes in follicle cells (Volkova et al., 2003). SuUR protein product interacts with heterochromatin protein 1 (HP1) which associates with heterochromatin, influences heterochromatin formation, gene silencing, regulation of gene expression and telomere stability (Fanti & Pimpinelli, 2008). In the absence of HP1, SuUR cannot bind appropriately to chromosomes, which suggests HP1 directs SuUR to the correct binding sites (Pindyurin et al., 2008).

Here, we examine the effect of the SuUR mutation on putatively underreplicated 4C nuclei in the majority of cells in the thorax of *Drosophila melanogaster*. Using flow cytometry to quantify DNA amounts in 2C unreplicated (G1) and 4C replicated (G2) nuclei, we ask if underreplication (and its many consequences) occurs in head and thoracic tissue and if SuUR mutations change the extent of that underreplication. We report that SuUR partially, but not completely, suppresses underreplication in 4C thoracic nuclei. Further, we show that the underreplication is suppressed in an age dependent manner - an SuUR effect not heretofore observed.

2. Material and Methods

Drosophila melanogaster strains Ore-R and SuUR (Bloomington/4445) were obtained from the Drosophila stock center and maintained at room temperatures on cornmeal, molasses, and yeast medium food in 6 oz plastic fly bottles with cotton stoppers. To score relative amounts of replicated DNA at different ploidy levels, flow cytometry was conducted following DeSalle et al. (2007). Nuclei were isolated from the head or thorax of a fly of a given age, sex and strain by grinding the appropriate tissue in Galbraith buffer (Galbraith et al., 1983) using 15 strokes of the "A" pestle in a Kontes 2 ml Dounce. The released nuclei were passed through a 50 micron filter, stained with 50 parts per million (ppm) of propidium iodide and run (after 30 minutes in the cold and dark) in a Beckman Coulter Epics Elete flow cytometer with the laser emitting an exciting light at 488 nm nanometers (nm). Red fluorescence from propidium iodide (intercalated into the DNA of the 2C and 4C nuclei of *Drosophila*) was detected using a high bandpass filter (615 nm). Doublet discrimination (based on the peak and total fluorescence) showed that clumped nuclei made up less than 1% of the total, which justified the inclusion of all fluorescent nuclei in the analysis.

The total number of nuclei and average channel numbers of 2C and 4C nuclei were calculated using software provided with the Epics Elete cytometer. The ratio of 4C to 2C counts and the ratio of average channel numbers for the 4C and 2C nuclei were calculated. The means of these ratios were compared using PROC GLM from SASS (SAS Institute, Inc., Cary, North Carolina), with Scheffe means comparisons and orthogonal contrasts. Both the untransformed ratios and the arcsin root transformed ratios were used in analyses. Because transformation did not change significance at any level of the GLM model, results are reported based on the untransformed data.

3. Results

The G2 and G1 nuclei, that we expected would correspond to 4C and 2C ploidy levels, were scored in the head and thorax of Ore-R and SuUR males and females using flow cytometry (Figure 1). The ratio of fluorescence averages for G2 and G1 nuclei scored from the head of Ore-R males and females of < 1, 1-10 and 10 days of age was 2.0080 +/- 0.0004 (N = 60) and not significantly different from the expected 2.0 ratio (Figure 1a, b). In contrast, the ratio of average fluorescence (amount of replicated DNA) for G2 and G1 nuclei from the thorax was 1.7830 +/- 0.0030, which is very significantly less than the expected 2.0 ratio observed for the head (P < 0.00001) (Table 1).

The reduced amount of fluorescence from 4C nuclei in the thorax suggests underreplication of heterochromatin - underreplication that is typical of the polytene chromosomes in dipteran salivary glands, follicle and nurse cells (Leach et al., 2000), but is not known to occur in the thorax of fully normal alates. Further, endoreduplication and associated underreplication appear to significantly alter the proportions of 2C, 4C, and 8C amounts of DNA in the thorax. As shown in Figure 1a, c the proportion of 4C nuclei in the thorax is fully 10 times that in the head (53.7% 4C in the thorax vs. 5.5% 4C in the head). Further, an 8C peak that was entirely missing in nuclei from the head was observed in nuclei from the thorax, an indication that the thoracic nuclei of the thorax are endoreduplicated through 1 round and (rarely) 2 rounds of endoreduplication.

Table 1. Underreplication of 4C DNA, represented as the ratio of 4C average fluorescence/2C average fluorescence, for males and females of SuUR, and Ore-R strains of D. melanogaster of different ages. The expected ratio is 2.0 for fully replicated 4C DNA. The effect of strain and age were very highly significant ($P < 0.0001$); the sex effect was significant only for $P < 0.5$. All interactions among these three main effects were nonsignificant ($0.37 < P < 0.87$)

Age	Strain***	Sex*	N	4C/2C Ave	SE
10 days	Ore-R	Female	19	1.759	.005657
10 days	Ore-R	Male	22	1.747	.009148
10 days	SuUR	Female	15	1.789	.007057
10 days	SuUR	Male	19	1.790	.009197
1-10 days	Ore-R	Female	29	1.776	.012989
1-10 days	Ore-R	Male	28	1.752	.010388
1-10 days	SuUR	Female	31	1.824	.013975
1-10 days	SuUR	Male	26	1.788	.015725
< 1 day***	Ore-R	Female	9	1.836	.009333
< 1 day***	Ore-R	Male	6	1.815	.011329
< 1 day***	SuUR	Female	8	1.888	.008178
< 1 day***	SuUR	Male	5	1.842	.016283

*** Overall means for Strain and for flies < 1 day age were very highly significantly different ($p < 0.0001$).

* Overall means for Sexes were significantly different for $p < 0.05$.

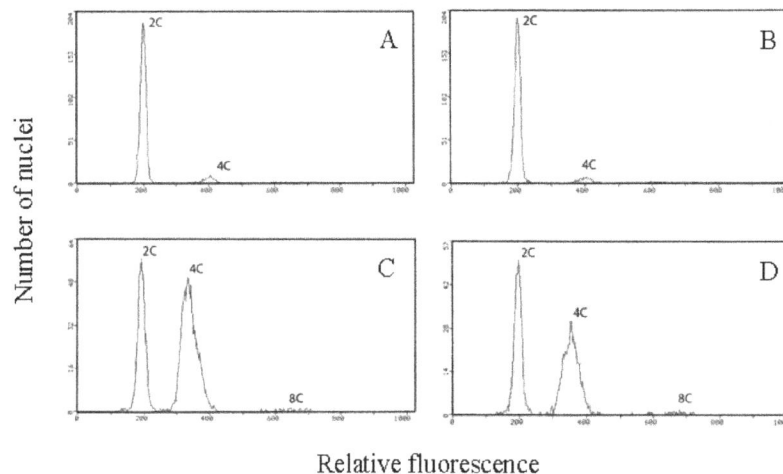

Figure 1. Fluorescent peaks for 2C and 4C nuclei of D. melanogaster extracted from 1a) head of Ore-R, 1b) head of SuUR. 1c) thorax of Ore-R, and, 1d) thorax of SuUR

Both the large proportion of 4C cells in the thorax and the relative underreplication of those 4C cells are evident by comparison of 1a,b with 1c,d. Considerable variation in amount of underreplication in 4C nuclei of both Ore-R and SuUR produces a broad 4C peak in 1c-d.

To test the hypothesis that the 4C cells of the thorax are underreplicated, we compared the result for Ore-R flies to those of a SuUR mutant strain known to suppress underreplication in the salivary gland. As observed for Ore-R, fluorescence from 4C and 2C nuclei from the head of SuUR flies produces a 2:1 ratio (2.0067 +/- 0.0001; N = 40) (Figure 1b), showing that DNA is fully replicated in the vast majority of 4C tissues in the head of SuUR strains. The interesting result was with the nuclei from the SuUR thorax. The ratio of fluorescence (relative amounts of DNA) in the 4C to 2C nuclei of the thorax was very significantly greater in SuUR than in Or-R (1.82 vs. 1.78; P < 0.0001) (Table 1). Given a 1C genome size of 175 Mb in Ore-R Drosophila, this change represents 28 million more nucleotides per nucleus in 4C SuUR thoracic nuclei than in comparable nuclei from the Ore-R strain. We conclude from this that underreplication of heterochromatin is found in a majority of cells in the thorax, with underreplication partially, but not completely suppressed in SuUR flies. Such underreplication has very important implications for the transcriptome from the thorax of the fly. Underreplication of heterochromatin changes the ratio of euchromatin to heterochromatin, creating possible position effects (Belyaeva et al., 2003), altered chromatin structure (Dorn et al., 1986) and modified expression levels for replicated and underreplicated (often heterochromatic) loci.

To test the possibility that the age of the adult fly influences the amount of underreplication, we compared the ratio of 4C:2C fluorescence in a subset of the above flies that had been scored as newly emerged flies or reared to 10 days of age. The difference was a highly significant 1.77 ratio of 4C: 2C fluorescence in 10 day old flies versus a ratio of 1.86 in newly emerged flies (P < 0.0001) (Table 1). Underreplication was greater in 10 day old flies than in those newly emerged, and both Ore-R and SuUR showed the same age dependent effect. A general linear statistical model showed no age by strain interaction for either SuUR or Ore-R flies. The ratio of DNA amount in 4C versus 2C cells in the thorax (Figure 1c, d) was less in older flies for both Ore-R (1.825 vs. 1.754) and SuUR (1.868 vs. 1.791) strains.

An age effect would be seen if 4C underreplication affects viability, and those cells that are least underreplicated in newly emerged flies undergo programmed cell death and autophagy, as observed for endoreduplicated cells of larvae (von Gaudecker & Schmale, 1974). However when the curve representing the population of 4C nuclei is compared for 10 day old flies and newly emerged flies, it is seen that the coefficient of variation of 4C fluorescence increases at the same time that the average amount of fluorescence decreases (CV = 8.1 vs. 7.4, P < 0.001). Additionally, there is a small, but significant increase, rather than loss, of 4C relative to 2C cells in older flies (P < .001). Thus, not programmed cell death/autophagy, but rather further endoreduplication from 2C to 4C nuclei (Pearson, 1974), with or without degradation of DNA associated with double strand breaks (Spradling, 2001), likely accounts for the age effect.

4. Discussion

Accurate measurement of DNA content in G1 and G2 cells of *D. melanogaster* was historically important, not only to settle old arguments about strandedness of DNA in chromatids (Rudkin, 1965; van de Flierdt, 1975) but also to show that metaphase cells from the brain of Drosophila are good standards for estimation of genomic DNA content (Zacharias et al., 1993). Here, we repeat these earlier reports showing that 4C (G2) cells of neural tissue have twice the DNA content of 2C (G1) cells; we then go on to examine the 4C cells that make up the majority of the *D. melanogaster* thorax. The 4C thoracic nuclei, unlike the 4C in the head, are underreplicated. The degree of underreplication is partially ameliorated in SuUR mutants, but not completely. This was anticipated by Umbetova et al. (1991) who noted that the degree of underreplication was proportional to the degree of polyteny, with the maximum suppression in the highly polytenized salivary gland cells, intermediate underreplication in fat body cells that have an intermediate degree of polyteny, and no underreplication in diploid cells of the head.

Zhimulev et al. (2003) showed that SuUR copy number can have dramatic effects on underreplication and expression levels in polytene tissues, and noted that SuUR proteins target genes that are repressed in embryos and then activated later in development (Pindyurin et al., 2007). We report here for the first time, that underreplication is typical of cells in the thorax and greater in 10 day old than in newly emerged flies. The observed increase in underreplication in 10 day old flies, coupled with an increase in the proportion of 4C nuclei, suggest that endoreduplication with underreplication continues after emergence of the imago, with or without a loss of heterochromatic DNA due to exonuclease activity (Spradling, 2001; Glaser et al., 1992).

These findings suggest a similarity of the underreplication control mechanism for single round endoreduplicated thoracic cell nuclei and highly endoreduplicated polytene salivary and nurse cell nuclei. In salivary polytene chromosomes, HP1 (heterochromatin protein 1) and SuUR interact and bind to underreplicated regions (Pindyurin et al., 2008), while a number of other genes are known to interact further with HP1 to modify position effects associated with changes in heterochromatin (Wallrath, 1998, review). The *Su(var)2-5* locus encodes the HP1 protein (Eissenberg et al., 1992) required for the assembly of the protein-modifying complex (Richards & Elgin, 2002, review). The *Su(var)3-6* locus encodes protein phosphatase PP1 (Baksa et al., 1993). The *Su(var)2-1* locus encodes histone acetylase (Dorn et al., 1986). And, the *Su(var)3-9* locus encodes the enzyme for methylation of histone H3 on lysine 9 (Rea et al., 2000; Czermin et al., 2001; Schotta et al., 2002).

Additionally, missense mutations at the histone deacetylase locus (HDAC1) are suppressors of position effect variagation (PEV) (Mottus et al., 2000); the product of the HisAv locus appears required for heterochromatin formation, as it interacts with HP1 and SuUR, directing the chromosomal distribution of the HP1 protein, of acetylated histone H4 at Lys12, and of methylated histone H3 at Lys9 (Swaminathan et al., 2005).

It is important to consider underreplication in the context of the endocycle. Our data indicate that the majority of thoracic cells in Drosophila have stalled at the mitotic/endocycle transition. In normal endocycling cells, many of the checkpoints that otherwise serve to regulate the cell cycle of dividing (mitotic) cells are lost. In the normal *Drosophila* endocycle, Cyclin E and its related cyclin-dependent kinase (*Cdk2*) has a conserved autoregulatory function: it is responsible for ensuring a stable G_1-S cycle (Lilly & Duronio, 2005). This is due to the conserved need for decreased Cdk activity in order to reset (or 'license') origins of replication in both endocycling and mitotic cells. During the endocycle, S-phase is truncated leading to incomplete DNA replication. Late replicating heterochromatin can be restored through transgenic overexpression of Cyclin E (Leach et al., 2000), and the continuous expression of Cyclin E throughout endocycling S-phase.

In normal endocycling cells, the DNA damage and/or presence of stalled replication forks that arise from underreplication do not typically halt progression of the cell cycle. In the thorax of Drosophila, we hypothesize that the majority of cells are stalled at the mitotic/endocycle transition, perhaps due to intra S-phase or G_2 checkpoint activity. The effect could be reduced in older flies because low E Cyclin activity in combination with incomplete DNA replication (truncation of S-phase) leads to the continued low-level licensing of new origins of replication, which over a period of days results in more complete DNA replication. This was anticipated in Lönn (1982) who reported in polytene nuclei of the midge *Chironomus*, "different populations of DNA replication intermediates are detectable in animals that have just passed a moult." Future studies must address how SuUR-mediated regulation of late origins of replication is linked to Cyclin E and S/G_2-phase checkpoint activity.

The consequences of underreplication in the majority of nuclei in the thorax are many. Position effects, transcription levels, chromatin condensation and genome architecture in general are altered by underreplication. Additionally double strand breaks have been shown to occur at all sites of underreplication and to increase in amount in direct proportion to the number of copies of SuUR[+] (Andreyeva et al., 2008). Understanding the molecular mechanisms that permit continuation or stalling of the endocycle in cells with truncated S-phase (and therefore underreplication) could offer significant insight into how checkpoints in normal mitotic cells work, and further, how cancer cells escape such controls. Overall our findings show the need for consideration of the proportion of 2C and 4C nuclei and the underreplication of the latter as it affects the transcriptome and genome architecture of the majority of cells in *D. melanogaster* thorax. It was thought that polyteny is only the rule among the specialized cells of larval and adult Diptera, where rapid growth is characteristic (Pearson, 1974). We find polyteny and endoreduplication are also characteristic of the majority of cells in the thorax; genomic studies will need to take this into account. Loci in underreplicated portions of the DNA are present in equal copy number (2C) in nuclei isolated from the head and thorax. In contrast, the proportion of DNA that is replicated (4C) is fully ten times greater in the thorax than in the head. Expression levels in the thorax can be 50% higher in the thorax than the head, simply as a consequence of different proportions of replication.

References

Andreyeva, E. N., Kolesnikova, T. D., Belyaeva, E. S., Glaser, R. L., & Zhimulev, I. F. (2008). Local DNA underreplication correlates with accumulation of phosphorylated H2Av in the *Drosophila melanogaster* polytene chromosomes. *Chromosome Res.* http://dx.doi.org/10.1007/s10577-008-1244-4

Ashburner, M. (1970). Function and structure of polytene chromosomes during insect development. *Adv. Insect Physiol., 7*, 1-95. http://dx.doi.org/10.1016/S0065-2806(08)60240-4

Baksa, K., Morawietz, H., Dombradi, V., Axton, M., Taubert, H., Szabo, G., ... Gausz, J. (1993). Mutations in the protein phosphatase 1 gene at 87B can differentially affect suppression of Position-Effect Variegation and mitosis in *Drosophila melanogaster*. *Genetics, 135*, 117-125.

Belyaeva, E. S., Boldyreva, L. V., Volkova, E. I., Nanayev, R. A., Alekseyenko, A. A., & Zhimulev, I. F. (2003). Effect of the Suppressor of Underreplication (SuUR) gene on position-effect variegation silencing in *Drosophila melanogaster*. *Genetics, 165*, 1209-1220.

Belyaeva, E. S., Demakov, S. A., Pokholkova, G. V., Alekseyenko, A. A., Kolesnikova, T. D., & Zhimulev, I. F. (2006). DNA underreplication in intercalary heterochromatin regions in polytene chromosomes of *Drosophila melanogaster* correlates with the formation of partial chromosomal aberrations and ectopic pairing. *Chromosoma, 115*, 355-366.

Belyaeva, E. S., Zhimulev, I. F., Volkova, E. I., Alekseyenko, A. A., Moshkin, Y. M., & Koryakov, D. E. (1998). Su(UR)ES: a gene suppressing DNA underreplication in intercalary and pericentric heterochromatin of *Drosophila melanogaster* polytene chromosomes. *Proc. Natl. Acad. Sci. USA, 95*, 7532-7537.

Bosco, G., Campbell, P., Leiva-Neto, J. T., & Markow, T. A. (2007). Analysis of *Drosophila* species genome size and satellite DNA content reveals significant differences among strains as well as between species. *Genetics, 177*, 1277-1290. http://dx.doi.org/10.1534/genetics.107.075069

Bridges, C. B. (1935). Salivary chromosome maps. *J. Hered, 26*, 60-64.

Czermin, B., Schotta, G., Hulsmann, B., Brehm, A., Becker, P., Reuter, G., & Imhof, A. (2001). Physical and functional association of SU(VAR)3-9 and HDAC1 in Drosophila. *EMBO Rep., 2*, 915-919. http://dx.doi.org/10.1093/embo-reports/kve210

DeSalle, R., & Templeton, A. R. (1996). The molecular through ecological genetics of *abnormal abdomen* in *Drosophila mercatorum*. III. Tissue-specific differential replication of ribosomal genes modulates the abnormal abdomen phenotype in *Drosophila mercatorum Genetics, 112*, 877-886.

DeSalle, R., Gregory, T. R., & Johnston, J. S. (2005). Preparation of samples for comparative studies of arthropod chromosomes: visualization, *in situ* hybridization, and genome size estimation. *Methods Enzymol, 395*, 460-488. http://dx.doi.org/10.1016/S0076-6879(05)95025-8

Dorn, R., Heymann, S., Lindigkeit, R., & Reuter, G. (1986). Suppressor mutation of position-effect variegation in *Drosophila melanogaster* affecting chromatin properties. *Chromosoma, 93*, 398-403. http://dx.doi.org/10.1007/BF00285820

Eissenberg, J. C., Ma, J., Gerber, M. A., Christensen, A., Kennison, J. A., & Shilatifard, A. (2002). dELL is an essential RNA polymerase II elongation factor with a general role in development. *Proc. Nat.l Acad. Sci. USA, 99*, 9894-9899. http://dx.doi.org/10.1073/pnas.152193699

Fanti, L., & Pimpinelli, S. (2008). HP1: a functionally multifaceted protein. *Curr. Opin. Genet. Dev., 18*, 169-174. http://dx.doi.org/10.1016/j.gde.2008.01.009

Galbraith, D. W., Harkins, K. R., Maddox, J. M., Ayres, N. M., Sharma, D. P., & Firoozabady, E. (1983). Rapid flow cytometric analysis of the cell cycle in intact plant tissues. *Science, 220*, 1049-1051. http://dx.doi.org/10.1126/science.220.4601.1049

Glaser R. L., Karpen G. H., & Spradling A. C. (1992). Replication forks are not found in a *Drosophila* minichromosome demonstrating a gradient of polytenization. *Chromosoma, 102*, 15-19. http://dx.doi.org/10.1007/BF00352285

Kolesnikova, T. D., Makunin, I. V., Volkova, E. I., Pirrotta, V., Belyaeva, E. S., & Zhimulev, I. F. (2005). Functional dissection of the Suppressor of Underreplication protein of *Drosophila melanogaster*: identification of domains influencing chromosome binding and DNA replication. *Genetica, 124*, 187-200. http://dx.doi.org/10.1007/s10709-005-1167-3

Laird, C. D. (1989). From Polytene Chromosomes to Human Embryology: Connections via the Human Fragile-X Syndrome. *Am. Zool, 29*, 569-591. http://www.jstor.org/stable/3883264

Leach, T. J., Chotkowski, H. L., Wotring, M. G., Dilwith, R. L., & Glaser, R. L. (2000). Replication of heterochromatin and structure of polytene chromosomes. *Mol. Cell. Biol., 20*, 6308-6316. http://dx.doi.org/10.1128/MCB.20.17.6308-6316.2000

Lamb, M. M., & Laird, C. D. (1987). Three euchromatic DNA sequences under-replicated in polytene chromosomes of *Drosohila* are localized in constrictions and ectopic fibers. *Chromosoma, 95*, 227-235. http://dx.doi.org/10.1007/BF00294779

Lilly, M. A., & Duronio, R. J. (2005). New Insights into cell cycle control from the *Drosophila* endocycle. *Oncogene, 24*, 2765-2775. http://dx.doi.org/10.1038/sj.onc.1208610

Lönn, U. (1982). DNA replication in polytene chromosomes. *Trends Biochem. Sci., 7*, 24-26. http://dx.doi.org/10.1016/0968-0004(82)90059-7

Makino, S. (1938). A morphological study of the nucleus in various kinds of somatic cells of *Drosophila virilis*. *Cytologia, 9*, 272-282. http://dx.doi.org/10.1508/cytologia.9.272

Malik, H. S., & Eickbush, T. H. (1999). Retrotransposable elements R1 and R2 in the rDNA units of *Drosophila mercatorum*: abnormal abdomen revisited. *Genetics, 151*, 653-665.

Moshkin, Y. M., Alekseyenko, A. A., Semeshin, V. F., Spierer, A., Spierer, P., Makarevich, G. F., ... Zhimulev, I. F. (2001). The Bithorax complex of *Drosophila melanogaster*: underreplication and morphology in polytene chromosomes. *Proc. Natl. Acad. Sci. USA, 98*, 570-574. http://dx.doi.org/10.1073/pnas.98.2.570

Mottus, R., Sobel, R. E., & Grigliatti, T. A. (2000). Mutational analysis of a histone deacetylase in Drosophila melanogaster: missense mutations suppress gene silencing associated with position effect variegation. *Genetics, 154*, 657-668.

Painter, T. S., & Reindorp, E. C. (1939). Endomitosis in the nurse cells of the ovary of *Drosophila melanogaster*. *Chromosoma, 1*, 276-283. http://dx.doi.org/10.1007/BF01271636

Pearson, M. J. (1974). Polyteny and the functional significance of the polytene cell cycle. *J. Cell Sci., 15*, 457-479.

Pindyurin, A. V., de Wit, E., Belyakin, S., Belyaeva, E., Christophides, G., Kafatos, F., ... Zhimulev, I. (2007). SUUR joins separate subsets of PcG, HP1, and B-type lamin targets in *Drosophila*. *J. Cell Sci., 120*, 2344-2351. http://dx.doi.org/10.1242/jcs.006007

Pindyurin, A. V., Boldyreva, L. V., Shloma, V. V., Kolenikova, T. D., Pokholkova, G. V., Andreyeva, E. N., ... Zhimulev, I. F. (2008). Interaction between the *Drosophila* heterochromatin proteins SUUR and HP1. *J. Cell Sci., 121*, 1693-1703. http://dx.doi.org/10.1242/jcs.018655

Rea, S., Elsenhaber, F., O'Carroll, D., Strahl, B. D., Sun, Z., Schmid, M., ... Jenuwein, T. (2000). Regulation of chromatin structure by site-specific histone H3 methyltransferases. *Nature (London), 406*, 593-599. http://dx.doi.org/10.1038/35020506

Richards, E., & Elgin, S. (2002). Epigenetic codes for heterochromatin formation and silencing: Rounding up the usual suspects. *Cell, 108*, 488. http://dx.doi.org/10.1016/S0092-8674(02)00644-X

Rudkin, G. T. (1969). Non-replicating DNA in *Drosophila*. *Genetics (Suppl.), 61*, 227-238.

Schotta, G., Ebert, A., Krauss, V., Fischer, A., Hoffmann, J., Rea, S., ... Reuter, G. (2002). Central role of Drosophila SU(VAR)3-9 in histone H3-K9 methylation and heterochromatic gene silencing. *EMBO J., 21*, 1121-1131. http://dx.doi.org/10.1093/emboj/21.5.1121

Spradling, A. C. (2001). Reflections on the *Drosophila* genome. *Funct. Integr. Genomics, 1*, 221-222. http://dx.doi.org/10.1007/s101420100034

Swaminathan, J., Baxter E. M., & Corces, V. G. (2005). The role of histone H2Av variant replacement and histone H4 acetylation in the establishment of *Drosophila* heterochromatin. *Genes Dev., 19*, 65-76. http://dx.doi.org/10.1101/gad.1259105

Templeton, A. R., Hollocher, H., & Johnston, J. S. (1993). The Molecular Through Ecological Genetics of *abnormal abdomen* in *Drosophila mercatorum*. V. Female Phenotypic Expression on Natural Genetic Backgrounds and in Natural Environments. *Genetics, 134*, 475-485.

Umbetova, G. H., Belyaeva, E. S., Baricheva, E. M., & Zhimulev, I. F. (1991). Cytogenetic and molecular aspects of position effect variegation in *Drosophila melanogaster*. IV. Underreplication of chromosomal material as a result of gene inactivation. *Chromosoma, 101*, 55-61. http://dx.doi.org/10.1007/BF00360687

van de Flierdt, K. (1975). No multistrandedness in mitotic chromosomes of *Drosophila melanogaster*. *Chromosoma, 50*, 431-434. http://dx.doi.org/10.1007/BF00327078

von Gaudecker, B., & Schmale E. M. (1974). Substrate-histochemical investigations and ultrahistochemical demonstrations of acid phosphatase in larval and prepupal salivary glands of *Drosophila melanogasater*. *Cell Tissue Res., 155*, 75-89.

Volkova, E. I., Yurlova, A. A., Kolesnikova, T. D., Makunin, I. V., & Zhimulev, I. F. (2003). Ectopic expression of the Suppressor of Underreplication gene inhibits endocycles but not the mitotic cell cycle in *Drosophila melanogaster*. *Mol. Genet. Genomics, 270*, 387-93. http://dx.doi.org/10.1007/s00438-003-0924-1

Wallrath, L. (1998). Unfolding the mysteries of heterochromatin. *Cur. Opin. Genet. Dev., 8*, 147-153. http://dx.doi.org/10.1016/S0959-437X(98)80135-4

Zacharias, H. (1993). Larger nuclei in the larval brain of *Drosophila nasutoides* often show underreplication, whereas metaphases provide a reliable DNA standard. *Genome, 36*, 294-301. http://dx.doi.org/10.1139/g93-041

Zhimulev, I. F., Belyaeva, E. S., Semeshin, V. F., Shloma, V. V., Makunin, I. V., & Volkova, E. I. (2003). Overexpression of the SuUR gene induces reversible modifications at pericentric, telomeric and intercalary heterochromatin of *Drosophila melanogaster* polytene chromosomes. *J. Cell Sci., 116*, 169-76. http://dx.doi.org/10.1242/jcs.00196

14

Osteonectin (SPARC) Is a Key Regulator of Malignant Pleural Mesothelioma

Fauzia Siddiq[1], Fulvio Lonardo[1], Harvey I. Pass[2], Arun K. Rishi[3] & Anil Wali[1,4]

[1] John D. Dingell VA Medical Center, Departments of Otolaryngology, Surgery, and Karmanos Cancer Institute, Wayne State University, Detroit, MI, USA

[2] Division of Cardiothoracic Surgery, New York University Cancer Center, New York, USA

[3] Department of Oncology, Wayne State University, John D. Dingell VA Medical Center, Research & Development Section, Detroit, MI, USA

[4] Center to Reduce Cancer Health Disparities, National Cancer Institute, National Institutes of Health, Rockville, MD, USA

Correspondence: Arun K. Rishi, Department of Oncology, Wayne State University, John D. Dingell VA Medical Center, Room B4334, Research & Development Section, 4646 John R. Street, Detroit, MI 48201, USA. E-mail: Rishia@Karmanos.org

Abstract

Malignant pleural mesothelioma (MPM) is a neoplasm arising from the mesothelial cells that line the thoracic pleura. Many factors such as asbestos, genetic predisposition, Simian Virus 40 (SV40) and radiation have been linked to MPM. Although, roles of osteonectin (SPARC) in various cancers including peritoneal MM have previously been studied, its involvement in development of MPM remains to be established. Our RT-PCR and Western blot analysis revealed elevated levels of osteonectin in archived MPM compared to normal peritoneum. We next generated Osteonectin-expressing stable cell lines derived from immortalized mesothelial Met5A cells. In addition, osteonectin expression was knocked-down by stable expression of osteonectin antisense in a patient-derived MPM cell line H2714. The osteonectin-expressing Met5A cells elicited faster migration rates when compared with their vector-expressing controls. Depletion of osteonectin, on the other hand, resulted in a slower migration rate of the MPM cells when compared with their vector-expressing counterparts. Expression of osteonectin also resulted in increased adhesion of the Met5A cells to collagen I, fibronectin and Laminin-coated plates while their adhesion to the collagen IV coated plates was decreased. Interestingly, knock-down of Osteonectin in MPM cells failed to influence their adhesion properties. Finally, we conducted in vitro invasion assays to determine whether osteonectin expression modulated invasive properties of these cells. Osteonectin expression in Met5A cells resulted in significant increase in their invasive properties while loss of osteonectin in MPM cells interfered with their abilities to invade. Taken together, our studies demonstrate that osteonectin regulates MPM cell adhesion, migration, and invasion properties, and therefore is an important regulator of MPM development and progression.

Keywords: malignant pleural mesothelioma, osteonectin, Met5A cells, RT-PCR

1. Introduction

Malignant pleural mesothelioma (MPM) is a neoplasm that arises from the mesothelial cells lining the pleura. About 2000-3000 new cases are diagnosed each year. Exposure to mineral fibers such as asbestos and erionite has been associated with increased incidence of MPM in men during the first half of the 20th century. However the MPM incidence rate in women remained almost unchanged even though women were also exposed to asbestos through their occupation and second hand exposure (Carbone et al., 2012). Some initial molecular evidence based studies cite the role of sex hormones and their receptors for this discrepancy (Chua et al., 2009; Pinton et al., 2009). Asbestos carcinogenesis is caused, in part, by chronic inflammation due to deposition of asbestos fibers and consequent release of pro-inflammatory molecules such as HMGB-1 (Yang et al., 2010). At the molecular level, asbestos causes DNA alteration by interfering with chromosomal segregation during mitosis (Olofsson & Mark, 1989) and by inducing mesothelial cells and macrophages to release mutagenic oxygen and nitrogen species (Xu, Zhou, Yu, & Hei, 2002), and thereby promoting survival of the cells with chromosomal mutations. Asbestos does not count for all the MPMs in men and women. About 50-80% of MPM in men and 20-30% of MPM in women are due to asbestos exposure. The association of other cofactors such as genetic predisposition, Simian Virus 40 (SV40) and radiation has been linked to MPM in addition to asbestos (Carbone et al., 2012).

Secreted protein acidic and rich in cysteine, SPARC (also known as osteonectin; or basement-membrane-40, BM-40) is a 32-kDa multifunctional matricellular glycoprotein. The gene encoding osteonectin is located on human chromosome 5q31-q33, and contains 10 exons spanning 34.6kb (Koukourakis et al., 2003). It is highly conserved through the evolution. There are a total of seven members within the osteonectin family. All members of this protein family share three similar domains (Tai & Tang, 2008). By binding to structural matrix proteins such as collagen and vitronectin, it functions in mediating cell-ECM interactions (Brekken & Sage, 2000). However, it does not have a role in structural stability of the ECM. It inhibits cell proliferation in a cell-type-dependent manner (Funk & Sage, 1993), regulates growth factor activity and inhibits cell adhesion and therefore it participates in physiological processes, where change in ECM and cell mobilization is required (Rivera, Bradshaw, & Brekken, 2011; Sweetwyne et al., 2004). Examples of such functions include bone mineralization (Termine et al., 1981), tissue remodeling (Tremble, Lane, Sage, & Werb, 1993; Alvarez et al., 2005) and morphogenesis (Motamed & Sage, 1998). Osteonectin plays a role in tumor initiation and progression through its counter adhesive properties (Sage & Bornstein, 1991; Tai & Tang, 2008) and ability to support migration (Campo McKnight, Sosnoski, Koblinski, & Gay, 2006). In this context, osteonectin has been found to influence matrix remodeling via metalloproteinases (Tremble et al., 1993), cells shape and proliferation through its interaction with various growth factors (Yan & Sage, 1999), and has an ability to settle a chronic low inflammatory state (Sangaletti et al., 2005; Alvarez et al., 2005; Sangaletti, Stoppacciaro, Guiducci, Torrisi, & Colombo, 2003). Although many studies have described the role of osteonectin in a variety of cancers, its role in tumorigenesis is complex and often dependent on the tissue or cell types involved. Osteonectin is differentially expressed in cancer tissue and surrounding stroma (Siddiq, Sarkar, Wali, Pass, & Lonardo, 2004) and consequently has different functions (Chlenski et al., 2006; Chlenski et al., 2007). High osteonectin expression in tumor cells has been correlated with metastasis, poor host survival and enhanced tumor growth in many cancers (F. Ledda et al., 1997; Schultz, Lemke, Ge, Golembieski, & Rempel, 2002; Kato et al., 2005; Thomas, True, Bassuk, Lange, & Vessella, 2000). Stromal osteonectin on the other hand has dual role in tumorigenesis. It inhibits angiogenesis in some cancers (Chlenski et al., 2006), while serving as an indicator of poor prognosis in others (N. Sato et al., 2003; Rempel, Ge, & Gutierrez, 1999; Derosa et al., 2012). Higher levels of osteonectin have been reported in melanoma (F. Ledda et al., 1997), glioblastoma (Rempel et al., 1998), Medulloblastoma (Bhoopathi, Gondi, Gujrati, Dinh, & Lakka, 2011), breast (Lien et al., 2007), colorectal cancer (Chew et al., 2011) and lung cancers (Siddiq et al., 2004). Lower levels of osteonectin due mainly to its promoter hypermethylation have been described in ovarian (Yiu et al., 2001), pancreatic (Puolakkainen, Brekken, Muneer, & Sage, 2004), acute mylogenous leukemia (DiMartino et al., 2006). Therefore osteonectin can support functions of tumor initiating or promoting genes in certain malignancies while it can also function as a tumor suppressor in other cancers. Many studies have described the role of osteonectin in different cancers including peritoneal MM. Its role in development of MPM is yet to be studied.

Due to the long latency period between first asbestos exposure and occurrence of the disease, the continued increase in MPM related mortality is predicted to last. The median survival time for MPM patients are between 8 and 18 months despite multimodality treatment options. Although recent clinical trials show some promise, MPM continues to be a serious health issue. Therefore, new and effective strategies for better management and treatment of this disease are urgently warranted. Such novel treatment strategies will provide hope for the medical community and patients diagnosed with MPM. New evidence based treatment options can be achieved if data explaining the disease mechanism is available. Here we investigated the molecular mechanisms by which osteonectin regulates MPM growth to better define its roles in biology and development of MPM.

2. Materials and Methods

2.1 Cell Lines and Reagents

Two cell lines, Met5A, generated from normal mesothelial cells and immortalized with SV40 large T-antigen and H2714 derived from a MPM patient, were used in our study for transfecting sense and anti-sense osteonectin respectively. Met5A was maintained in LHCMM (Biofluid) and H2714 was maintained in RPMI 1640 (HyClone) supplemented with 100 units/ml penicillin, 100 μg/ml streptomycine, 4mM L-glutamine and 10% fetal calf serum (HyClone) under 7.5% CO_2 at 37°C. The cultures were passaged weekly. Antibody against SPARC was purchased from Haematologic Technologies, Inc., Essex Jct., and VT. Secondary antimouse IgG-conjugated with horseradish peroxidase was bought from Santa Cruz Biotechnology, Santa Cruz, CA.

2.2 RNA Preparation, Reverse Transcription, and Cloning of Osteonectin cDNA

RNA was isolated using Trizol reagent from Life technologies Inc (Rockville, MD). Cells grown in monolayer at sub-confluency level and frozen tumor samples were lysed in Trizol, extracted in chloroform and precipitated with isopropanol. 3 μg poly (A)+ RNA was reverse transcribed with superscriptII reverse transcriptase (Life Technologies Inc, Rockville, MD) using oligo(dt)$_{12-18}$ primers (Life Technologies, Rockville, MD) into first-strand cDNA according to the manufacturer's instructions. The first strand cDNA was then utilized as template for a PCR reaction using 5' and 3' primers (Forward Primer 'GAGGGTTCCCAGCACCATG 3', Reverse Primer 5'AGGGGAGGGGTTAAAGAGAG 3'). PCR product was analyzed by agarose gel electrophoresis, the amplified DNA purified, and subcloned in PCRII vector with the advantage of TA cloning system. A number of recombinant plasmids were isolated and cDNA inserts were sequenced to ascertain validity of osteonectin-encoding open reading frame. Osteonectin cDNA was then excised from recombinant PCRII using BstXI and then sub-cloned into BstXI digested pcDNA3.1 expression vector (InVitrogen, Inc.). A number of recombinant pcDNA3.1 plasmids were isolated followed by restriction mapping to obtain constructs harboring sense and anti-sense orientations of osteonectin cDNA.

2.3 Generation of Osteonectin Stable Transfactants

We used Lipofectamine (Life Technologies Inc., Rockville, MD) to introduce recombinant pcDNA3.1 plasmids into human mesothelial Met5A as well as MPM cells. Transfections were performed according to the manufacturer's protocol. Briefly, equal numbers ($1.5x10^5$) of cells were seeded 48h prior to transfection, the medium was replaced with fresh serum–free medium and the cells were transfected with 1 to 2 μg of respective recombinant plasmid DNAs. Following 3 h incubation with the transfection solution, cells were washed in PBS and replenished with serum containing medium (10% fetal calf Serum). Forty eight hour later G418 (Calbiochem) was added at concentration of 200 μg/ml, and selection for stable transfectants continued for 3-4 weeks. Multiple, single colonies were isolated using cloning cylinders and expanded under G418. Only one clone each from sense and anti-sense transfection experiment was then used for subsequent experiments. For negative controls cell were transfected with pcDNA3.1 plasmid only.

2.4 Genomic DNA isolation and Genomic PCR

To confirm the integration of osteonectin in the genome of transfected cells, genomic DNA was isolated from the transfected cells and integrated vector plasmid DNA was amplified using PCR primers that span segment of the vector. DNA was isolated using puregene ^{TM} DNA purification kit from Gentra systems (Minneapolis, MN). Briefly, 1 to 2 million cells was lysed in lysis buffer. RNA was removed from the cell lysates by treating with RNAse A. Contaminant proteins were removed by adding protein precipitation solution. DNA was precipitated with 100% isopropanol. DNA pellet was hydrated overnight at room temperature and stored at -80°C. 100ng genomic DNA was PCR amplified with primer pairs 5'CGGGACTTTCCAAAATGTCG3' and 5'CAACAGATGGCTGGCAACTA3' that span pcDNA3.1. DNA was amplified using 2.5U Taq Gold from Applied Biosystems (Fostercity, CA). The 50 μl amplification reaction mixture contained the template, 400 ng primer, 10 mM dNTP and 25 mM MgCl$_2$ and the manufacturer's recommended buffer. Before amplification, Taq Gold was activated by heating at 95°C for 10min. DNA was then amplified with each amplification cycle consisting of a denaturation step at 95°C for 1min, an annealing step at 55°C for 1min and an extension step at 72°C for 1 min. Each amplification cycle was repeated 35 times with a final extension at 72°C for 5 min. Amplification products were then analyzed by agarose gel electrophoresis using a 2% agarose gel containing ethidium bromide.

2.5 Real-Time PCR

For quantitative real time PCR analysis the ABI prism 7700 sequence detector (Applied Biosystems Fostercity, Ca) was used. Briefly 1 μl c-DNA was amplified using SYBR green PCR master mix in a 50 μl reaction (Applied Biosystems Fostercity, Ca). The 50 μl reaction contained 400 ng each of forward and reverse primers specific each for sense and anti-sense osteonectin. To ensure the fidelity of mRNA extraction and reverse transcription all samples were amplified with 18S rRNA specific primers and normalized to 18S rRNA levels. Each mRNA and control samples were amplified in triplicate. The amplification conditions were 2 min at 50°C, 10min at 95°C and then 40 cycles of 95°C 15 sec, 60°C 1 min. Data were analyzed using v.1.7 software provided by PE Biosystems.

2.6 Protein Isolation and Western Blot Analyses

Cells that were grown in monolayer at 70% confluency or the frozen tumor samples were lysed in RIPA buffer (50 mM Tris-Hcl pH 7.5, 150 mM Nacl, 0.1% SDS, 1.0% nonidet P40, 0.5% sodiumdeoxycholate, 4 µg/ml aprotinin, 100 µg/ml leupeptin, 1 mM Na-orthovanadate, 0.1 mg/ml PMSF) on ice. The lysed suspensions were then centrifuged at 12,000 rpm for 10min at 4°C and the supernatants were collected, vortexed, aliquoted and stored at -80°C until used. The protein content was determined using Bio-Rad Dc protein assay kit (BioRad Laboratories Inc., Hercules, CA). For western blot analyses, 50 µg total protein from each specimen was resolved on a 15% SDS-PAGE gel under reducing conditions, transferred to Hybond-C super membrane (Amersham Life Science, Piscataway, NJ) The membranes were blocked for 1hr in 5% milk, followed by incubation with mouse anti-human osteonectin primary antibodies (Haematologic Technologies, Inc., Essex Jct., VT) for one hour. The membranes were then incubated with secondary antimouse IgG-conjugated with horseradish peroxidase (1:1000) dilution, Santa Cruz Biotechnology, Santa Cruz, CA) for additional one hour. The membranes were then subjected to ECL (Amersham Life Science, Piscataway, NJ), and autoradiographic signals were quantified by scanning followed by analysis with NIH image software.

2.7 Cell Cycle Analysis

Cells transfected with vector plasmid or plasmid encoding antisense-osteonectin were plated at equal density. 24 hours later, the cells were washed in PBS and allowed to synchronize in serum-free media for 72 h. The serum-starved cultures were replenished with 10% serum containing RPMI media and cells were allowed to re-enter the proliferation cycle. Cells were harvested following 24, 48 and 72 h of serum addition, and fixed in 70% ice cold ethanol overnight at 4°C. For flow cytometric analysis, the fixed cells were resuspended in 500 µl PBS with 0.5% triton X-100. To remove aggregates, fixed cells were treated with RNAseI at a final concentration of 40 µg/ml for 30 min at 37°C. Cells were then stained with 25 µg/ml propidium iodide for 1 h at room temperature. Stained cells were counted on a Becton Dickinson FACScan. About 15×10^3 cells were analyzed for each time interval. The DNA content profiles were analyzed with Cellquest from Becton Dickinson.

2.8 Migration Assay

Migration assay was performed essentially as described before (Dhanesuan, Sharp, Blick, Price, & Thompson, 2002) with minor modifications using the stable cell lines that express sense or antisense osteonectin. Briefly 5×10^4 cells were seeded in complete RPMI media in 6 well dishes. The cells were grown for 72 h under 5% CO_2 at 37°C to reach 70% confluency. After 72 h an incision was then made through the middle of the well with a sterile 1 ml blue pipette tip. The detached cells produced by the incision were removed by washing the plates twice with PBS. Fresh complete RPMI was then added and cells were allowed to grow. The widths of the incisions were photographed at 0, 48 and 72 h for cells expressing osteonectin cDNA in sense orientation, while the cells expressing anti-sense orientation of the osteonectin cDNA were photographed at 0, 18, and 36 h time periods. Marking was made at the bottom of the plate at 0 h to measure incision width at the same spot every time.

2.9 Adhesion Assay

Adhesion assay was performed in 6 well tissue culture plates according to previously published methods (Said & Motamed, 2005) with minor modification. Briefly plates were pre-coated with collagen I, collagen IV, Laminin and Fibronectin plates. A 5×10^4 cells/well were seeded in the coated plates and incubated at 37°C/5%CO_2. After an hour of incubation, plates were gently washed in PBS. Attached cells were stained and fixed in 0.2% crystal violet in MetOH for 15min. Stained plates were washed under running water, dried and photographed. Studies were done in triplicates per experimental condition. Attachment was quantified by solubilizing the attached cells in 1% SDS and taking absorbance at 540 nm.

2.10 Invasion Assay

Invasion assay was performed using Cell Invasion Assay kit (Chemicon International, Temecula, CA). Cells were first grown to subconfluence as monolayer, harvested and suspended at a concentration of 0.5×10^6 cells/ml in serum free RPMI medium. A 300 µl cell suspension was then placed in the upper chamber of a polycarbonate insert that contained an 8 µm pore size polycarbonate membrane, pre-coated with a thin layer of extracellular matrix (ECM). Before the cells were placed, the ECM inside the inserts were pre-equilibrated with the serum free RPMI for an hour. The inserts were then placed in 24 well tissue culture plates. As a chemo attractant, 500 µl RPMI containing 10% serum was then added to the outside of the insert. The cells were incubated under 5% CO_2 at 37°C for 72 h. Invasive cells migrated through the ECM layer and remained attached to the bottom of the polycarbonate membrane. Using a cotton-tipped swab, non-invading cells as well as the ECM matrix gel were removed from the interior of the inserts. Invasive cells attached to the underside of the inserts were then stained by dipping the bottom part of the inserts in the staining solution provided with the kit. Stained cells were counted with an inverted microscope equipped with camera. All assays were done in triplicates and the results were expressed as the mean ± SEM.

3. Results

3.1 MPM Tumor Tissues Express Elevated Levels of Osteonectin RNA and Protein

RT-PCR analysis showed increased osteonectin RNA expression in tumor tissue compared to normal pleura (Figure 1A). The tumor specimens were next analyzed for presence of osteonectin protein by western blotting as detailed in methods. Consistent with the RT-PCR analysis, all the MPM tumors showed presence of osteonectin while minimal to absent levels of osteonectin was noted in the normal pleura (Figure 1B). Quantitative analysis of the blot with NIH image soft-ware revealed up to 7.6 fold increase in osteonectin protein in certain tumor tissues when compared with osteonectin levels in the sample from normal pleura (Figure 1C).

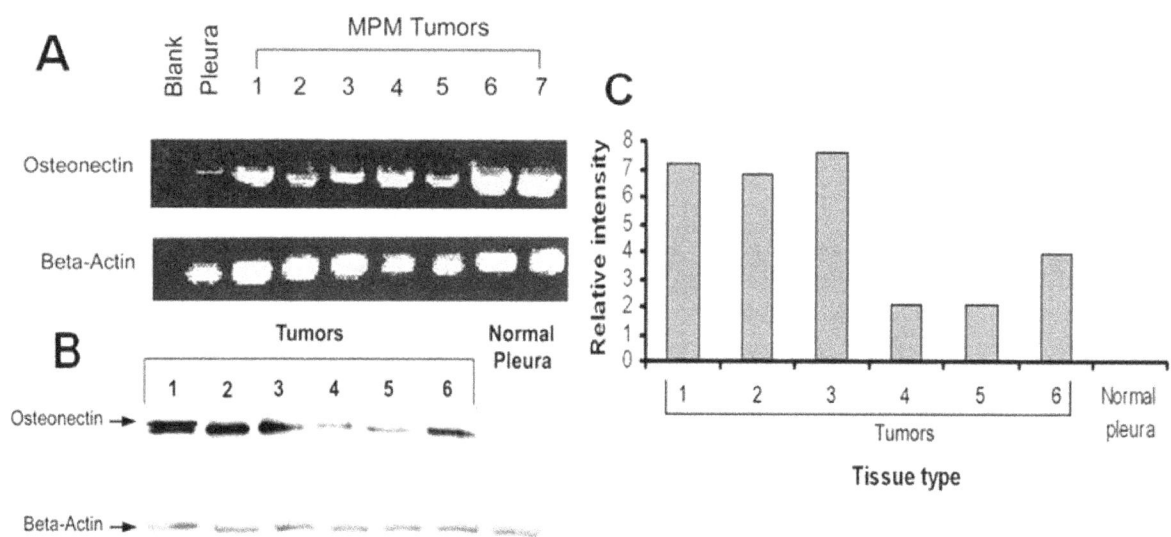

Figure 1. Osteonectin RNA and protein levels in different MPM tumors

(A) Gel showing osteonectin RNA expression in MPM tumors. The data represents one of the three independent experiment replicates. Equal amount of RNA from tumors were reverse transcribed and amplified with osteonectin and Beta-actin specific primers. An aliquot of the PCR products were run on agarose gel and visualized with ethidium bromide. From left to right, blank (no cDNA was used for PCR), pleura, Lane 1-7, different MPM tumors. (B) Representative (one of three replicates) western blot shows protein levels in different MPM tumors. The proteins were transferred to Hybond-C super membrane and osteonectin specific antibody was used to probe the blots. From left to right, Lane 1-6, protein levels in different MPM tumors, normal pleura. The blot was reprobed with Beta-actin specific antibody to assess the loading accuracy and used as an internal control. (C) Differential expression of osteonectin protein in MPM tumors and normal pleura. The autoradiographic signals from blot in panel B were quantified by NIH image software. All protein levels were adjusted relative to beta-actin level. From left to right, Lane 1 to 6, MPM tumors, normal pleura.

3.2 Ectopic Expression or Knock-Down of Osteonectin Alters Biological Properties of MPM Cells

To test the extent perturbation of osteonectin levels altered biological properties of the immortalized mesothelial as well as the MPM cells, we first generated and characterized several stable cell lines that harbored expression plasmids encoding osteonectin sense or antisense cDNAs. The osteonectin cDNA was PCR amplified, and subcloned in the pcDNA3.1 vector as described in methods. For osteonectin overexpression, the recombinant plasmid having osteonectin cDNA in the sense orientation was transfected into to the SV40 large T-antigen-immortalized normal mesothelial Met5A cells followed by isolation of stable cell lines as detailed in methods. The levels of osteonectin were analyzed by western blotting and its expression in a representative subline is shown in Figure 2A. Expression of osteonectin cDNA resulted in a 1.8-fold increase in osteonectin levels. For knock-down of the oesteonectin, the recombinant plasmid having anti-sense orientation of the osteonectin cDNA was transfected into H2714 MPM cells. Stable sublines expressing osteonectin antisense were isolated and characterized as in methods. Figure 2B shows integration of the antisense construct of a representative stable subline as revealed by PCR amplification and subsequent sequencing of a 1.4 kb chimeric fragment containing part of vector plasmid and full-length osteonectin cDNA. The vector control transfected cell line however yielded an expected 300bp PCR product. The levels of osteonectin were further analyzed by western blotting and were found to be significantly reduced in the subline stably expressing anti-sense construct (Figure 2C). Stable expression of recombinant plasmid encoding osteonectin antisense resulted in ~8.2-fold lower levels of osteonectin when compared with the vector-transfected control.

We next utilized the cells having either elevated or depleted levels of osteonectin to study the role of this molecule in cellular remodeling in vitro. As a first step, we performed wound healing assay. The ability of the cells to close monolayer wound was monitored over a 72 h period in the case of osteonectin-overexpressing cells while anti-sense expressing H2714 MPM cells were monitored over a period of 36 h. Results of the representative migration assays are shown in Figure 3. Overexpression of osteonectin resulted in accelerated migration of the cells in to the area of the incision (Figure 3A). Knock-down of osteonectin on the other hand caused a slower migration rate compared to the vector expressing cells.

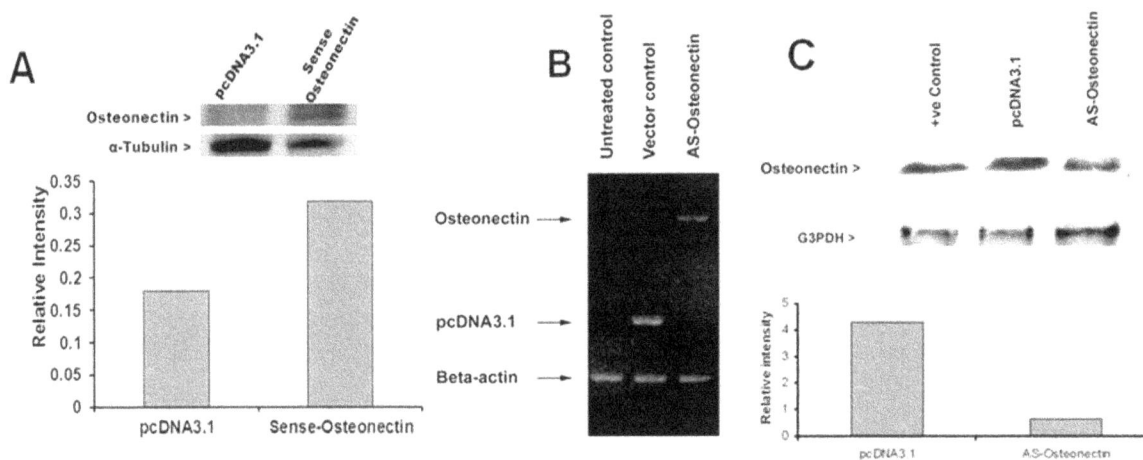

Figure 2. Alteration of osteonectin levels in transfected cells

(A) Real time PCR analysis of osteonectin levels in pcDNA 3.1 and sense osteonectin transfected cells. cDNA was amplified using SYBER green PCR master mix. Alpha-tubulin was used as an internal control. Data was analyzed using v.1.7 provided by the PE systems and relative intensity plotted. (B) Integration of antisense-osteonectin in transfected cells. Isolation of the genomic DNAs from the vector or antisense-transfected cells, and subsequent PCR analysis was carried out as detailed in methods. In antisensense-osteonectin transfected cells a 1.4 kb band specific for osteonectin (1.1 kb)+ surrounding pcDNA 3.1 (300bp) was amplified. To assess loading, genomic DNA was amplified with beta-actin specific primers. The integration of sense construct was similarly verified (data not shown). (C) Real-Time PCR analysis reveal decrease in osteonectin level in antisense osteonectin transfected cells compared to pcDNA 3.1 transfected counterparts. The isolation of genomic DNA and the real time PCR analysis were carried out essentially as in panel A except that G3PDH was used as an internal control. RNA from a tumor sample was used as a positive control. Data was analyzed using v.1.7 provided by the PE systems and relative intensity plotted in the histogram.

A

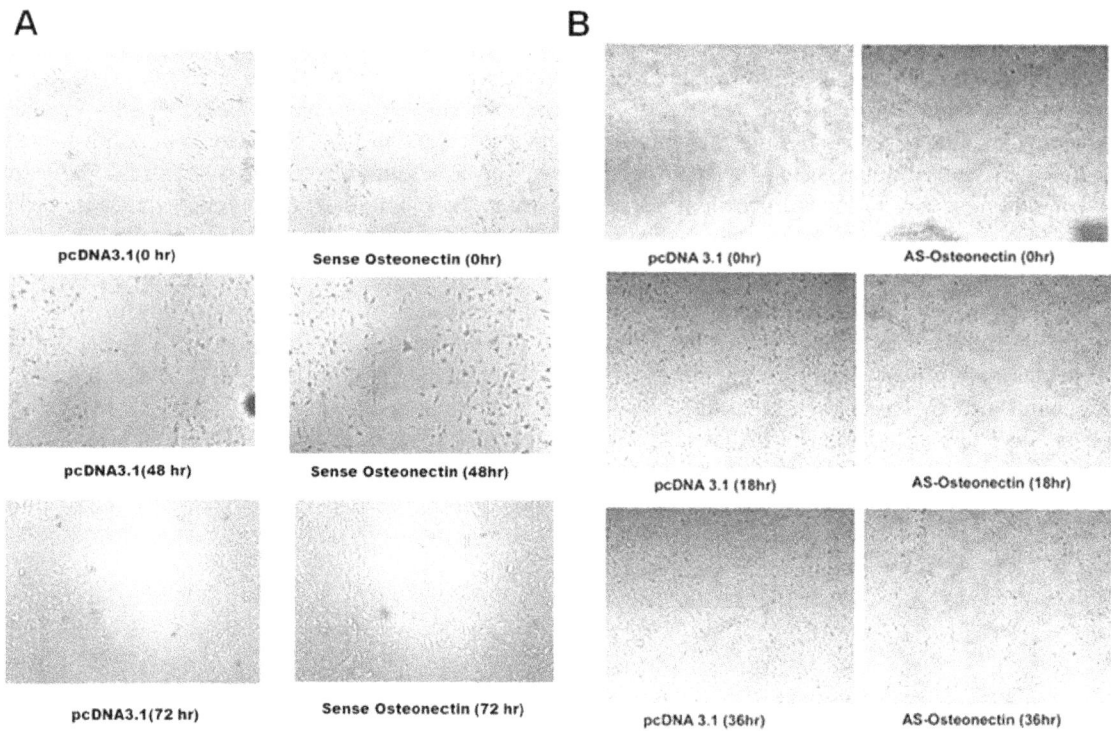

pcDNA3.1(0 hr)

Sense Osteonectin (0hr)

pcDNA3.1(48 hr)

Sense Osteonectin (48hr)

pcDNA3.1(72 hr)

Sense Osteonectin (72 hr)

B

pcDNA 3.1 (0hr)

AS-Osteonectin (0hr)

pcDNA 3.1 (18hr)

AS-Osteonectin (18hr)

pcDNA 3.1 (36hr)

AS-Osteonectin (36hr)

Figure 3. Osteonectin expression alters cell migration

(A) Migration assay of sense osteonectin tansfected cells and their respective vector transfected counterparts. A wound was created on a subconfluent culture of transfected cells and wound closure was measured as in methods. (B) Migration assay of antisense osteonectin tansfected cell lines. Similar to the experiment in panel A, a wound was created on a subconfluent culture of antisense transfected cells and migration of cells into the wound were monitored as above.

Since wound closure often depends on both proliferation and migration, to delineate the effect of proliferation on wound healing, we then performed cell cycle analysis on antisense osteonectin trasfected H2714 cells. Flow cytometry analysis at various time intervals revealed a relatively higher percentage of cells in the S and G2/M phase in the antisense transfected cells when compared with their vector-transfected counterparts (Figure 4). Although osteonectin depletion induced significant alterations in the cell cycle phase distribution of the cells, it appears unlikely that the slower migration of these cells following knock-down of osteonectin involved altered cell proliferation. Rather, our current data would suggest that osteonectin likely regulates biological properties of MPM cell migration.

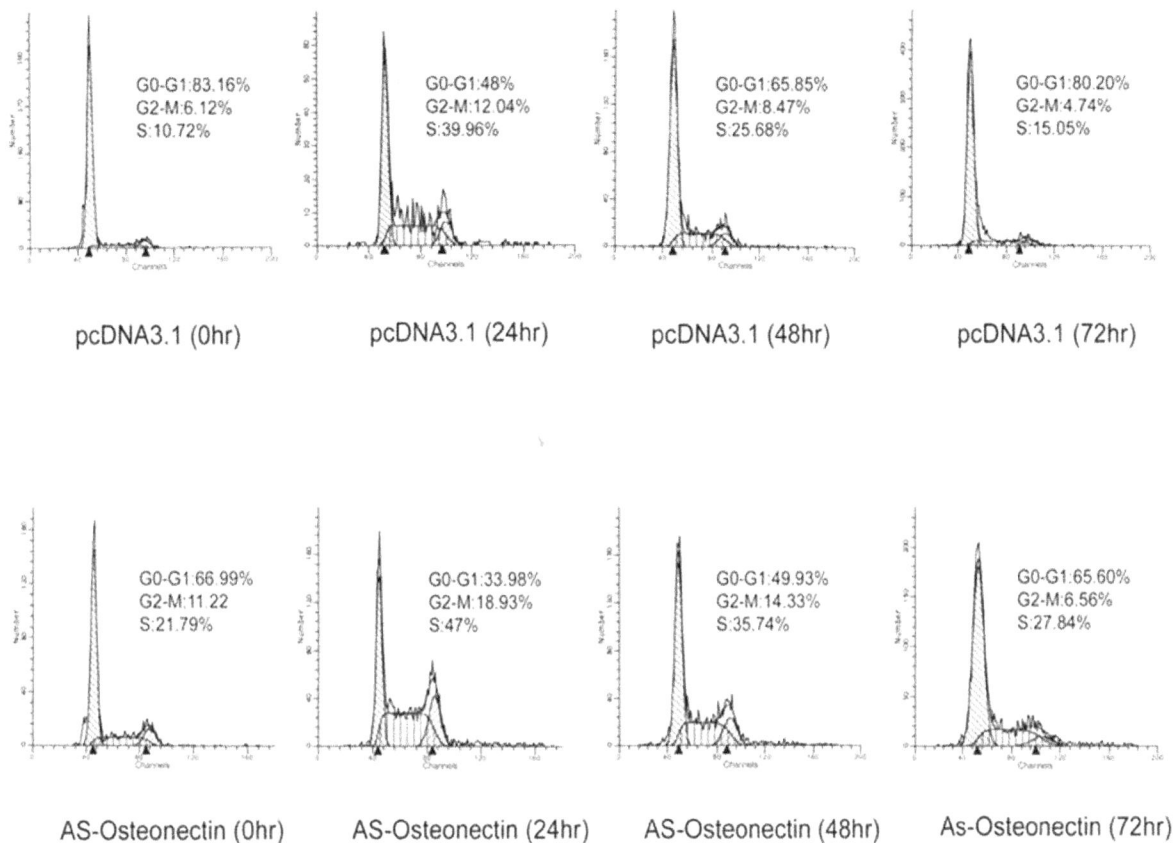

Figure 4. Osteonectin alters cellular migration properties independent of cell cycle alterations

Representative FACS analysis of pcDNA 3.1 and antisense osteonectin transfected cell line is shown. The various cell cycle population corresponding to individual cytofluorimetry peaks are indicated as inserts in each plot. Even though more cells are present in S-phase in anti-sense transfected cell line at different time intervals, the cells displayed attenuated migration in the wound healing assay suggesting that their migration properties were unlikely influenced by the alterations in their cell cycle phase distribution/cell division.

In light of the fact that loss of adhesion is often a pre-requisite for tumor cell invasion, and the fact that a number of earlier reports have suggested an ability/property of osteonectin to influence adhesion (Said & Motamed, 2005; Campo McKnight et al., 2006), we then investigated the extent osteonectin altered MPM tumor cell adhesion to the various ECM constituents (Collagen I, Fibronectin, Laminin and Collagen IV). Our study revealed, in comparison to the vector transfected cell line, osteonectin over-expressing cells displayed increased adhesion towards various ECM constituents. The maximum adhesion was observed towards collagen I and fibronectin (p = 0.0001) (Figure 5A) followed by laminin (p = 0.0053). However, elevated osteonectin resulted in reduced adhesion to collagen IV (not shown). Increasing the level of osteonectin in H2417 cells did not alter their adhesion properties in a statistically significant manner when compared with the vector expressing H2714 cells.

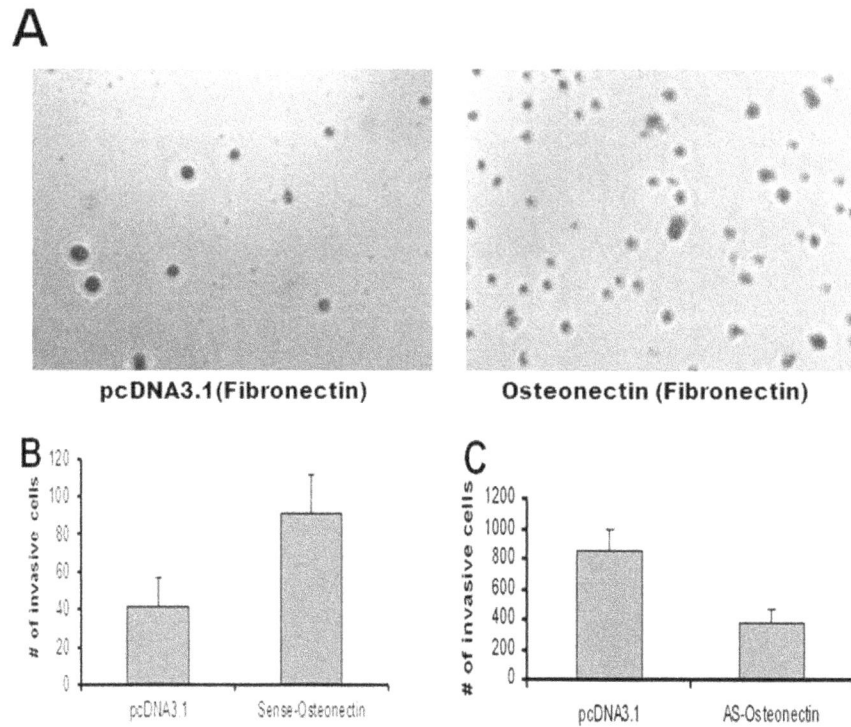

Figure 5. Osteonectin expression alters cell adhesion

(A) Representative image of the differences in adhesion of the osteonectin-overexpressing cells compared with the vector expressing cells. (B) Histogram showing number of invasive cells in osteonectin overexpressing cells and their vector transfected counterpart. (C) Histogram showing number of invasive cells in antisense osteonectin transfected cells and their vector transfected counterpart.

To further study the role of osteonectin in MPM cell invasion, in vitro assays were performed in matrigel coated inserts as described in materials and methods. Results of the invasion assays are shown in Figures 5B and C. Our data revealed up regulation of osteonectin expression in Met-5A cells resulted in 2.2 fold increase in invasion of cells when compared with their vector-transfected counterparts (Figure 5B). However, loss of osteonectin in H2714 cells resulted in 2.2 fold reduction in invasive properties of these cells in comparison with their vector expressing controls (Figure 5C).

4. Discussion

Osteonectin is matrix associated glycoprotein which is abundant in tissues undergoing remodeling. Several studies so far revealed up-regulated osteonectin expression in malignancies of both mesenchymal and epithelial origin. Osteonectin is over-expressed in many cell lines that originated from different malignancies such as glioblastoma (Rempel et al., 1998), melanoma (F. Ledda et al., 1997), prostate and breast cancers (Thomas et al., 2000), as well as the cells that acquire mesenchymal properties (Gilles et al., 1998). Osteonectin has dramatic effect on cell behavior in vitro. The major known functions of osteonectin include anti-adhesive and anti-proliferative effects. The counter-adhesive properties of osteonectin in non-malignant cells led to further exploration of its potential role in tumor invasion and metastasis. The ability of cancer cells to disseminate though requires that the malignant cells first adhere to ECM matrices followed by their invasion of the basement membrane. However, the role osteonectin plays in regulating adhesion and invasion properties is also cell type dependent. Our interest in mesothelioma prompted us to study the genes involved in mesothelioma initiation and progression. Our preliminary data from gene expression array (data not shown) revealed osteonectin was up-regulated in MPM that is further supported by increased osteonectin levels in MPM tumors noted in our current study. Although inconsistent pattern exists in terms of both expression and biological activity of osteonectin, we speculated that better understanding of the mechanism(s) of action of osteonectin may shed light into its role(s) in the processes of tumorigenesis. This prompted us to first clone osteonectin in the sense or antisense orientations and then generates various cell lines that either overexpress or have reduced levels of osteonectin. Following transfections, we generated stable cell lines, and characterized them for levels of osteonectin by RT-PCR and western immunoblotting.

As expected, over-expression of osteonectin caused increased migration and invasion *in vitro*. In a similar assay, knock-down of osteonectin resulted in decreased migration and invasion. Although over-expression of osteonectin interestingly promoted cell interactions with collagen I, fibronectin and laminin, knock-down of osteonectin intriguingly failed to cause significant changes in adhesive properties of the MPM cells. This could be due to the fact that the matrices involved in our experiments do not represent the entire repertoire of "stromal-tumor microenvironment" milieu under which tumor initiation, progression and local spread of MPM takes place within thoracic cavity. Based on the compensatory mechanisms with "osteopontin" being the predominant factor involved in conjunction with "osteonectin", it is likely that differences in the critical threshold levels of endogenous, constitutively active osteonectin within affected pleural space, and the released form of osteonectin in the extracellular space, besides other secretary molecules present in the region, influence interactions with regards to differences observed in adhesion and migratory properties of osteonectin. The extent absence of changes in the adhesive properties of osteonectin-depleted MPM cells is due to altered cellular "compensatory mechanisms" remains to be clarified. Nevertheless, a thorough further investigation will be necessary to explain this discrepancy. Although, overexpression of osteonectin was noted in the case of melanoma (Ledda et al., 1997) and breast cancer cell lines (Gilles et al., 1998), suppression of osteonectin in melanoma cells resulted in significant decrease in *in vitro* adhesive and invasive capacity of the cells, interfering with the ability of osteonectin-depleted cells to form tumors in vivo. In contrast, ovarian cancer cells that had elevated levels of osteonectin were significantly inhibited in their ability to adhere and invade a number of matrices such as collagen I, collagen IV, vitronectin, fibronectin, hyaluronic acid and laminin (Said & Motamed, 2005). This was supported by lower rate of intra-peritoneal dissemination of ovarian cancer in osteonectin+/+ mice. Similar effect was shown by pancreatic cancer cell grown in osteonectin-/- mice. In case of pancreatic cancer, tumor grown in osteonectin null mice was larger and more prone to metastasize compared to tumor grown in osteonectin+/+ mice (Puolakkainen et al., 2004; Arnold et al., 2008). Later the effect of osteonectin on ovarian cancer was shown to be caused by lowering the cell surface expression of α_V-integrin and $\beta 1$- subunits (Said, Najwer, & Motamed, 2007). Osteonectin was also shown to be capable of inhibiting the VEGF-induced integrin activation and down-regulation of metalloproteinases such as MMP2 and MMP9 (Said & Motamed, 2005; Socha et al., 2007). MMP2 can cleave Collagen type IV, which results in degradation of basement membrane and can help in metastasis and tumor progression (Hornebeck, Emonard, Monboisse, & Bellon, 2002). MMP2 is secreted as an inactive pro-enzyme. It becomes activated after the proteolytic cleavage of its NH2-terminal pro-fragment. (Atkinson et al., 1995; Stetler-Stevenson, 1994). MT1-MMP is another MMP that contributes to MMP2 activation (Lohi & Keski-Oja, 1995). *In vivo* major source of MMP2 and MT1-MMP are fibroblasts (Polette, Nawrocki-Raby, Gilles, Clavel, & Birembaut, 2004). However, up-regulated MMP2 and MT1-MMP have been observed in cases of invasive tumors (Sato & Seiki, 1996), (Gilles, Polette, Seiki, Birembaut, & Thompson, 1997). Elevetaed levels of MMP9 have also been found to be associated with increased invasiveness of both the murine and human MPM cells as well as the orthotopic MPM tumors in mice (Servias et al., 2012). In contrast to ovarian cancer cells where osteonectin reduced MMP activity, overexpression of MMP9 in pancreatic cancer cells resulted in a greater cell migration and invasion in vitro. This effect was completely abolished by the addition of osteonectin (Arnold et al., 2008). In cases of prostate cancer, osteonectin induced MMP2 activity and aided prostate cancer cells to metastasize in the bone (Jacob, Webber, Benayahu, & Kleinman, 1999), while in the ovarian and breast cancer cells, osteonectin and smaller peptide fragments containing its N-terminal region, induced MMP2 activation and their invasive phenotype. The role of osteonectin in invasion was further strengthened by its regulation/promotion of EMT transition (Girotti et al., 2011; Conant, Peng, Evans, Naud, & Cooper, 2011). Cell cycle analysis using osteonectin over-expressing breast cancer cells showed slower progression to S phase (Dhanesuan et al., 2002). This is consistent with our current observations where knock-down of osteonectin in MPM cells resulted in a greater number of cells in the G2/M and S phases.

In conclusion we have demonstrated for the first time the role of osteonectin in mesothelioma progression *in vitro*. We have found that osteonectin is a positive regulator of adhesion to various ECM constituents, invasion and migration, and thus an important transducer of MPM cancer progression. Our proof-of-concept studies therefore warrant further research to elucidate potential of osteonectin as a target for therapeutic intervention in MPM.

References

Alvarez, M. J., Prada, F., Salvatierra, E., Bravo, A. I., Lutzky, V. P., Carbone, C., ... Podhajcer, O. L. (2005). Secreted protein acidic and rich in cysteine produced by human melanoma cells modulates polymorphonuclear leukocyte recruitment and antitumor cytotoxic capacity. *Cancer Res., 65*, 5123-32. http://dx.doi.org/10.1158/0008-5472.CAN-04-1102

Arnold, S., Mira, E., Muneer, S., Korpanty, G., Beck, A. W., Holloway, S. E., ... Brekken, R. A. (2008). Forced expression of MMP9 rescues the loss of angiogenesis and abrogates metastasis of pancreatic tumors triggered by the absence of host SPARC. *Exp Biol Med (Maywood), 233*, 860-73. http://dx.doi.org/10.3181/0801-RM-12

Atkinson, S. J., Crabbe, T., Cowell, S., Ward, R. V., Butler, M. J., Sato, H., ... Murphy, G. (1995). Intermolecular autolytic cleavage can contribute to the activation of progelatinase A by cell membranes. *J. Biol. Chem., 270*, 30479-85. http://dx.doi.org/10.1074/jbc.270.51.30479

Bhoopathi, P., Gondi, C. S., Gujrati, M., Dinh, D. H., & Lakka, S. S. (2011). SPARC mediates Src-induced disruption of actin cytoskeleton via inactivation of small GTPases Rho-Rac-Cdc42. *Cell Signal, 23*, 1978-87. http://dx.doi.org/10.1016/j.cellsig.2011.07.008

Brekken, R. A., & Sage, E. H. (2000). SPARC, a matricellular protein: at the crossroads of cell-matrix. *Matrix Biol., 19*, 569-80. http://dx.doi.org/10.1016/S0945-053X(00)00105-0

Campo McKnight, D. A., Sosnoski, D. M., Koblinski, J. E., & Gay, C. V. (2006). Roles of osteonectin in the migration of breast cancer cells into bone. *J. Cell Biochem, 97*, 288-302. http://dx.doi.org/10.1002/jcb.20644

Carbone, M., Ly, B. H., Dodson, R. F., Pagano, I., Morris, P. T., Dogan, U. A., ... Yang, H. (2012). Malignant mesothelioma: facts, myths, and hypotheses. *J. Cell Physiol, 227*, 44-58. http://dx.doi.org/10.1002/jcp.22724

Chew, A., Salama, P., Robbshaw, A., Klopcic, B., Zeps, N., Platell, C., & Lawrance, I. C. (2011). SPARC, FOXP3, CD8 and CD45 correlation with disease recurrence and long-term disease-free survival in colorectal cancer. *PLoS One, 6*, e22047. http://dx.doi.org/10.1371/journal.pone.0022047

Chlenski, A., Guerrero, L. J., Yang, Q., Tian, Y., Peddinti, R., Salwen, H. R., & Cohn, S. L. (2007). SPARC enhances tumor stroma formation and prevents fibroblast activation. *Oncogene, 26*, 4513-22. http://dx.doi.org/10.1038/sj.onc.1210247

Chlenski, A., Liu, S., Guerrero, L. J., Yang, Q., Tian, Y., Salwen, H. R., ... Cohn, S. L. (2006). SPARC expression is associated with impaired tumor growth, inhibited angiogenesis and changes in the extracellular matrix. *Int. J. Cancer, 118*, 310-6. http://dx.doi.org/10.1002/ijc.21357

Chua, T. C., Yao, P., Akther, J., Young, L., Bao, S., Samaraweera, U., ... Morris, D. L. (2009). Differential expression of Ki-67 and sex steroid hormone receptors between genders in peritoneal mesothelioma. *Pathol. Oncol. Res., 15*, 671-8. http://dx.doi.org/10.1007/s12253-009-9170-0

Conant, J. L., Peng, Z., Evans, M. F., Naud, S., & Cooper, K. (2011). Sarcomatoid renal cell carcinoma is an example of epithelial--mesenchymal transition. *J. Clin. Pathol., 64*, 1088-92. http://dx.doi.org/10.1136/jclinpath-2011-200216

Derosa, C. A., Furusato, B., Shaheduzzaman, S., Srikantan, V., Wang, Z., Chen, Y., ... Petrovics, G. (2012). Elevated osteonectin/SPARC expression in primary prostate cancer predicts metastatic progression. *Prostate Cancer Prostatic Dis., 15*, 150-6. http://dx.doi.org/10.1038/pcan.2011.61

Dhanesuan, N., Sharp, J. A., Blick, T., Price, J. T., & Thompson, E. W. (2002). Doxycycline-inducible expression of SPARC/Osteonectin/BM40 in MDA-MB-231 human breast cancer cells results in growth inhibition. *Breast Cancer Res. Treat., 75*, 73-85. http://dx.doi.org/10.1023/A:1016536725958

DiMartino, J. F., Lacayo, N. J., Varadi, M., Li, L., Saraiya, C., Ravindranath, Y., ... Dahl, G. V. (2006). Low or absent SPARC expression in acute myeloid leukemia with MLL rearrangements is associated with sensitivity to growth inhibition by exogenous SPARC protein. *Leukemia, 20*, 426-32. http://dx.doi.org/10.1038/sj.leu.2404102

Funk, S. E., & Sage, E. H. (1993). Differential effects of SPARC and cationic SPARC peptides on DNA synthesis by endothelial cells and fibroblasts. *J. Cell Physiol., 154*, 53-63. http://dx.doi.org/10.1002/jcp.1041540108

Gilles, C., Bassuk, J. A., Pulyaeva, H., Sage, E. H., Foidart, J. M., & Thompson, E. W. (1998). SPARC/osteonectin induces matrix metalloproteinase 2 activation in human breast cancer cell lines. *Cancer Res., 58*, 5529-36.

Gilles, C., Polette, M., Seiki, M., Birembaut, P., & Thompson, E. W. (1997). Implication of collagen type I-induced membrane-type 1-matrix metalloproteinase expression and matrix metalloproteinase-2 activation in the metastatic progression of breast carcinoma. *Lab Invest., 76*, 651-60.

Girotti, M. R., Fernandez, M., Lopez, J. A., Camafeita, E., Fernandez, E. A., Albar, J. P., ... Llera, A. S. (2011). SPARC promotes cathepsin B-mediated melanoma invasiveness through a collagen I/alpha2beta1 integrin axis. *J. Invest. Dermatol, 131*, 2438-47. http://dx.doi.org/10.1038/jid.2011.239

Hornebeck, W., Emonard, H., Monboisse, J. C., & Bellon, G. (2002). Matrix-directed regulation of pericellular proteolysis and tumor progression. *Semin. Cancer Biol., 12*, 231-41. http://dx.doi.org/10.1016/S1044-579X(02)00026-3

Jacob, K., Webber, M., Benayahu, D., & Kleinman, H. K. (1999). Osteonectin promotes prostate cancer cell migration and invasion: a possible mechanism for metastasis to bone. *Cancer Res., 59*, 4453-7.

Kato, Y., Nagashima, Y., Baba, Y., Kawano, T., Furukawa, M., Kubota, A., ... Tsukuda, M. (2005). Expression of SPARC in tongue carcinoma of stage II is associated with poor prognosis: an immunohistochemical study of 86 cases. *Int. J. Mol. Med., 16*, 263-8.

Koukourakis, M. I., Giatromanolaki, A., Brekken, R. A., Sivridis, E., Gatter, K. C., Harris, A. L., & Sage, E. H. (2003). Enhanced expression of SPARC/osteonectin in the tumor-associated stroma of non-small cell lung cancer is correlated with markers of hypoxia/acidity and with poor prognosis of patients. *Cancer Res., 63*, 5376-80.

Ledda, F., Bravo, A. I., Adris, S., Bover, L., Mordoh, J., & Podhajcer, O. L. (1997). The expression of the secreted protein acidic and rich in cysteine (SPARC) is associated with the neoplastic progression of human melanoma. *J. Invest. Dermatol, 108*, 210-4. http://dx.doi.org/10.1111/1523-1747.ep12334263

Ledda, M. F., Adris, S., Bravo, A. I., Kairiyama, C., Bover, L., Chernajovsky, Y., ... Podhajcer, O. L. (1997). Suppression of SPARC expression by antisense RNA abrogates the tumorigenicity of human melanoma cells. *Nat. Med., 3*, 171-6. http://dx.doi.org/10.1038/nm0297-171

Lien, H. C., Hsiao, Y. H., Lin, Y. S., Yao, Y. T., Juan, H. F., Kuo, W. H., ... Hsieh, F. J. (2007). Molecular signatures of metaplastic carcinoma of the breast by large-scale transcriptional profiling: identification of genes potentially related to epithelial-mesenchymal transition. *Oncogene, 26*, 7859-71. http://dx.doi.org/10.1038/sj.onc.1210593

Lohi, J., & Keski-Oja, J. (1995). Calcium ionophores decrease pericellular gelatinolytic activity via inhibition of 92-kDa gelatinase expression and decrease of 72-kDa gelatinase activation. *J. Biol. Chem., 270*, 17602-9. http://dx.doi.org/10.1074/jbc.270.29.17602

Motamed, K., & Sage, E. H. (1998). SPARC inhibits endothelial cell adhesion but not proliferation through a tyrosine phosphorylation-dependent pathway. *J. Cell Biochem, 70*, 543-52. http://dx.doi.org/10.1002/(SICI)1097-4644(19980915)70:4<543::AID-JCB10>3.0.CO;2-I

Olofsson, K., & Mark, J. (1989). Specificity of asbestos-induced chromosomal aberrations in short-term cultured human mesothelial cells. *Cancer Genet Cytogenet, 41*, 33-9. http://dx.doi.org/10.1016/0165-4608(89)90105-2

Pinton, G., Brunelli, E., Murer, B., Puntoni, R., Puntoni, M., Fennell, D. A., ... Moro, L. (2009). Estrogen receptor-beta affects the prognosis of human malignant mesothelioma. *Cancer Res., 69*, 4598-604. http://dx.doi.org/10.1158/0008-5472.CAN-08-4523

Polette, M., Nawrocki-Raby, B., Gilles, C., Clavel, C., & Birembaut, P. (2004). Tumour invasion and matrix metalloproteinases. *Crit Rev Oncol Hematol, 49*, 179-86. http://dx.doi.org/10.1016/j.critrevonc.2003.10.008

Puolakkainen, P. A., Brekken, R. A., Muneer, S., & Sage, E. H. (2004). Enhanced growth of pancreatic tumors in SPARC-null mice is associated with decreased deposition of extracellular matrix and reduced tumor cell apoptosis. *Mol. Cancer Res., 2*, 215-24.

Rempel, S. A., Ge, S., & Gutierrez, J. A. (1999). SPARC: a potential diagnostic marker of invasive meningiomas. *Clin. Cancer Res., 5*, 237-41. http://dx.doi.org/10.1097/00005072-199812000-00002

Rempel, S. A., Golembieski, W. A., Ge, S., Lemke, N., Elisevich, K., Mikkelsen, T., & Gutierrez, J. A. (1998). SPARC: a signal of astrocytic neoplastic transformation and reactive response in human primary and xenograft gliomas. *J. Neuropathol Exp. Neurol., 57*, 1112-21.

Rivera, L. B., Bradshaw, A. D., & Brekken, R. A. (2011). The regulatory function of SPARC in vascular biology. *Cell Mol Life Sci., 68*, 3165-73. http://dx.doi.org/10.1007/s00018-011-0781-8

Sage, E. H., & Bornstein, P. (1991). Extracellular proteins that modulate cell-matrix interactions. SPARC, tenascin, and thrombospondin. *J. Biol. Chem., 266*, 14831-4.

Said, N., & Motamed, K. (2005). Absence of host-secreted protein acidic and rich in cysteine (SPARC) augments peritoneal ovarian carcinomatosis. *Am. J. Pathol., 167*, 1739-52. http://dx.doi.org/10.1016/S0002-9440(10)61255-2

Said, N., Najwer, I., & Motamed, K. (2007). Secreted protein acidic and rich in cysteine (SPARC) inhibits integrin-mediated adhesion and growth factor-dependent survival signaling in ovarian cancer. *Am. J. Pathol, 170*, 1054-63. http://dx.doi.org/10.2353/ajpath.2007.060903

Said, N., Socha, M. J., Olearczyk, J. J., Elmarakby, A. A., Imig, J. D., & Motamed, K. (2007). Normalization of the ovarian cancer microenvironment by SPARC. *Mol. Cancer Res., 5*, 1015-30. http://dx.doi.org/10.1158/1541-7786.MCR-07-0001

Sangaletti, S., Gioiosa, L., Guiducci, C., Rotta, G., Rescigno, M., Stoppacciaro, A., Chiodoni, C., & Colombo, M. P. (2005). Accelerated dendritic-cell migration and T-cell priming in SPARC-deficient mice. *J. Cell Sci., 118*, 3685-94. http://dx.doi.org/10.1242/jcs.02474

Sangaletti, S., Stoppacciaro, A., Guiducci, C., Torrisi, M. R., & Colombo, M. P. (2003). Leukocyte, rather than tumor-produced SPARC, determines stroma and collagen type IV deposition in mammary carcinoma. *J. Exp. Med., 198*, 1475-85. http://dx.doi.org/10.1084/jem.20030202

Sato, H., & Seiki, M. (1996). Membrane-type matrix metalloproteinases (MT-MMPs) in tumor metastasis. *J. Biochem, 119*, 209-15. http://dx.doi.org/10.1093/oxfordjournals.jbchem.a021223

Sato, N., Fukushima, N., Maehara, N., Matsubayashi, H., Koopmann, J., Su, G. H., ... Goggins, M. (2003). SPARC/osteonectin is a frequent target for aberrant methylation in pancreatic adenocarcinoma and a mediator of tumor-stromal interactions. *Oncogene, 22*, 5021-30. http://dx.doi.org/10.1038/sj.onc.1206807

Schultz, C., Lemke, N., Ge, S., Golembieski, W. A., & Rempel, S. A. (2002). Secreted protein acidic and rich in cysteine promotes glioma invasion and delays tumor growth in vivo. *Cancer Res., 62*, 6270-7.

Servais, E. L., Colovos, C., Rodriguez, L., Bograd, A. J., Nitadori, J., Sima, C., ... Adusumilli, P. S. (2012). Mesothelin overexpression promotes mesothelioma cell invasion and MMP-9 secretion in an orthotopic mouse model and in epithelioid pleural mesothelioma patients. *Clin Cancer Res*earch, 18, 2478-89.

Siddiq, F., Sarkar, F. H., Wali, A., Pass, H. I., & Lonardo, F. (2004). Increased osteonectin expression is associated with malignant transformation and tumor associated fibrosis in the lung. *Lung Cancer, 45*, 197-205. http://dx.doi.org/10.1016/j.lungcan.2004.01.020

Stetler-Stevenson, W. G. (1994). Progelatinase A activation during tumor cell invasion. *Invasion Metastasis, 14*, 259-68.

Sweetwyne, M. T., Brekken, R. A., Workman, G., Bradshaw, A. D., Carbon, J., Siadak, A. W., ... Sage, E. H. (2004). Functional analysis of the matricellular protein SPARC with novel monoclonal antibodies. *J. Histochem Cytochem, 52*, 723-33. http://dx.doi.org/10.1369/jhc.3A6153.2004

Tai, I. T., & Tang, M. J. (2008). SPARC in cancer biology: its role in cancer progression and potential for therapy. *Drug Resist Updat., 11*, 231-46. http://dx.doi.org/10.1016/j.drup.2008.08.005

Termine, J. D., Kleinman, H. K., Whitson, S. W., Conn, K. M., McGarvey, M. L., & Martin, G. R. (1981). Osteonectin, a bone-specific protein linking mineral to collagen. *Cell, 26*, 99-105. http://dx.doi.org/10.1016/0092-8674(81)90037-4

Thomas, R., True, L. D., Bassuk, J. A., Lange, P. H., & Vessella, R. L. (2000). Differential expression of osteonectin/SPARC during human prostate cancer progression. *Clin. Cancer Res., 6*, 1140-9.

Tremble, P. M., Lane, T. F., Sage, E. H., & Werb, Z. (1993). SPARC, a secreted protein associated with morphogenesis and tissue remodeling, induces expression of metalloproteinases in fibroblasts through a novel extracellular matrix-dependent pathway. *J. Cell Biol., 121*, 1433-44. http://dx.doi.org/10.1083/jcb.121.6.1433

Xu, A., Zhou, H., Yu, D. Z., & Hei, T. K. (2002). Mechanisms of the genotoxicity of crocidolite asbestos in mammalian cells: implication from mutation patterns induced by reactive oxygen species. *Environ Health Perspect, 110*, 1003-8. http://dx.doi.org/10.1289/ehp.021101003

Yan, Q., & Sage, E. H. (1999). SPARC, a matricellular glycoprotein with important biological functions. *J. Histochem Cytochem, 47*, 1495-506. http://dx.doi.org/10.1177/002215549904701201

Yang, H., Rivera, Z., Jube, S., Nasu, M., Bertino, P., Goparaju, C., … Carbone, M. (2010). Programmed necrosis induced by asbestos in human mesothelial cells causes high-mobility group box 1 protein release and resultant inflammation. *Proc. Natl. Acad. Sci. USA, 107*, 12611-6. http://dx.doi.org/10.1073/pnas.1006542107

Yiu, G. K., Chan, W. Y., Ng, S. W., Chan, P. S., Cheung, K. K., Berkowitz, R. S., & Mok, S. C. (2001). SPARC (secreted protein acidic and rich in cysteine) induces apoptosis in ovarian cancer cells. *Am. J. Pathol., 159*, 609-22. http://dx.doi.org/10.1016/S0002-9440(10)61732-4

15

Extended Spectrum Beta- Lactamase- Producing Uropathogenic *Escherichia coli* in Pregnant Women Diagnosed With Urinary Tract Infections in South-Western Nigeria

Oluduro Anthonia Olufunke[1], Aregbesola Oladipupo Abiodun[1] & Fashina Christina Dunah[1]

[1] Department of Microbiology, Faculty of Science, Obafemi Awolowo University, Ile-Ife 220005, Nigeria

Correspondence: Oluduro Anthonia Olufunke, Department of Microbiology, Faculty of Science, Obafemi Awolowo University, Ile-Ife 220005, Nigeria, E-mail: aoluduro2003@yahoo.co.uk

Abstract

Extended spectrum beta-lactamase (ESBL)-producing uropathogenic *Escherichia coli* (UPEC) in symptomatic pregnant women with confirmed urinary tract infections in Southwest. Nigeria was reported. Susceptibility of UPEC isolates to β-lactam and other classes of antibiotics was determined by the Kirby-Bauer's disc diffusion method on Mueller-Hinton agar plate. Detection of plasmid DNA in multiple antibiotic resistant isolates was carried out by alkali lysis (TENS) method. Extended-spectrum-β-lactamase (ESBL) production was determined by double disk synergy test (DDST). Isolates that were positive to ESBL were mated with non-ESBL- producing *E. coli* and other enterics in a conjugation experiment. Transfer of ESBL-enzyme and other resistance phenotypes in the transconjugants was investigated by DDST. Data obtained were statistically analyzed using SPSS 17. Greater percentage of the isolates were multiple antibiotic resistant (MAR). Sixty-nine (26.1 %) of UPEC were ESBL producers. Some of the ESBL producers transferred ESBL- enzyme and other resistance determinants to the recipients. Large size plasmid DNA of molecular weight (23.13-33.04 kb) was detected in some representative MAR isolates.

Keywords: ESBL, UPEC, pregnant women, UTI, multiple antibiotic resistance

1. Introduction

Escherichia coli have been reported to be common causes of hospital acquired infections which can have severe clinical implications with corresponding multiple antibiotic resistance (Aibinu et al., 2002). Extended spectrum β–lactamases (*ESBLs*) are plasmid mediated enzymes that confer resistance to penicillin, third generation cephalosporins and aztreonam but are inhibited by clavulanic acid (Paterson and Bonomo, 2005). In Africa, *ESBLs*-producing bacteria have been reported in Egypt, Morocco, Tunisia, Senegal and South Africa (Bloomberg et al., 2005).

Extended-spectrum β-lactamase (ESBL)-producing enterobacteriaceae have become widespread in hospitals and are increasing in community settings where they cause a variety of infections. In addition to hydrolyzing most β-lactam agents, bacteria harboring these enzymes display resistance to other unrelated antimicrobial agents and thus often pose a therapeutic dilemma (Maina et al., 2013). Increase in ESBL-producing enteric Gram-negative bacteria has led to the choice of inappropriate therapy; as a result, the rate of resistance has increased. Antibiotic therapy of infections due to ESBL-producing pathogens is still a clinical challenge. In most cases, carbapenems and fluoroquinolones have been used (Ramphal & Ambrose, 2006).

Horizontal gene transfer by plasmid exchange between *E. coli* strains is a recognized source of rapid spread of antimicrobial resistance phenotypes (Fang et al., 2006). The significance of plasmids in disseminating antimicrobial resistance genes is further enhanced by the association of plasmids with mobile genetic elements, such as transposons, integrons and insertion elements (Pitout et al., 2007). Resistance to third generation cephalosporins, ciprofloxacin, trimethoprim-sulfamethoxazole, gentamicin and amikacin was detected in the ESBL- producer group. Selective pressure of the antimicrobials selects those strains that are resistant to the applied antimicrobials causing the resistant strains to multiply and spread. In the last decade, CTX-M enzymes have replaced TEM and SHV mutants as the most prevalent ESBLs worldwide, with *E. coli* being the major host (Livermore et al., 2007; Paterson & Bonomo, 2005). Extended spectrum beta-lactamases production have been reported among *E. coli* in both hospital and community settings ((Pitout et al., 2007)). They have also been detected in pets and farm animals, products of the food chain and sewage (Warren et al., 2008). The resistance rate of *E. coli* to third generation cephalosporins (3GCs) is a broad indicator of the occurrence of ESBLs. It has been discovered that, travellers to countries with high rates of ESBLPCs (e.g. Egypt, India, etc) readily acquire asymptomatic faecal carriage.

Extended spectrum β-lactamase-producing *E. coli* and other enterobacteriaceae, particularly those producing CTX-M, have spread rapidly among humans and there is evidence of spread among animal populations. Factors that influence the spread of resistance genes as well as resistant bacteria include, antimicrobial usage, co-selection for resistance genes

A number of risk factors of acquiring ESBL-producing bacteria have been identified in hospitalized patients, most of which also apply to other multi-resistant Gram-negative bacilli. These risk factors include: prolonged hospital stay; prior hospitalization; previous use of 3GCs, aminoglycosides and quinolones; presence of medical devices such as urinary catheters and mechanical ventilation (Rodriguez-Bano et al., 2006). In the case of community acquired ESBL infections, older age, female gender, recurrent UTIs/prior invasive procedures (e.g. catheterisation), known faecal carriage, contact with healthcare facilities/residents in care homes and previous antimicrobial treatment are the risk factors (Moor et al., 2008).

Travellers to areas of the world such as India where very high rates of *ESBL* are present, have been noted to become readily colonized, asymptomatically, with CTX-M-producing *ESBL* strains (Tham et al., 2010). The present study reports the prevalence of ESBL- producing UPEC in pregnant women with confirmed UTIs in Ondo and Ekiti States, south-western Nigeria.

2. Methodology

2.1 Study Design and Sample Collection

The study population include symptomatic UTIs pregnant women in Ekiti and Ondo States, South-western Nigeria. Ekiti State is located between longitudes 40°51′ and 50°451′ east of the Greenwich meridian and latitudes 70°151′ and 80°51′ north of the Equator while Ondo State lies between longitudes 4"30" and 6" East of the Greenwich Meridian, 5" 45" and 8" 15" North of the equators

The preliminary identification of *E. coli* recovered from 400 voided mid-stream urine samples of symptomatic pregnant women with confirmed cases of UTIs was based on colonial morphology by a characteristic green metallic sheen on EMB agar (Oxoid Ltd., Hampshire, England) plate. The identity of isolate was confirmed by various conventional biochemical tests with reference to Bergey's Manual of Determinative Bacteriology (Holt et al., 1994).

2.2 Antibiotic Susceptibility Testing

Antibiotic susceptibility testing of isolate was performed by Kirby-Bauer's disc diffusion method. The antibiotics tested and their concentrations (μg) include; cefadroxil (30), ampicillin (10), nalidixic acid (30), cefepime (30), augumentin (30), cefuroxime (30), ceftazidime (30), cefotaxime (30) (Oxoid, Basingstoke, Hampshire, UK); amoxicillin (30), gentamicin (10), ofloxacin (5), ciprofloxacin (30), tetracycline (25), augmentin (30), ceftriazole (30), nitrofurantoin (300), cotrimoxazole (30), and pefloxacin (Fondos, Nigeria). The standardized inoculum (0.5 McFarland Barium Sulphate turbidity standard- 10^6 CFU/ml) was seeded on Mueller-Hinton Agar (MHA) (Hi Media, Vadhani, India) plates. The antibiotic disks were firmly placed on the sterile MHA plates (Oxoid, England) and incubated at 37°C for 24 h. Diameter of zones of inhibition was measured to the nearest millimeter using a transparent calibrated ruler and compared to the Clinical and Laboratory Standards Institute (2012). *Escherichia coli* ATCC 25922 was used as reference.

2.3 Detection of β-lactamase Enzymes Producing E. coli

Extended spectrum β-lactamase production among the isolates was detected following the double-disk synergy technique (DDST) (Clinical and Laboratory Standards Institute (CLSI), 2012). The standardized isolates were subjected to double-disk synergy tests (DDST) on sterile Mueller Hinton agar plates. *Escherichia coli* ATCC 25922 was used as reference. The test was performed by placing ceftazidime (30 μg) and cefotaxime (30 μg) at 20 mm (center to center) from a centre disk containing augmentin (30 μg) (amoxicillin (20 μg) plus clavulanate (10 μg). Enhancement or potentiation in the zone of inhibition of any of the antibiotic disks toward the center disk containing clavulanic acid indicates the presence of ESBLs (Clinical and Laboratory Standards Institute (CLSI), 2012).

2.4 In Vitro Conjugation Experiment

Minimum inhibitory concentrations of the antibiotics (augmentin, amoxicillin, cefadroxil, Cefotaxime, cefepime, ceftazidime, cefuroxime, alidixicAcid, ciprofloxacin, ofloxacin, pefloxacin, gentamicin, tetracycline, cotrimoxazole) used in the conjugation experiments were determined by agar dilution method as prescribed by CLSI.[6] Conjugation experiment was performed by the mating assay using the method of Aibinu et al. (2003) in tryptic soy broth for six of the ESBL-producing *E. coli* isolates as donors and non- ESBL-producing *E. coli* ATCC 25922 and other enterobacteriaceae (*Klebsiella* sp, *Proteus* sp., *Salmonella* sp. and *Shigella* sp) as recipients. The donors were tested and confirmed to be sensitive to the antibiotic resistance markers on MHA plates. A suspension of each organism was made in the sterile double strength nutrient broth at 37°C and adjusted to 0.5 McFarland Standard. The donor and the recipient were then mixed in a ratio 1:9 (50 µl of the donor to 450 µl of the recipient) and incubated at room temperature for 3 h for conjugation to take place.

One milliliter of each conjugated samples was serially diluted (10-folds) and 0.1 ml from 10^{-3} and 10^{-4} dilution fractions was spread inoculated onto the surface MacConkey, eosin methylene blue agar and *Salmonella-Shigella* agar plates supplemented with appropriate minimum inhibitory concentration of antibiotics used as the recipients' markers. Transfer of resistance was read by observation of recovery of the recipient colonies on the agar plates containing the corresponding antibiotics. The transconjugants were subjected to DDST to confirm the transfer of ESBL enzymes and co-transfer of other resistance determinants present in the donor isolates.

2.5 Plasmid Profiling

Plasmid DNA extraction of selected multiple antibiotic resistant ESBL-producing isolates was performed using the alkaline lysis 'TENS' (Tris 25 mM, EDTA 10 mM, 0.1 N NaOH and 0.5% SDS- all Sigma products) method of Kraft et al. (1988) and Lech et al. (1987). The extracted plasmid DNA was separated on 0.8% agarose gel (Oxoid, Basingstoke, England) in a 20-40 µl of TE (Tris-EDTA) buffer and a 100 bp ladder (Promega, Madison, USA) was used as standard. The electrophoretic products were viewed using ultraviolet trans-illuminator and the plasmid size was compared to the reference marker.

2.6 Statistical Analysis

Significant differences and relationship between the prevalence of ESBL- producing UPEC strains in pregnant women in the study areas were compared using analysis of variance (SPSS 17 version). A value of $p < 0.05$ was set as significant

3. Results

Table 1. Prevalence of antibiotic resistance among ESBL-producing uropathogenic *E. coli* in pregnant women with UTI in Ondo and Ekiti States

Antibiotics tested		\% Resistance of the (MAR) isolates (n=264)					
		ESBL (%) (n=69)			Non ESBL (%) (n=195)		
Classes of antibiotics tested	Specific antibiotics tested	Ondo	Ekiti	Total (%)	Ondo	Ekiti	Total (%)
β-Lactams	Augmentin	34	30	64 (92.8)	57	57	114 (58.5)
	Amoxicillin	33	31	64 (92.8)	57	68	125 (64.1)
	Ampicillin	35	33	68 (98.6)	75	77	152 (77.9)
	Ceftriaxone	30	32	62 (89.6)	79	87	166 (85.1)
	Cefadroxil	36	32	68 (98.6)	93	93	186 (95.4)
	Cefotaxime	31	29	60 (87.0)	54	66	120 (61.5)
	Cefepime	14	20	34 (49.3)	16	26	42 (21.5)
	Ceftazidime	36	31	67 (97.1)	92	92	184 (94.4)
	Cefuroxime	34	33	67 (97.1)	87	88	175 (89.7)
Quinolones	Nalidixic acid	28	28	56 (81.2)	57	65	122 (62.6)
	Ciprofloxacin	28	24	52 (75.4)	50	66	116 (59.5)
	Ofloxacin	28	22	50 (72.5)	50	58	108 (55.4)
	Pefloxacin	34	30	64 (92.8)	60	71	131 (67.2)
Aminoglycosides	Gentamicin	28	30	58 (84.1)	34	72	106 (54.4)
Nitrofurantoins	Nitrofurantoin	31	27	58 (84.1)	58	68	126 (64.6)
Tetracyclines	Tetracycline	36	33	69 (100)	90	91	181 (92.8)
Sulphonamides/Trimethroprim	Cotrimoxazole	32	32	64 (92.8)	68	88	156 (80.0)

Key: MAR= Multiple antibiotic resistant; ESBL= Extended spectrum β-lactamase.

Table 2. Multiple antibiotic resistance (MAR) among ESBL-producing uropathogenic *E. coli* in pregnant women with UTIs in Ondo and Ekiti States

Number of classes of antibiotics tested	ESBL producer (n=69)		Total	Percentage (%)
	Ondo State	Ekiti State		
6	15	21	36	52.2
5	10	8	18	26.1
4	7	4	11	15.9
3	4	0	4	5.8

Key: MAR = Multiple antibiotic resistant; ESBL= Extended spectrum β-lactamase.

Table 1 shows the prevalence of antibiotic resistance among ESBL- producing UPEC isolates in Ondo and Ekiti States. Sixty-nine (26.1%) of the UPEC were ESBL- producing strains with concomitant multiple antibiotic resistance profiles. Isolates that produced ESBL-enzymes showed high rate of resistance to the extended spectrum β-lactam and other classes of antibiotics as well. There was a significant statistical difference in the incidence of resistance between ESBL- and non- ESBL- producing *E. coli* isolates in Ondo and Ekiti States ($P < 0.05$) (Table 1).

Each of the ESBL producers was resistant to more than one class of antibiotic. The ESBL-producing UPEC (52.2%) were resistant to all the six classes of antibiotics tested, 26.1% to five, 15.9% to four, and 5.8% to three classes (Table2).

Figure 1 shows the percentage rate of transfer of resistance phenotypes to the transconjugants. Extended spectrum β-lactamase enzyme and other antibiotic resistance phenotypes were transferred to non-ESBL- producing *E. coli, Salmonella* sp., *Shigella* sp., *Proteus vulgaris* and *Klebsiella* sp. recipients. Consequently, resistance to other antibiotics was also transferred in the same trend, augmentin, cefotaxime and ceftazidime were 100% transferred to the transconjugants, followed by amoxicillin (98.0%), tetracycline (94.0%) and cotrimoxazole (90.0%). However, pefloxacin, ciprofloxacin, ofloxacin and nalidixic acid resistance traits were not transferred (Figure 1).

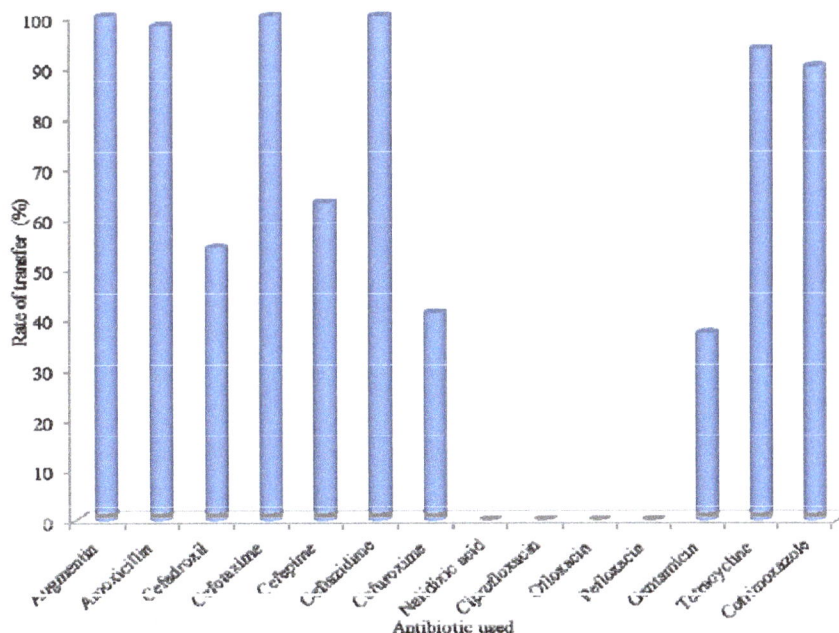

Figure 1. Rate of transfer of resistance phenotypes to the transconjugants (%)

AUG=Augmentin,   AMX=Amoxicillin, CFR=Cefadroxil,   CTX=Cefotaxime,   FEP=Cefepime,
CAZ=Ceftazidime, CXM=Cefuroxime,   NAL=NalidixicAcid,   CPX=Ciprofloxacin,   OFL=Ofloxacin,
PFX=Pefloxacin,   GEN=Gentamicin,   TET=Tetracycline,   COT=Cotrimoxazole.

Molecular weights of plasmid DNA in the representative MAR -UPEC isolates are presented in table 4 and the gel electrophoresis of the amplified plasmid DNA is shown in Figure 2. Large plasmid size of molecular weights ranging from 23.13– 33.04 kb were harboured by the selected MAR isolates, and 62.5% of the isolates were ESBL- producing strains. Some of the non-β-lactam antibiotic resistant isolates had same plasmid DNA size as those of ESBL-producing UPEC (Table 3).

Table 3. The molecular weight of plasmid DNA in the representative MAR uropathogenic *E. coli* in pregnant women with UTIs in Ondo and Ekiti States

Isolate code	Antibiotics to which isolates were resistant	Plasmid (Estimated) Mol. Wt. [kb]
EKNG028	AUG, AMX, AMP, CRO, CFR, CTX, CAZ, CXM, GEN, NIT, TET, COT	25.70
EKNG114	AUG, AMX, AMP, CRO, CFR, CTX, FEP, CMX, GEN, NIT, TET, COT	25.70
ODNG051	AUG, AMX, CRO, CFR, CAZ, CXM, GEN, NIT, TET, COT	25.70
ODNG172	AUG, AMX, AMP, CFR, FEP, CAZ, CXM, GEN, NIT, TET, COT	25.70
EKNG099	AUG, AMP, CRO, CFR, FEP, CAZ, CXM, NIT, TET, COT	25.70
EKNG004	AUG, AMX, AMP, CRO, CRO, CFR, CTX, FEP, CAZ, CXM, GEN, TET, COT	25.70
EKNG060	AUG, AMX, AMP, CRO, CFR, CTX, FEP, CAZ, CXM, GEN, NIT, TET, COT	25.70
ODNG124	AUG, AMX, AMP, CRO, CFR, CTX, CAZ, CXM, GEN, NIT, COT	25.70
ODNG059	AUG, AMX, AMP, CRO, CFR, CTX, FEP, CAZ, CXM, GEN, NIT, TET, COT	33.04
0DNG024	AUG, AMX, AMP, CRO, CFR, CTX, CAZ, CXM, GEN, NIT, TET, COT	33.04
EKNG007	AUG, AMX, AMP, CRO, CFR, CTX, FEP, CAZ, CXM, GEN, NIT, TET, COT	33.04
EKNG080	AUG, AMX, AMP, CRO, CFR, CTX, CAZ, CXM, GEN, NIT, TET, COT	33.04
ODNG021	AUG, AMX, AMP, CRO, CFR, CTX, CAZ, CXM, GEN, NIT, TET, COT	23.13
EKNG022	AUG, AMP, CRO, CFR, CTX, CAZ, CXM, GEN, NIT, TET, COT	23.13

EKNG= Isolates from Ekiti State, Nigeria; ODNG= Isolates from Ondo State, Nigeria.

AUG= Augmentin   AMX=Amoxicillin   AMP=Ampicillin   CRO=Ceftriaxone   CFR=Cefadroxil
CTX=Cefotaxime   FEP=Cefepime   CAZ=Ceftazidime   CXM=Cefuroxime   GEN=Gentamicin
NIT=Nitrofurantoin TET=Tetracycline   COT=Cotrimoxazole   AMC=Amoxicillin/Clavulanic Acid

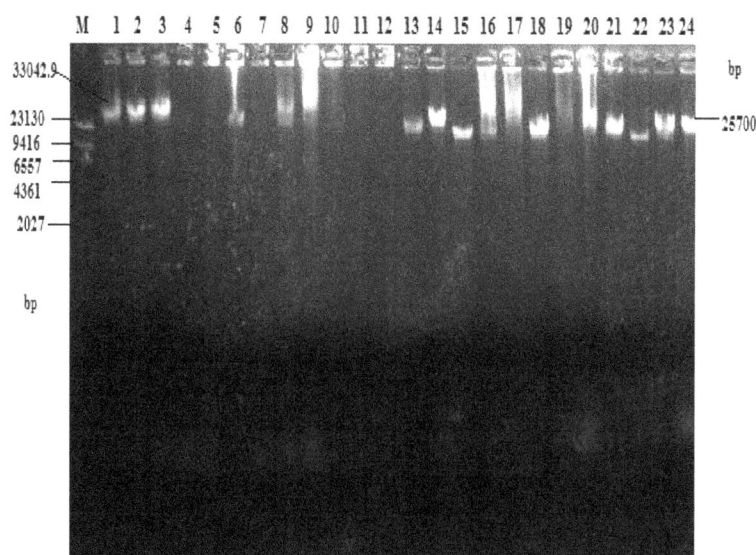

Figure 2. Gel electrophoresis of the amplified plasmid DNA of selected multiple antibiotic resistant uropathogenic *E. coli* in Ondo and Ekiti Sates

Lane M: DNA marker, Lanes 1-24: the test isolates.

4. Discussion

Present study reveals 26.1% prevalence of ESBL-enzyme production among UPEC in symptomatic UTIs pregnant women in Ondo and Ekiti States, Nigeria. This contradicts the earlier reports of 75% by Padmavathy et al. (2014), 52.4% by Yusuf et al. (2005), 52.4% by Iroha et al. (2008), and 37.3% by Aibinu et al. (2003) on prevalence of ESBL enzymes detected in *E. coli* from UTI patients in the East, South and Northern Nigeria, respectively. Prevalence of ESBL-producing UPEC in this study is higher than reports from other countries and other parts of Nigeria. For instance, in Cameroun, Gangoue-Pieboji (2005) reported 12.0% prevalence of *E. coli* with ESBL, El-Khizzi and Bakheshwain (2006) obtained 15.8% prevalence in Riyadh, Saudi Arabia and Yushau et al. (2007) reported 9.3% prevalence in Kano, Nigeria. In general, prevalence of ESBLs in *E. coli* in UTIs cases varies from country to country and from health institution to another (Iroha et al., 2010). The Pan European Antimicrobial Resistance Using Local Surveillance (PEARIS) study between (2001- 2002) showed that, the percentage of ESBL producer among *E. coli* in UTIs cases was 5.8% for all the study sites (Bouchillon et al., 2004). In Egypt, a high rate of 38.5% was observed, 27.4% were reported from Greece, 2.0% from the Netherlands and 2.6% in Germany. In Japan, Korea and Hong Kong, the percentage of ESBL production among *E. coli* was also low (Ho et al., 2000).

In South and Eastern Nigeria, prevalence of ESBL among *E. coli* isolated from pregnant patients from 2003-2007 was very high (52.4%) (Aibinu et al., 2003; Iroha et al., 2008) and is at variance with the findings of the present study. This may likely due to geographical location and antimicrobial usage. The implication of ESBL-producing UPEC in pregnant women may include prolonged stay in the hospital, cost, treatment failure and relapsed cases. A number of risk factors of acquiring ESBL-producing bacteria have been identified in hospitalized patients, most of which also apply to other multi-resistant Gram-negative bacilli. These risk factors include: prolonged hospital stay; prior hospitalization; previous use of 3GCs, aminoglycosides and quinolones; presence of medical devices such as urinary catheters and mechanical ventilation (Rooney et al., 2009). In the case of community acquired ESBL infections, older age, female gender, recurrent UTIs/prior invasive procedures (e.g. catheterization), known faecal carriage, contact with healthcare facilities/residents in care homes and previous antimicrobial treatment are the risk factors (Moor et al., 2008). Transfer of ESBL enzyme and other antibiotic resistance phenotypes in some of the isolates is plasmid-linked. This may be due to the fact that the ESBL genes were located on transposable element or integron thus resulting in transfer function (Akortha et al., 2011). Inter-generic transfer rate in this study is lower than intra species transfer rate. This has been linked to fertility inhibition, incompatibility, inability to synthesize adhesion or narrow host range (Akortha, 2009). Transmission in the community is probably quite complicated. Individuals in long-term care homes where high carriage rates of CTX-M producing enterobacteriaceae have been observed may spread strains (Rooney et al., 2009) to other noncare-home residents. The evidence of a significant spread amongst household contacts has been presented in a Spanish study which showed that 70% of index cases of patients with ESBL-producing strains causing UTI in the community had positive contacts with 16.7% of their household members. The finding of this study explains co-selection of resistance as resistance to other non-β-lactam antibiotics is located on the same plasmid as ESBL factor.

The presence of plasmid-mediated ESBL resistance among the isolates in the present study is an evidence of its transfer capability of ESBL and other resistance phenotypes between its species and other genera. This implies that under favourable condition, horizontal gene transfer of resistance plasmid by conjugation could occur *in vivo* (Yah et al., 2008). This indicates that plasmid carrying ESBL gene in one bacterium can spread rapidly to members of the same or organisms of different species in the same or different individual. Greater percentage of UPEC in the study area MAR capable of transferring ESBL enzymes and other resistance phenotypes.

5. Conclusion

In conclusion, antibiotic resistance remains one of the nature's never ending process whereby organisms develop tolerance to new environmental condition. The development and spread of ESBL- producing UPEC and horizontal transfer of resistance in the study is of great concern especially in therapeutic management of UPEC-induced urinary tract infections.

Acknowledgements

Authors acknowledge the Chief Medical Director and the Laboratory Scientists of the various hospitals selected for the study.

References

Aibinu, I. E., Ohagbulam, V. C., Adenipekun, E. A., Ogunsola, F. T., Odugbemi, T. O., & Mee, B. J. (2003). Extended–spectrum beta-lactamase enzymes in clinical isolates of Enterobacter species from Lagos, Nigeria. *J Clin Microbiol, 41*(5), 2197-2200. http://dx.doi.org/10.1128/JCM.41.5.2197–2200.2003

Akortha, E. E., Filgona, J. (2009). Transfer of gentamicin resistance genes among enterobacteriaeceae isolated from the outpatients with urinary tract infections attending 3 hospitals in Mubi, Adamawa State. *Sci Resear Essay*, 4(8), 745–752. Retrieved from http://www.academicjournals.org/SRE.

Akortha, E. E., Aluyi, H. A. S., Enerrijiofi, K. E. (2011). Transfer of amoxicillin resistance gene among bacterial isolates from sputum of pneumonia patients attending the University of Benin Teaching Hospital, Benin City, Nigeria. *Shiraz E-Medical J*, 12(4), 1-11. http://semj.sums.ac.ir/vol12/apr2011/89048.htm.

Bloomberg, B., Jureen, R., Manji, K. P, Tamim, B. S., Mwakagile, D. S. M., Urassa, W. K., Fataki, M., Msangi, V. et al. (2005). High rate of fatal cases of pediatric septicemia caused by Gram-negative bacteria with extended-spectrum beta-lactamases in Dar es Salaam, Tanzania. *J Clin Microbiol*, 43, 745-749. http://dx.doi.org/10.1186/1471-2334-7-43.

Bouchillon, S. K., Jonson, B. M., Hoban, D. J., Johnson, J. L., Dowzicky, M. J., Wu, D. H., Visalli, M. A., Bradford, A. (2004). Determining incidence of extended spectrum β-lactamase producing Enterobacteriaceae, vancomycin-resistant *Enterococcus faecium* and methicillin-resistant *Staphylococcus aureus* in 38 centers from 17 countries: the PEARLS Study 2001–2002. *Intern J Antimicro Agents*, 24, 119-124. http://dx.doi.org/10.1016/j.ijantimicag.2004.01.010.

Clinical and Laboratory Standards Institute (CLSI) (2012). *Performance standards for antimicrobial susceptibility testing*. 22nd informational supplement M100-S22. Wayne, PA.

El-Khizzi, N. A. and Bakheshwain, S. M. (2006). Prevalence of extended-spectrum beta-lactamases among Enterobacteriaceae isolated from blood culture in a tertiary care hospital. *Saudi Medical J*, 27(1), 37-40.

Fang, H., Ferda, A., Göran, H., Dornbusch, K. (2008). Molecular epidemiology of extended-spectrum β-lactamases among *Escherichia coli* Isolates collected in a Swedish hospital and its associated health care facilities from 2001 to 2006, *J Clin Microbiol*, 46(2), 707-712. http://dx.doi.org/10.1128/JCM.01943-07

Gangoue-Pieboji, J. (2005). Extended spectrum-beta-lactamase-producing enterobacteriaceae in Yaounde, Cameroon. *J Clin Microbiol*, 43, 3273 – 3277. http://dx.doi.org/10.1128/JCM.43.7.3273–3277.2005.

Ho, P. L., Tsang, D. N. C., Que, T. L., Ho, M., Yuen, K. Y. (2000). Comparison of screening methods for detection of extended-spectrum β-lactamase and their prevalence among *Escherichia coli* and *Klebsiella* sp. in Hong Kong. *Acta Pathol, Microbiol et Immunol Scandinavica*, 108, 237-240. http://dx.doi.org/10.1093/jac/dki010.

Holt, J. G., Krieg, N. R., Sneath, P. H., Staley, J. T., Williams, T., Hensyl, W. R. (1994). *Bergy's manual of determination bacteriological*. 9th ed., Williams and Wilkins, Baltimore, Meryland.

Iroha, I. R., Adikwu, M. U., Amadi, E. S., Aibinu, I., Esimone, C. O. (2008). Characterization of extended spectrum β-lactamase producing *E. coli* from secondary and tertiary hospital in South Eastern Nigeria. *Resear J Microbiol*, 3, 514-519. http://dx.doi.org/10.3923/jm.2008.514.519.

Iroha, I. R., Adikwu, M. U., Esimone, C. O., Aibinu, I., Amadi, E. S. (2009). Extended spectrum beta-lactamase (ESBL) in *E. coli* isolated from a tertiary hospital in Enugu State, Nigeria. *Pakistan J Medicin Sci*, 25, 279-288.

Iroha, I. R., Amadi, E. S., Nwazo, A. C., Ejike-ugwu, P. C. (2010). Detection of plasmid borne ESBLs from blood and urine isolates of Gram negative bacteria from University Teaching Hospital in Nigeria. *Current Resear Bacter*, 3, 77-83. http://dx.doi.org/10.3923/crb.2010.77.83.

Kraft, R., Tardiff, J., Krauter, K. S., Leinwand, L. A. (1988). Using mini-prep plasmid DNA for sequencing double stranded template with sequences. *Biotechnol*, 6, 544.

Lech, K. and Brent, R. (1987). Mini-prep of plasmid DNA. P. 1.6.1. 1.6.4. In Ausubel, F. M., Brent, R., Kingston, R. E., Moore, D. D., Seidman, J. G., Smith, J. A. (Eds). *Current Protocol in Molecular Biology*. John Wiley and Sons, NY.

Livermore, D. M., Canton, R., Gniadkowski, M., Nordmann, P., Rossolini, G. M., Arlet, G., Ayala Coque, T. M. et al. (2007). Changing the face of ESBLs in Europe. *J Antimicrob Chemoth*, 59(2), 165–174. DOI:10.1093/jac/dkl483.

Maina, D., Makau, P., Nyerere, A. and Revathi, G. (2013). Antimicrobial resistance patterns in extended-spectrum β-lactamase producing *Escherichia coli* and *Klebsiella pneumoniae* isolates in a private tertiary hospital, Kenya. *Microbiology Discovery*, 1(5), 1-4. http://dx.doi.org/10.7243/2052-6180-1-5.

Moor, C. T., Roberts, S. A., Simmons, G., Briggs, S., Morris, A. J., Smith, J., Heffernan, H. (2008). Extended-spectrum beta-lactamase producing enterobacteria: factors associated with infection in the community setting, Auckland, New Zealand. *J Hospital Infect,* 68, 355–362. http://dx.doi.org/10.1016/j.jhin.2008.02.003.

Padmavathy, K., Krishnan, P., Rajasekaran, S. (2014). Fluoroquinolone resistance among CTX-M producing uropathogenic *Escherichia coli* from HIV and non-HIV patients in South India. *BMC Infect Dis*, 14 (3), 63. http://dx.doi.org/10.1186/1471-2334-14-S3-P63.

Paterson, D. L and Bonomo, R. A. (2005). Extended-spectrum β-lactamases: a clinical update. *Clin Microbiol Rev*, 18, 657–86. http://dx.doi.org/10.1128/CMR.18.4.657-686.2005.

Pitout, J. D. D., Daniel, B. G., Lorraine, C., Kevin, B. L. (2009). Molecular characteristics of extended-spectrum-β-lactamase-producing *Escherichia coli* isolates causing bacteremia in the Calgary health region from 2000 to 2007: Emergence of clone ST131 as a cause of community-acquired infections. *Antimicr Agents Chemotherap*, 53(7), 2846-2851. http://dx.doi.org/10.1128/AAC.00247-09.

Pitout, J. D. D., Nordmann, P., Laupland, K. B., Poirel, L. (2005). Emergence of Enterobacteriaceae producing extended-spectrum β-lactamases (ESBLs) in the community. *J Antimicrob Chemother* 5; 56: 52–59. http://dx.doi.org/10.1093/jac/dki166.

Ramphal, R. and Ambrose, P. G. (2006). Extended-spectrum beta-lactamases and clinical outcomes: current data. *Clin Infect Dis*, 42(4): 164-172. http://dx.doi.org/10.1086/500663.

Rodriguez-Bano, J., Navarro, M. D., Romero, L., Muniain, M. A., Perea, E. J., Perez-Cano, R., Hernandez, J. R. and Pascual, A. (2006). Clinical and molecular epidemiology of extended-spectrum beta-lactamase-producing *Escherichia coli* as a cause of nosocomial infection or colonization: implications for control. *Clin Infect Dis*, 42, 37–45. http://dx.doi.org/10.1086/498519.

Rooney, P. J., O'Leary, M. C., Loughrey, A. C., McCalmont, M., Smyth, B., Donaghy, P., Badri, M.et al. (2009). Nursing homes as a reservoir of extended-spectrum β-lactamase (ESBL)-producing ciprofloxacin-resistant *Escherichia coli. J Antimicrob Chemother,* 64, 635–4. http://dx.doi.org/10.1093/jac/dkp220.

Tham, J., Odenholt, I., Walder, M., Melander, E. and Odenholt, I. (2010). Extended-spectrum beta-lactamase-producing *Escherichia coli* in patients with travellers' diarrhoea. *Scandinavian J Infect Dis*, 42, 275–280. http://dx.doi.org/10.2147/IDR.S34941.

Warren, R., Butler, V. N., O'Neill, K., Taylor, J., Ensor, V. and Harvey, P. (2008). Imported chicken meat as a potential source of quinolone-resistant *Escherichia coli* producing extended-spectrum β-lactamases in the UK. *Antimicrob Chemother*, 61(3), 504-508. http://dx.doi.org/10.1093/jac/dkm517.

Yah, S. C. Yusuf, O. E., Eghafona, N. O. (2008). Pattern of antibiotic usage by adult populations in the city of Benin, Nigeria. *Scientific Resear Essays*, 3(3), 081-085. Retrieved from http://www.academicjournals.org/SRE

Yushau, M., Olonitola, S. O., Aliyu, B. S. (2007). Prevalence of extended – spectrum beta lactamases (ESBLs) among members of the enterobacteriaceae isolates obtained from Mohammed Abdullahi Wase specialist hospital, Kano, Nigeria. *Inter J Pure Appl Sci,* 1(3), 42–48. http://dx.doi.org/10.4314/bajopas.v4i2.37.

Yusuf, I., Arzai, A. H., Umar, A., Magaji, N., Salisu, N., Tukur, A., Haruna, M., Hamid, K. M. (2011). Prevalence of extended spectrum beta-lactamases (ESBL) producing *Escherichia coli* and *Klebsiella pneumoniae* in tuberculosis patients in Kano, Nigeria. *Bayero J Pure Appl Sci*, 4(2), 182-185. http://dx.doi.org/10.4314/bajopas.v4i2.37

16

Expression of Human Chloride Channels ClC1 or ClC2 Revert the Petite Phenotype of a *Saccharomyces cerevisiae* GEF1 Mutant

Fernando Rosas-Sanchez[1], Angélica López-Rodríguez[1,2], Carlos Saldaña[3], Lenin Ochoa-de la Paz[1], Ricardo Miledi[1] & Ataúlfo Martínez-Torres[1]

[1] Instituto de Neurobiología, Campus Juriquilla - Universidad Nacional Autónoma de México, Querétaro, México

[2] National Institute of Neurological Disorders and Stroke, National Institutes of Health, Bethesda, USA

[3] Facultad de Ciencias Naturales, Universidad Autónoma de Querétaro, Querétaro, México

Correspondence: Ataúlfo Martínez-Torres, Departamento de Neurobiología Molecular y Celular, Laboratorio de Neurobiología Molecular y Celular II, Instituto de Neurobiología, Campus Juriquilla-Universidad Nacional Autónoma de México. Boulevard Universitario 3001, C.P. 76230 Juriquilla, Querétaro, México. E-mail: ataulfo@unam.mx

Abstract

The mechanism of activation of the yeast ClC chloride channel/transporter *GEF1* is unknown, and in this study we tested the ability of human ClC1 and ClC2, two channels with different activation kinetics, to revert the *petite* phenotype of a strain whose *GEF1* gene was deleted. We found that when the human channels are expressed in a low-copy plasmid, the reversion of the phenotype does not occur; in contrast, when the channels are over expressed by means of a strong transcriptional promoter in a multiple-copy plasmid, the cells reach the normal size, and show a normal membrane surface and oxygen consumption. To determine the size variationsof individual cells, we employed flow-cytometry as a quantitative tool to evaluate the *petite* phenotype.

These results suggest that the human ClC channels, when abundantly present in the cells, can support the metabolism disrupted in the knock-out strain. We also observed that the fluorescence emitted by GFP-tagged channels was found mostly towards the periphery of the *wt* yeast, whereas in the *GEF1* knock-out it was detected in intracellular clusters. GFP-tagged channels expressed in *X. laevis* oocytes produced robust currents and did not show any evident difference with respect to the normal ClCs, whereas Gep1p did not show voltage-dependent activation.

Keywords: chloride channel, functional complementation, voltage-clamp, *Xenopus laevis* oocytes

1. Introduction

Chloride channels/transporters (ClCs) are members of a large family present in a wide variety of organisms from bacteria to higher eukaryotes. ClCs carry out multiple physiological roles, from plasma membrane and cell volume modulation to the control of vesicular pH (Fahlke, 2001; Jentsch, Stein, Winreich & Zdebik, 2002; Sardini et al., 2003; Soleimani & Xu, 2006; Jentsch, 2008). A clear example of this functional diversification is illustrated by comparing the properties of mammalian ClC1 and ClC2. They are both located in the plasma membrane; however, whereas ClC1 is activated by plasma membrane depolarization and thus is responsible for the repolarization current in muscle fibers, ClC2 is activated by hyperpolarization, as well as by other mechanisms such as changes in pH and cell volume (Conte, De Luca, Mamrini, & Vrbovà, 1989; Steinmeyer, Ortland, & Jentsch, 1991; Klocke, Steinmeyer, Jentsch, & Jockusch, 1994; Jordt & Jentsch, 1997).

The mechanism of activation of the *Saccharomyces cerevisiae* Gef1p, the sole ClC found in this species of yeast, is still not clearly understood. Gef1p plays a critical role in yeast iron metabolism and is found mainly in the *trans*-Golgi (Greene, Brown, DiDomenico, Kaplan & Eide, 1993; Schwappach, Stobrawa, Hechenberger, Steinmeyer & Jentsch, 1998). Mutations of the GEF1 gene lead to an iron requirement for growth on non-fermentable carbon sources due to a failure to load copper onto the iron uptake system; thus, knocking down the expression of GEF1 leads to *petite* (*pet*) colonies when grown in these conditions (Gaxiola et al., 1998). Gef1p forms a Cl⁻ transporter/channel in the plasma membrane of the yeast that does not show voltage-dependent activation when expressed in heterologous systems (López-Rodríguez et al., 2007). Interestingly, several ClC genes from plants, fungi, and vertebrates functionally complement the *pet* phenotype of yeast whose gene *GEF1* had been deleted, whereas others such as the mammalian ClC7 gene, which codes for a protein of the lysosomal membrane, does not revert the mutation (Hechenberger et al., 1996; Gaxiola, Yuan, Klausner & Fink, 1998; Miyazaki et al., 1999; Kida, Uchida, Miyazaki, Sasaki, & Mauro, 2001; Marmagne et al., 2007).

To determine if human ClC1 and ClC2 complement the *pet* phenotype of Gef1p⁻ yeast, we expressed these two genes in a *GEF1* knock-out strain. This paper describes the results of complementation assays and some details of the yeast phenotype revealed by scanning electron microscopy (SEM). To quantify the reversion of the *pet* phenotype, the colony size assay was supported with an analysis of cell volume by flow cytometry, which allowed us to measure the size and estimate the cell surface complexity of up to 5,000 individual cells per second. The results suggest that overexpression of ClC1 or ClC2 rescue the *pet* phenotype of a Gef1p⁻ strain, whereas expression of the same channels in a single-copy plasmid and under a constitutive promoter do not rescue the mutant phenotype.

2. Methods

2.1 Yeast Strains and Media

The *S. cerevisiae* strains used in this study were RGY30 (*wt*) MATa ura3-1; leu2-3,112 trp1-1; his3-11, 15, and RGY 192 (Gef1p⁻) MATa*gef1: HIS3*, leu2-3,112; trp1-1; his15. Both strains were donated by Dr. R. Gaxiola (Gaxiola et al., 1998). The cells were propagated on standard YPD media that contained 1% bacto-yeast extract, 2% peptone, and 2% dextrose; they were made competent for transformation by the LiAc method (Greene et al., 1993; Geitz, Schiestls, Willems & Woods, 1995). For auxotroph selection we used a medium without uracil (SC-U), containing 0.67% yeast nitrogen base without amino acids. Induction and analysis of the *pet* phenotype for cells transformed with pYES was made in SC-U supplemented with 0.01% leucine, 0.01% tryptophan, 0.005% histidine, 2% galactose, 2% ethanol, 2% glycerol, and 1mM ferrozine. For the cells transformed with pUG35, we also included 2% galactose to eliminate the possibility that this carbohydrate couldrevert the phenotype, as observed previously (Greene et al., 1993).

2.2 Plasmid Construction

Plasmids carrying either hClC1 (pRc/CMV_hClC1) or hClC2 (pBK/RSV_hClC2) were donated by Dr. Al George and Dr. Gary Cutting, respectively. The ClCs were amplified with Platinum®Taq DNA polymerase using the following conditions: for hClC-1 (35 cycles): 94°C/30s, 55°C/30s, and 68°C/3min, and for hClC-2 (35 cycles): 94°C/30s, 63°C/30s, and 68°C/3min. The translation stop codon was eliminated in both cases.

Two expression plasmids were used for complementation experiments: 1) the multicopy vector pYES2.1/V5-His-TOPO (pYES2.1 TOPO®, Invitrogen) and 2) the centromeric pUG35 (Donated by Dr. J.H.Hegemann), which permitted tagging the hClCs with the green fluorescent protein (GFP) at the carboxy-terminus. RNA was isolated from transformed yeast using the acid guanidinium thiocyanate-phenol-chloroform extraction method (Chomczynski & Sacchi, 1987). In addition, RNA was isolated from yeast induced to overexpress *GEF1* with the plasmid pYEX-BX-GEF1 (López-Rodríguez et al., 2007).

2.3 Yeast Complementation Assays

Strains RGY30 (*wt*) and RGY192 (Gef1p⁻) were transformed with plasmids derived from pYES (pYES-hClC1 or pYES-hClC2) and plated on YPD; after selection in restrictive media, positive colonies were transferred to SC-U supplemented with 2% galactose to induce the GAL promoter. The size of the transformed yeast was visualized and measured under the light microscope. Cell diameters were measured from ten different ocular fields (100X), and statistical analysis was performed with one-way ANOVA and Tukey *post hoc* tests; in order to have a more accurate measure of the cell diameter and membrane complexity, flow cytometry was used, as indicated below.

Plasmids derived from pUG35 (pUG-hClC1 and pUG-hClC2) were also introduced into RGY30 and RGY192; galactose was added to as above this carbohydrate to revert the phenotype, eventhough the MET promoter allowed the constitutive expression of the transgenes (Mumberg, Müller & Funk, 1994).

2.4 Flow Cytometry

Cells grown in liquid YNB were collected from samples of three independent transformations, optic density measured in a spectrophotometer (λ 480 nm) after 4 h of induction with galactose, and sorted using a

Fluorescence Activated Cell Sorter apparatus (FACS calibur; Becton Dickinson). Acquisition and analysis of the FACS data were performed using CELLQUEST software (Becton Dickinson) and SUMMIT V4.3 (DAKO Colorado, Inc.). Frontal light dispersion was a direct indication of cell volume whereas lateral dispersion suggests the complexity of the cell surface. Data analysis was performed with Windows Multiple Document Interface Flow Cytometry Application, Version 2.9 (WinMDI V2.9).

2.5 O_2 Consumption Rates

Yeast strains were grown at 30°C in SC-U medium with 2% dextrose to an OD_{600} of 3 and then arrested in M phase by adding 1.5 mg/mL of nocodazole in 1% DMSO. After 4 h, yeast were collected by centrifugation for 5 min at 1000 g at 4°C and resuspended in SC-U supplemented with 2% galactose. After 4 h of incubation, cells were counted in a Nuebauer chamber (Optyk Labor), and the culture was diluted in 3 mL of fresh SC-U set at 30°C. The rate of O_2 consumption was determined using a Clark-type oxygen electrode and YSI Benchtop Biological Oxygen Monitor (5300 model) as reported (López-Rodríguez et al., 2007; Hernandez-Muñoz, Díaz-Muñoz, & Chagoya de Sanchez, 1992).

2.6 Scanning Electron Microscopy (SEM)

Yeast were transformed and fixed in 3% glutaraldehyde in H_2O for 2 h. Then the cells were covered with a thin coat of gold using an Ion Sputter FC 1100 (Jeol) operating at 1200kV and 5 mA for 10 min. Samples were observed and photographed under an electron microscope (Jeol, JSM-54110LV) at a 10,000 X.

2.7 Electrophysiology

Isolation of oocytes and recordings were as described previously (Miledi, 1982). Briefly, *X. leavis* frogs were anesthetized with 0.17% 3-aminobenzoic acid methylester (MS-222) for 20-30 min. The oocytes were manually removed, enzymatically defolliculated with 0.5mg/ml collagenase type I (Sigma) at room temperature for 45 min, and then kept at 16°C in Barth's medium (in mM): 88, NaCl; 1, KCl; 0.33, Ca(NO3)2; 0.41, CaCl; 0.82, MgSO4; 2.4, NaHCO3; 5, HEPES; pH 7.4, containing 0.1 mg/mL gentamycin sulfate. The next day 50 nL of RNA (1 µg/µl) from yeast that were induced to express hClC1, hClC1-GFP, hClC2, hClC2-GFP, or *GEF1* were injected into the vegetal hemisphere, and electrophysiological recordings were obtained 4-5 days later.

To obtain the membrane currents generated by the hClCs we used the two-microelectrode voltage-clamp technique (Miledi, 1982). The oocytes were placed in a 100-µL chamber, impaled with two microelectrodes filled with 3 M KCl (0.5-1.5 MΩ), and clamped at 0 mV. For induction of voltage-activated currents, the oocyte's membrane potential was stepped from 0 to -120 to +40 in 20-mV steps. All recordings were made at room temperature (20-23°C) in a chamber continually perfused with Ringer's solution (in mM): 115, NaCl; 2, KCl; 1.8, CaCl2; 5, HEPES; pH 7.4.

3. Results

3.1 The Gef1p⁻ Yeast Phenotype is Reverted by Overexpression of hClC1 and hClC2

The first indication showing the complementation of the *pet* phenotype by expressing hClC1 or hClC2 was provided by a simple drop assay. Figure 1A illustrates that the mammalian genes revert the size of spotted cells when expressed in a Gef1p⁻ strain growing in low-iron and non-fermentable carbon sources. This complementation was found when the ClCs were introduced and induced to express under the GAL1 promoter contained in the plasmid pYES. Flis et al. (2005) reported that expression of the mouse ClC2 was not capable of complementing a Gef1p⁻ strain when using a single-copy plasmid and thus, we decided to see if this was also true with our strains and the human ClCs.

The ClCs clones derived from pUG35 were grown in restrictive media supplemented with galactose to discard any ability of this carbohydrate to revert the phenotype. Consistent with Flis' findings, neither hClC1 nor hClC2 reverted the *pet* phenotype (Figure 1B). The results above suggested that the reversion of the *pet* phenotype observed in the first series of experiments was due to a dose effect, since the expression derived from pYES is expected to be higher than that driven by the MET25 promoter of the pUG35 vector.

A visual inspection of the cells transformed with pYES under the light microscope revealed that the cell diameter correlated well with the size of the colony (Figure 1A and 2A). The cell diameter of Gef1p⁻ (5.91µ ± 0.06) and the strain transformed with the multicopy vector (5.95 µ ± 0.03) differed from that of the *wt* strain (6.88µ ± 0.03). Cells expressing hClC1 were clearly larger (6.77 µ ± 0.01) than the knock-out mutant but did not reach the full size of the *wt*, whereas those expressing hClC2 (6.82 µ ± 0.03) were undistinguishable from the *wt* yeast (Figure 2A). The yeast surface was analyzed in samples of these cells under the SEM (Figure 2B and 3B), but other than changes in cells diameter we did not observe any major difference among the samples.

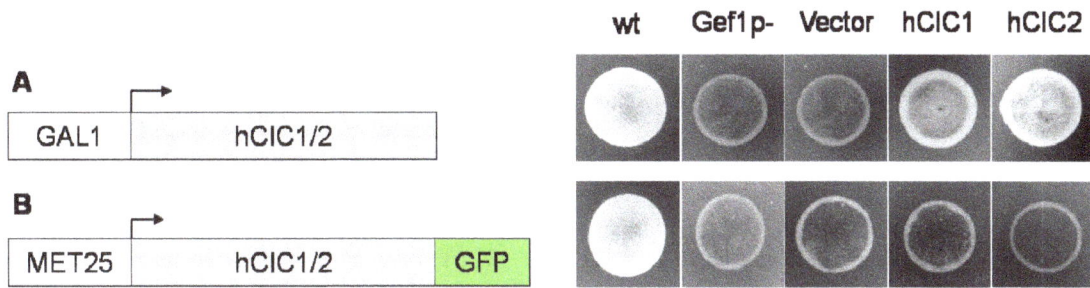

Figure 1. Overexpression of hClCs revert the *pet* phenotype

A. The human ClCs expressed under the GAL1 promoter revert the *pet* phenotype of a Gef1p⁻ strain. B. In contrast, a centomeric plasmid where the ClCs were expressed under the direction of MET25 did not revert the phenotype of the strain.

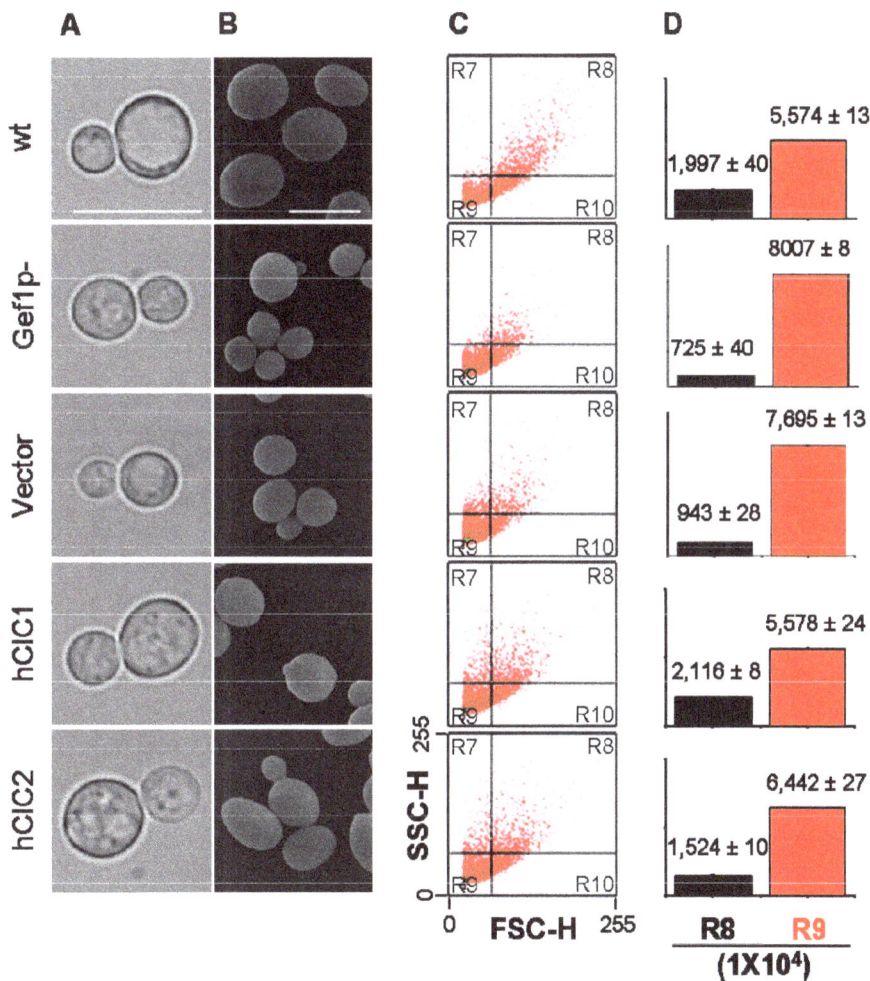

Figure 2. Phenotype of reverted cells

A. Sample images of *wt* and reverted cells seen under the light microscope. Bar = 10 μm. B. Sample images of the yeast under the SEM; notice the diversity of cell sizes within and among samples. Bar = 5 μm. C. Distribution of the cells resulting from the flow cytometry; FSC-H (cell size) and SSC-H (surface complexity), comparative data was subtracted from quadrant R8. D. Distribution of cells (events) in R8 and R9 in three independent experiments (means ± SE).

Observation of the cells under the light microscope showed a wide diversity of cell diameters among samples; thus flow cytometry, a standard easy and quick methodology, was used to analyze a large population of yeast samples to have a better idea of the variants within and among the samples.

The results of flow cytometry were plotted in Figure 2C, which shows the wide variability of cell size regardless of the cell sample, and this was consistent with the diversity of cell diameters observed under the light microscope. Nevertheless, comparing the distribution of cells in quadrants R8 and R9 revealed some difference in volume and complexity of the cell surface between the *wt* and the *GEF1* knock-out; these parameters allowed us to establish a clear *quantitative* reference to determine whether the reversion of the *pet* phenotype occurred or not. The number of events recorded in quadrant R8 for the *GEF1* mutant and for cells transformed with the core vector (pYES) were slightly different (725 ± 4 and 943 ± 28, respectively), whereas the number of cells expressing ClC1 (2116 ± 8) in R8 approached that of the *wt* (1998 ± 14), and the number of ClC2-expressing cells in R8 was intermediate (1524 ± 10), suggesting a partial reversion (Figure 2C-D).

Cells transformed with pUG35 (472 ± 1) showed an unexpected distribution that was different from the *GEF1* mutant (725 ± 4); in addition, the number of small cells was minimally changed by expressing either hClC1 (938 ± 4) or hClC2 (767 ± 8) (Figure 3C-D).

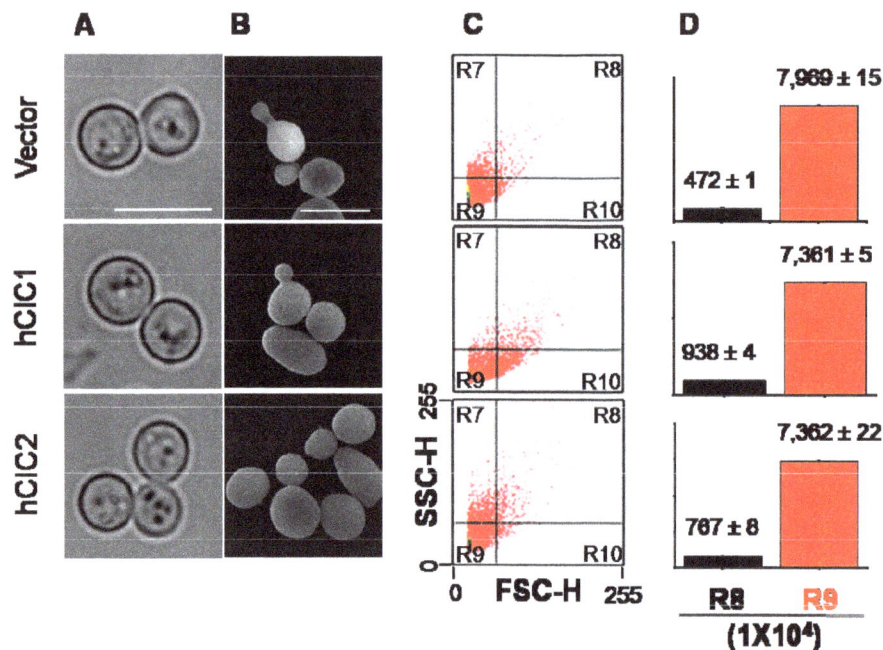

Figure 3. Expression of ClC derived from pUG-35 does not revert the pet phenotype

A. Images of cells under the light microscope. Bar = 10μm. B. Cells seen under the SEM Bar = 5μm. C. Distribution of cells in the flow cytometry assay. D. Distribution of the cells in R8 and R9 in three independent experiments (means ± SE).

3.2 Oxymetry

Respiratory metabolism is significantly diminished in strains that lack the *GEF1* gene (Gaxiola et al., 1998); thus, we determined if the level of oxygen consumption was normal in the reverted strains that expressed the hClCs. In three independent experiments the *wt* strain presented a higher respiratory rate (3.65 ± 0.68 nAO$_2$/min per 10^6 cells) when compared to Gef1p⁻ transformed with the core plasmid (3.04 ± 0.10 nAO$_2$/min per 10^6 cells, Table 1). However, when the plasmid expressed either hClC1 or hClC2, the strain exhibited the normal level of oxygen consumption: 3.62 ± 0.05 or 3.59 ± 0.05, respectively. (In three independent experiments, the expression of hClC2 showed a slightly lower rate of O_2 consumption; however, it was not statistically significant. Expression of the hClCs in the *wt* strain did not increase the O_2 consumption (Table 1, RGY30 strain), size of the colony, or microscopic characteristics (not shown).

Table 1. O_2 consumption rates in *wt* (RGY30) and mutant (RGY192) yeast strains

Plasmid/Strain	$(nAO_2/min/10^6 \text{ cells})$
pYES-RGY30	3.658 ± 0.068
pYES-hClC1-RGY30	3.609 ± 0.086
pYES-hClC2-RGY30	3.626 ± 0.063
pYES-RGY192	3.048 ± 0.101
pYES-hClC1-RGY192	3.625 ± 0.052
pYES-hClC2-RGY192	3.596 ± 0.058

3.3 Expression of hClCs Tagged with GFP

In sharp contrast to the results described above, expression of hClC1 and hClC2 fused to GFP using a centromeric plasmid (pUG35) did not completely revert the *pet* phenotype. The spot assay correlated well with the images taken under the light microscope (Figure 1A and 3A), whereas flow cytometry revealed that the characteristics of the mutant were not totally reverted by expressing the ClCs (Figure 3C and D). This was also evidenced when observing the cell structure under the SEM, which showed that cells with the *pet* phenotype remained in the population of yeast transformed with either hClC1 or hClC2, and only a few cells among the population appeared to have reverted to the *wt* phenotype (Figure 3B).

A previous report describing similar results suggested that for the proper expression of ClC2, several codons have to be switched to those more frequently found in yeast. In addition, the overexpression of the Kah1 transporter is also needed to suppress the *pet* phenotype (Flis et al., 2005). In our results, fluorescence derived from the hClCs tagged with GFP and expressed in the *wt* strain was observed in intracellular compartments but mainly distributed around the periphery of the cells (Figure 4); thus,it is not necessary to introduce the yeast preferred codons for proper expression of the hClCs. When those plasmids were used to transform the Gef1p⁻ strain, roughly 20-25% of the cells expressed the gene, and the fluorescence was found in intracellular compartments (Figure 4); however, as indicated above, this level of expression did not suffice to rescue the *pet* phenotype. A remaining question was if the GFP-tagged channels were functional; to probe that, we used *X. laevis* oocytes to test the electrophysiological properties of these channels.

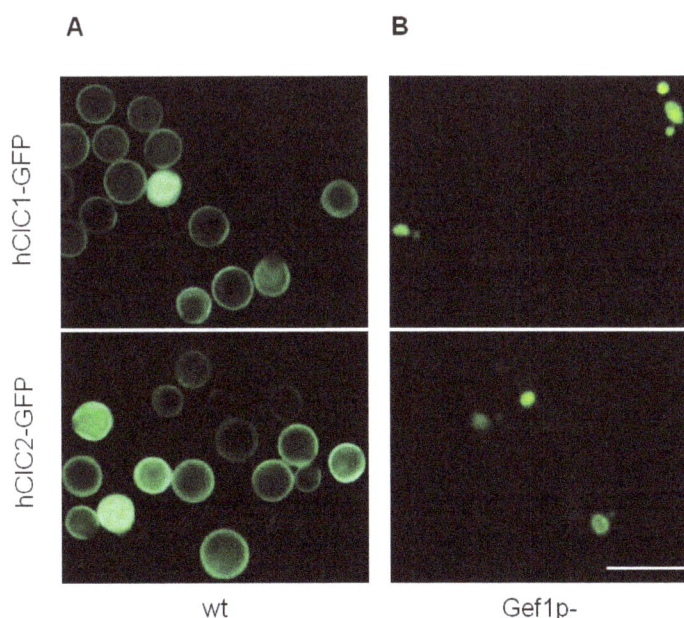

Figure 4. Distribution of GFP-tagged channels

A. fluorescence of ClC1 and ClC2 expressed in the *wt* strain was observed in the periphery of the cells. B. The same plasmids did not revert the *pet* phenotype, and the fluorescence was localized in intracellular compartments. Bar = 10μm.

3.4 Functional Expression of hClC1 and hClC2 Channels in X. laevis Oocytes

Injection of 50 nl of RNA (1 µg/µl) isolated from yeast expressing either hClC1 or hClC2 into frog oocytes induced the expression of functional channels; in contrast, *GEF1* did not present a voltage-dependent current. The resting membrane potential of oocytes injected with the hClC1 was usually around -25 mV, while the uninjected oocytes as well as those expressing hClC2 oscillated between -35 and -40 mV. Voltage stepping the oocytes from 0 to -120 to +40 in 20-mV steps elicited currents derived from the expressed channels. Sample currents generated by hClC1, hClC2, and the GFP-tagged channels are shown in Figure 5. hClC1 and hClC1-GFP showed a fast activation and a pronounced deactivation at voltages more negative than -100 mV, as previously demonstrated (Lorenz, Pusch, & Jentsch, 1996; Pusch, Steinmeyer, & Jentsch, 1994). hClC2 and hClC2-GFP showed a slow activation upon hyperpolarization of the plasma membrane, similar to that previously reported (Gründer, Thiemann, Pusch & Jentsch, 1992; Thiemann, Gründer, Pusch & Jentsch, 1992). This indirect assessment of the channels expressed in yeast gives no indication that the GFP tag alters the properties of the channel.

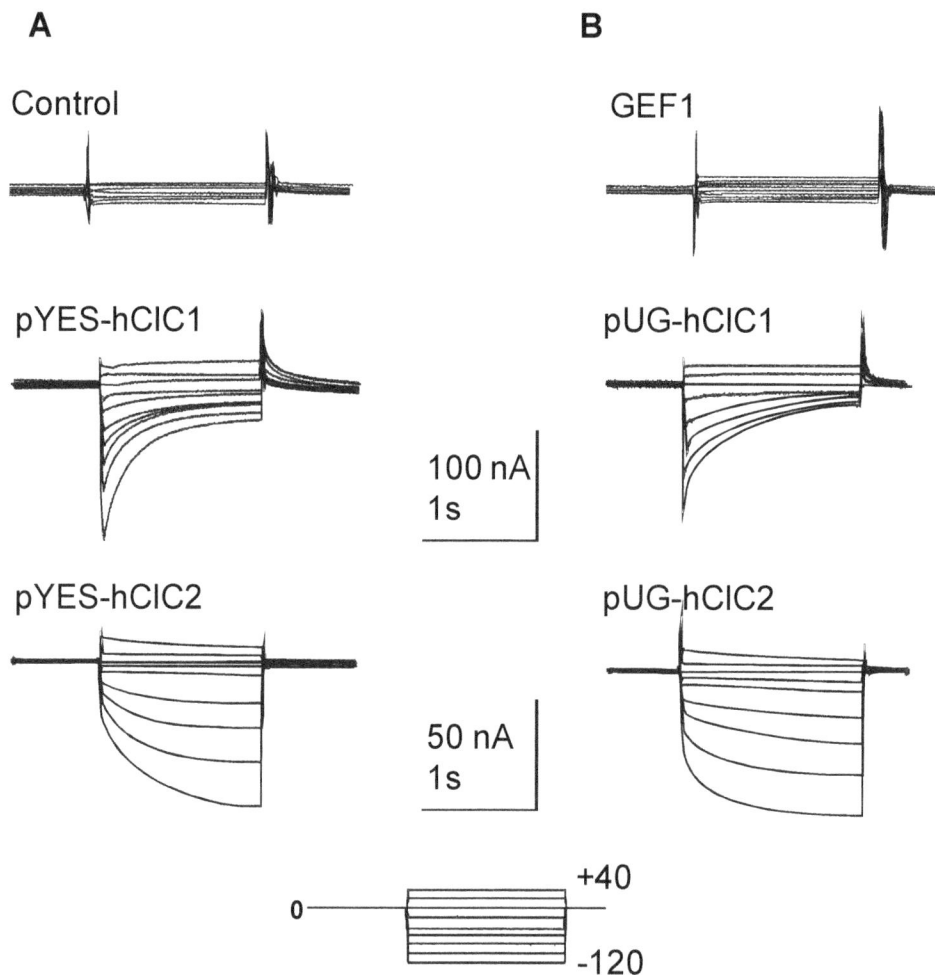

Figure 5. Functional expression in frog oocytes

Neither control nor GEF-1-injected oocytes generated a voltage-activated current, whereas oocytes injected with RNA from yeast induced to express ClC1 and ClC2, whether tagged or not with GFP, generated voltage-gated currents.

4. Discussion

The aim of this study was to determine whether the opposite activation kinetics of hClC1 and hClC2, *i.e.* either slow or fast activation, respectively, as well as other functional differences were related to their ability to revert the *pet* phenotype of a Gef1p⁻ strain of *S. cerevisiae*. Initially, we observed that both hClCs were able to revert the *pet* phenotype of the colonies formed by *GEF1* mutant cells; however, a previous report by Flis et al. (2005) contrasted with our observations. Thus, we repeated our experiments using a centromeric expression plasmid as reported by Flis et al. (2005); in this case our results were consistent with those of Flis et al. (2005); that is to say: the expression of hClC1 or hClC2 derived from pUG35 did not rescue the mutant phenotype. Therefore, we can explain our initial results by a dose effect: overexpression of the hClCs under the GAL1 promoter in pYES permits many channels to be properly located in the cell. In contrast, expression of hClCs under the MET25 promoter and in a centromeric plasmid did not induce the expression of sufficient, properly located protein to complement the functions lost in the *GEF1* mutant.

A second possibility to explain the inability of our pUG35-derived plasmids to revert the *pet* phenotype is the presence of GFP at the carboxy-terminus of the receptor. However, the membrane currents generated by oocytes injected with the ClCs showed no evident differences between the channels that were tagged or not with GFP. The fluorescence detected in yeast that were induced to express the GFP-tagged human channels indicates that it is not absolutely necessary to change the codons to those prefered by *S. cerevisiae*, as reported by Flis et al. (2005). This may reflect differences in the nucleotide sequence between the murine cDNAs used in their studies and the human genes that we used in our experiments. Furthermore, the *wt* strain expressing ClC1 or ClC2-GFP presented fluorescence at the cell periphery.

Considering that hClC1 and hClC2 show obvious differences in their activation mechanism and kinetics, we had aimed to correlate their ability to revert the *pet* phenotype with the specific properties of one of the channels; unexpectedly, both human ClCs induced the reversion. *GEF1* does not show voltage dependence either in HEK cells or in *X. laevis* oocytes heterologously expressing the protein (López-Rodríguez, 2007) (Figure 5) for what is considered mainly as an intracellular chloride transporter. There is also some evidence showing the functional role of ClC1 and ClC2 in intracellular compartments as well as their active role in transporting protons and their modulation by pH (Steinmeyer et al., 1991; Bösl et al., 2001). This last functional property may explain the ability of both channels to compensate for the absence of Gef1p in the-knockout yeast.

Acknowledgements

This work was supported by grants from CONACYT (101851) and UNAM-PAPIIT to RM and AM-T. FR-S and AL-R were supported by fellowships from CONACYT-Mexico and Posgrado en Ciencias Biomédicas-UNAM. Dr. R. Gaxiola donated yeast strains. Dr. A. George and Dr. G. Cutting donated plasmids carrying ClCs. We thank Dr. D.D. Pless for editing the final version of this manuscript.

References

Bösl, M., Stain, V., Hübner, C., Zdebik, A., Jordt, S. E., Mukhopadhyay, A., ... Jentsch, T. J. (2001). Male germ cells and photoreceptors, both dependent on close cell-cell interactions, degenerate upon ClC-2 Cl⁻ channel disruption. *EMBO J., 20*, 1289-1299. http://dx.doi.org/10.1093/emboj/20.6.1289

Chomczynski, P., & Sacchi, N. (1987). Single-step method of RNA isolation by acid guanidiniumthiocyanate-phenol-chloroform extraction. *Anal. Biochem., 162*, 156-159. http://dx.doi.org/10.1016/0003-2697(87)90021-2

Conte, C. D., De Luca, A., Mamrini, M., & Vrbovà, G. (1989). Membrane ionic conductances in normal and denervated skeletal muscle of the rat during development. *Pflugers. Arch., 413*, 568-570. http://dx.doi.org/10.1007/BF00594192

Fahlke, C. (2001). Ion permeation and selectivity in ClC-type chloride channels. *Am. J. Physiol. Renal Physiol., 280*, 748-757.

Flis, K., Hinzpeter, A., Edelman, A., & Kurlandzka, A. (2005). The functioning of mammalian ClC-2 chloride channel in Saccharomyces cerevisiae cells requires an increased level of Kha1p. *Biochem. J., 390*, 655-664. http://dx.doi.org/10.1042/BJ20050480

Gaxiola, R. A., Yuan, D. S., Klausner, R. D., & Fink, G. R. (1998). The yeast CLC chloride channel functions in cation homeostasis. *Proc. Natl. Acad. Sci. U.S.A., 95*, 4046-4050. http://dx.doi.org/10.1073/pnas.95.7.4046

Gietz, D. R., Schiestls, R., Willems, A. R., & Woods, R. (1995). Studies on the Transformation of Intact Yeast Cells by the LiAc/SS-DNA/PEG Procedure. *Yeast, 11*, 355-360. http://dx.doi.org/10.1002/yea.320110408

Greene, J. R., Brown, N. H., DiDomenico, B. J., Kaplan, J., & Eide, D. J. (1993). The *GEF1* gene of Saccharomyces cerevisiae encodes an integral membrane protein; mutations in which have effects on respiration and iron-limited growth. *Mol. Gen. Genet., 241*, 542-553. http://dx.doi.org/10.1007/BF00279896

Gründer, S., Thiemann, A., Pusch, M., & Jentsch, T. J. (1992). Regions involved in the opening of CIC-2 chloride channel by voltage and cell volume. *Nature, 360*, 759-762. http://dx.doi.org/10.1038/360759a0

Hechenberger, M., Schwappach, B., Fischer, W., Frommer, W., Jentsch, T. J., & Steinmeyer, K. (1996). A family of putative chloride channels from Arabidopsis and functional complementation of a yeast strain with a CLC gene disruption. *J. Biol. Chem., 271*, 33632-33638. http://dx.doi.org/10.1074/jbc.271.52.33632

Hernandez-Muñoz, R., Díaz-Muñoz, M., & Chagoya de Sanchez, V. (1992). Effects of adenosine administration on the function and membrane composition of liver mitochondria in carbon tetrachloride-induced cirrhosis. *Arch. Biochem. Biophys., 294*, 160-167. http://dx.doi.org/10.1016/0003-9861(92)90151-L

Jentsch, T. J., Stein, V., Weinreich. F., & Zdebik, A. A. (2002). Molecular structure and physiological function of chloride channels.*Physiol. Rev., 82*, 503-56. http://dx.doi.org/10.1080/10409230701829110

Jentsch, T. J. (2008). CLC chloride channels and transporters: from genes to protein structure, pathology and physiology. *Crit. Rev. Biochem. Mol. Biol., 43*, 3-36.

Jordt, S. E., & Jentsch, T. J. (1997). Molecular dissection of gating in the ClC2 chloride channel. *EMBO J., 16*, 1582-1592. http://dx.doi.org/10.1093/emboj/16.7.1582

Kida, Y., Uchida, S., Miyazaki, H., Sasaki, S., & Marumo, F. (2001). Localization of mouse CLC-6 and CLC-7 mRNA and their functional complementation of yeast CLC gene mutant. *Histochem. Cell Biol., 115*, 189-194.

Klocke, R., Steinmeyer, K., Jentsch, T. J., & Jockusch, H. (1994). Role of innervation, excitability and myogenic factors in the expression of the muscular chloride channel ClC1.A study on normal and myogenic muscle. *J. Biol. Chem., 269*, 27635-27639.

López-Rodríguez, A., Carabez-Trejo, A., Coney, L., Halliwell, R. F., Miledi, R., & Martínez-Torres, A. (2007). The product of the gene *GEF1* of Saccharomyces cerevisiae transports Cl⁻ across the plasma membrane. *FEMS Yeast Res., 7*, 1218-1229. http://dx.doi.org/10.1111/j.1567-1364.2007.00279.x

Lorenz, C., Pusch, M., & Jentsch, T. J. (1996). Heteromultimeric CLC chloride channelswith novel properties. *Proc. Natl. Acad. Sci. U.S.A., 93*, 13362-13366. http://dx.doi.org/10.1073/pnas.93.23.13362

Marmagne, A., Vinauger-Douard, M., Monachelo, D., de Longevialle, A. F., Charon, C., Allot, M., ... Ephritikhine, G. (2007). Two members of the Arabidopsis CLC (chloride channel) family, AtCLCe and AtCLCf, are associated with thylakoid and Golgi membranes, respectively. *J. Exp. Bot., 58*, 3385-3393. http://dx.doi.org/10.1093/jxb/erm187

Miledi, R. (1982). A calcium-dependent transient outward current in Xenopuslaevis oocytes. *Proc. R. Soc. Lond., B, Biol. Sci., 215*, 491-497. http://dx.doi.org/10.1098/rspb.1982.0056

Miyazaki, H., Uchida, S., Takei, Y., Hirano, T., Marumo, F., & Sasaki, S. (1999). Molecular Cloning of CLC Chloride Channels in OreochromisMossambicus and Their Functional Complementation of Yeast CLC Gene Mutant. *Biochem.Biophys. Res. Commun., 255*, 175-181. http://dx.doi.org/10.1006/bbrc.1999.0166

Mumberg, D., Müller, R., & Funk, M. (1994). Regulable promoters of *Saccharomyces cerevisiae*: comparison of transcriptional activity and their use of heterologous expression. *Nucleic Acids Res., 22*, 5767-5768. http://dx.doi.org/10.1093/nar/22.25.5767

Pusch, M., Steinmeyer, K., & Jentsch, T. J. (1994). Low single channel conductance of the major skeletal muscle chloride channel, ClC-1. *Biophys. J., 66*, 149-152. http://dx.doi.org/10.1016/S0006-3495(94)80753-2

Sardini, A., Ameya, J. S., Weylandt, K-H., Nobles, M., Valverde, M. A., & Higgins, C. F. (2003). Cell volume regulation and swelling-activated chloride channels. *Biochim. Biophys.Acta, 1618*, 153-162. http://dx.doi.org/10.1016/j.bbamem.2003.10.008

Schwappach, B., Stobrawa, S., Hechenberger, M., Steinmeyer, K., & Jentsch, T. J. (1998). Golgi localization and functional important domains in the NH2 and COOH terminus of the yeast ClC putative chloride channel Gef1p. *J. Biol. Chem., 273*, 15110-15118. http://dx.doi.org/10.1074/jbc.273.24.15110

Soleimani, M., & Xu, J. (2006). SLC26 chloride/base exchangers in the kidney in health and disease.*Semin.Nephrol., 26*, 375-385. http://dx.doi.org/10.1016/j.semnephrol.2006.07.005

Steinmeyer, K., Ortland, C., & Jentsch, T. J. (1991). Primary structure and functional expression of a developmentally regulated skeletal muscle chloride channel. *Nature, 354*, 301-304. http://dx.doi.org/10.1038/354301a0

Thiemann, A., Gründer, S., Pusch, M., & Jentsch, T. J. (1992). A Chloride channel widely expressed in epithelial and nonepithelial cells. *Nature, 356*, 57-60. http://dx.doi.org/10.1038/356057a0

17

Study on the Heterosis of the First Generation of Hybrid between Chinese and Korean Populations of *Scapharca broughtonii* using Methylation-Sensitive Amplification Polymorphism (MSAP)

Hailin Sun[1], Yanxin Zheng[2], Chunnuan Zhao[2], Tao Yu[2] & Jianguo Lin[2]

[1] Changdao Fisheries Research Institute, Changdao, China

[2] Changdao Enhancement and Experiment Station, Chinese Academy of Fishery Sciences, Changdao, China

Correspondence: Tao Yu, Changdao Enhancement and Experiment Station, Chinese Academy of Fishery Sciences, Changdao, China. Email: cdyutao@126.com

Abstract

DNA methylation is known to play an important role in the regulation of gene expression in eukaryotes. In this study, the author assessed the extent and pattern of cytosine methylation in the *Scapharca broughtonii* genome using the technique of methylation-sensitive amplified polymorphism (MSAP).The results showed that, DNA methylation rate was negatively related to the shell length, the gross weight and the weight of soft body, but positively related to the shell broadness and the shell height; there was significantly different between the parents and the offspring: 31.6% of 5'-CCGG sites in the *Patinopecten yessoensis* of Korean populations genome were cytosine methylated, and in the *Patinopecten yessoensis* of Chinese populations were 33%, the methylation rates of F1 was 29.98%; four classes of patterns were identified in a comparative assay of cytosine methylation in the parents and hybrid, increased methylation was detected in the hybrid compared to the parents at some of the recognition sites, while decreased methylation in the hybrid was detected at other sites. It indicated that the alteration of methylation resulted from cross-breeding, and the inbreeding did not change the methylation ratio and patterns; The DNA cytosine methylation has a relationship with the heterosis.

Keywords: *Scapharca broughtonii*, DNA methylation, methylation-sensitive amplified polymorphism (MSAP), heterosis

1. Introduction

Scapharca broughtonii (Mollusca, Bivalve, Arcoida), one of the most important marine commerical bivalve species, mainly distributes in the coasts of BoHai Sea and North of Yellow Sea, China. Because of its large body, rapid growth rate, delicious tastes and high protein and vitamin contents, the export in the exchange rates of ark shell is higher in the aquatic products, and *Scapharca broughtonii* has become one of the most popular farming mollusks in North China due to its high economic value in recent years. However, with the deterioration of marine ecological environment resulted from the extended farming scale and frequency human activities in coastal waters, and over-fishing, the wild resources decreased. The mass mortality has become a major constraint for the development of the *Scapharca broughtonii* culture. It is imperative for us to actively manage the resource and turn to breed high adversity resistance, fast-growing variety using traditional and new breeding methods. The hybridization of different populations proved to be a good way of breeding.

The genetic basis of heterosis has been debated for decades, dominance, pseudo-overdominance, real overdominance, and epistasis are the major genetic models proposed to explain heterosis (Crow, 2000; Lamkey & Edwards, 1997; Lippman & Zamir, 2007; Reif et al., 2006), but there is still a striking discordance between an extensive use of heterosis in variety development and our understanding of the basis of heterosis (Birchler, Auger, & Riddle, 2003; Reif et al., 2006). In recent years, many research thought that the molecular basis of heterosis may be attributed to the increased gene expression level in the hybrid or to the altered regulation of gene expression in the hybrid either at the global level or for specific classes of genes (Leonardi, Damerval, Hebert, Gallais, & Vienne, 1991; Romagnoli, Maddaloni, Livini, & Motto, 1990; A. Tsaftaris, Kafka, Polidoros, & Tani, 1997; S. Tsaftaris, 2006). Two different alleles brought together in the hybrid may create a combined allelic expression pattern in the hybrids. Alternatively, at some loci, allelic interaction or a change in the spectrum of trans-acting factors causes gene expression in the hybrid to deviate from simple additive allelic expression patterns of the parents(Birchler et al., 2003; Gibson & Weir, 2005). Considering effects of DNA methylation on gene expression, there may be a relationship between DNA methylation and the expression of heerosis (Finnegan, Peacock, & Dennis, 2000; Rangwala & Richards, 2004).

DNA cytosine methylation is the most common covalent modification of DNA in eukaryotes, in recent years, DNA methylation has received considerable attention in eukaryotic organisms (Xiong, Xu, Saghai Maroof, & Zhang, 1999), which plays an important role in many aspects of biology, including differential gene expression, cell differentiation, genomic imprinting, chromatin inactivation, transposable elements and gene silencing, and so on(Finnegan et al., 2000; Paszkowski & Whitham, 2001; Tariq & Paszkowski, 2004).

DNA methylation analysis has been approached either by studying global levels of cytosines methylated or by analyzing specific gene sequences(Jacobsen, Sakai, Finnegan, Cao, & Meyerowitz, 2000; Luff, Pawlowski, & Bender, 1999; Riddle & Richards, 2002; Soppe et al., 2000). There are several methods used for detecting DNA methylation, such as bisulfite conversion, methylation-sensitive restriction enzymes, methyl-binding proteins, methylation-sensitive amplified polymorphism (MSAP), and anti -methylation cytosine antibodies(Zilberman & Henikoff, 2007). Among these, two methods are routinely used for the detection of DNA methylation in the tissues of eukaryotic organisms. These depend on the application of bisulfites or methylation-sensitive restriction enzymes. Bisulfites convert unmethylated cytosine into thymine, thus allowing the detection of cytosine methylation. Some restriction enzymes (isoschizomers) share the same recognition sites but show differential sensitivity to DNA methylation. Thus, polymorphic DNA fragments can be generated after digestion of methylated genomic DNA with isoschizomers(Xu, Li, & Korban, 2000). Methylation sensitive amplified polymorphism (MSAP) analysis is based on the use of isoschizomers for detection of DNA methylation. It is an adaptation of the amplified fragment length polymorphism (AFLP) technique(Reyna-Lopez, Simpson, & Ruiz-Herrera, 1997), in which the isoschizomers HpaII and MspI are employed as 'frequent-cutter' enzymes for AFLP, instead of the usual MseI. HpaII and MspI recognize the same tetranucleotide sequence (5'-CCGG-3'), but display differential sensitivity to DNA methylation. HpaII is inactive when either of the two cytosines is fully methylated, but cleaves hemi-methylated 5'-CCGG-3' at a lower rate than the unmethylated sequence. MspI cleaves 5'-C^{5m}CGG-3', but not 5'-^{5m}CCGG-3'. MSAP allows for detection of genetic diversity throughout the genome without any prior knowledge of the nucleotide sequence(Vos et al., 1995) and has been successfully applied in various studies.

In this study, the MSAP technique was firstly used to analyze the *Scapharca broughtonii* genome DNA cytosine methylation. We discussed the differences in the level of cytosine methylation among the parents and the offspring, the differences of methylation patterns between the parents and offspring, the correlation between the methylation and phenotypic traits, and emphasized the discussion on the molecular basis of heterosis in terms of the DNA methylation.

2. Materials and Methods

2.1 Sampling

The *Scapharca broughtonii*, Chinese populations(6.9±0.53 cm of shell length), were collected randomly from Penglai sea area (Shandong province, China) in April, 2014. Korean populations (7.2±0.47 cm)were collected randomly from Incheon sea area in April，2014. The heterozygous **F1** (6.7±0.62 cm) was a cross between the *Scapharca broughtonii* of Chinese and Korean populations.

2.2 DNA Extraction

The sample DNA was extracted from adductor muscle by traditional phenol-chloroform method. Approximately 100 mg of adductor muscle was dissected out and transferred to an Eppendorf tube containing 500μL of a lysis solution (50 mmolL-1 Tris–Cl pH8.0, 10 mmolL-1 EDTA, 10% sodium dodecylsulfate[SDS] and 200 μgmL-1 proteinase K) at 55 ℃ for 3 h. DNA was extracted with phenol, phenol/chloroform/isoamyl alcohol (25:24:1), and chloroform, and then precipitated with two volumes of ice-chilled absolute ethanol and 1/10 volume of 3 mol L-1 sodium chloride at -20 ℃ for 1h. The rough extraction was washed with 70% ethanol for three times, natural dried and resuspended in 50μL autoclaved ddH$_2$O. Extracted DNA was stored at-20 ℃.

2.3 MSAP Analysis

The MSAP protocol was adapted from Xiong et al(Xiong et al., 1999). Briefly, DNA was double-digested with one of the methylation sensitive enzymes HpaII or MspI, which cuts at the CCGG site, and then with the methylation insensitive EcoRI. Two digestion reactions were set up at the same time for each genomic DNA sample, each containing 400 ng of DNA with 3 Units of isoschizomers either HpaII or MspI (Fermentas) and 2μL

10×Buffer Tango™ in a final volume of 20 μL for 6 h at 37°C, and then add in 3 Units of EcoRI (Fermentas) and 4μL 10×Buffer Tango™ in a final volume of 30 μL for 6 h at 37 ℃.

Subsequently, the digested DNA fragments from the two reactions were ligated separately with an equal volume of the ligation solution containing 5μL digested fragments with 5 U of T4 DNA Ligase (Trans), 5 pmolL-1 EcoRI adapter, 50 pmolL-1 HpaII/MspI adapter, and 4μL 5×T4 DNA Ligase Buffer in a final volume of 20 μL at 16℃ for overnight. The reactions were stopped by incubating at 65°C for 10 min and diluted to 200μL for PCR amplification.

Preamplification was conducted by using 5μL of the above ligation product with E0/HM0 primers in a volume of 20μL containing 2 μL PCR buffer, 0.1 mmolL-1 each dNTPs, 20 ng of each primer and 0.1 U Taq polymerase (TIANGEN). The reaction involved 27 cycles of 94°C for 30 s, 56°C for 1min, 72°C for 1 min, with a final extension at 72°C for 10 min. The preamplified products were then diluted to 600μL and stored at -20°C before use.

Ingredients of the selective amplification were the same as described above using 2.5μL of diluted preamplification mixture DNA. The selective amplification was performed by the touchdown program using amplification primers. The PCR conditions were as follows: 13 cycles at 94℃ for 30 s, 0.7℃ per cycle from 65 to 56℃ for 30 s and 72℃ for 1 min; and another 23 cycles of PCR amplification were used following the touchdown program. The denaturing step was done at 94℃ for 30 s, annealing at 56℃ for 30 s, extension at 72℃ for 1 min; and a final extension at 72℃ for 10 min.

The final selective amplification products were denatured, separated on a 6% polyacrylamide sequencing gel, and visualized by silver staining.

All reactions were performed in triplicate to avoid false positive results. If results were reproducible, the sample was used for further analysis. Only clear and reproducible bands that appeared in four independent PCR amplifications were scored.

2.4 Restriction of Isolated and Re-Amplified Fragments

The special bands were excised directly from the polyacrylamide gels on the plate using a razor blade. The bands were rehydrated with 50μL of sterile distilled water, heated at 98℃ for 5 min and let cool slowly to room temperature for the night. The tubes were centrifuged at 12,000 g for 10 min and the supernatant transferred into a fresh tube. Aliquots of 5μL were used as template for re-amplification in a total PCR reaction volume as the selective amplification with the same primer combinations. The products were checked on 1.5% agarose gel for the presence of the bands.

Then, two sets of digestion reactions were carried out simultaneously, in the first reaction, 5μL of re-amplification PCR product was added to 15μL of the digestion system above described, the reamplified fragment was excised from the H lane. The second digestion reaction was carried out in the same way, except that MspI was used in place of HpaII, and the reamplified fragment was excised from the M lane. The digestion products were checked on 1.5% agarose gel with the re-amplification PCR product.

3. Results

The isoschizomers HpaII and MspI recognize and digest 5'-CCGG-3 sites, but display differential sensitivity to DNA methylation. HpaII is inactive if one or both cytosines are fully methylated (both strands methylated), but cleaves hemimethylated sequences (only a single DNA strand is methylated) or no methylation sequences; whereas, MspI digests inner methylation of double-stranded DNA or no methylation. Hemimethylation of either of the two cytosines would lead to the appearance of a fragment in the amplification product from the EcoRI+HpaII digest but not the EcoRI+MspI digest; on the contrary, it is a full methylation site; and if the fragments appeared in the products of the two digestions, the cytosine were not methylated(Lu et al., 2006).

3.1 The Correlation between Methylation Rates and Phenotypic Traits

The methylation may affected *Scapharca broughtonii* phenotypic traits were studied with adductor muscle DNA samples of *Scapharca broughtonii* using 9 pairs of primers (Table 1). And Figure 1 show the correlation between methylation rates and phenotypic traits, results showed that DNA methylation rate was negatively related to the shell length, the gross weight and the weight of soft body, but positively related to the shell broadness and the shell height, there was great correlativity between the DNA methylation and the gross weight (from Korean, the correlation coefficient is -0.59) (Figure 1). It indicated that DNA methylation affected on the shell length, the gross weight and the weight of soft body, DNA methylation may play an important part during the growth and development of the organisms, in the aspect of survival rate it has a different effect on Chinese and Korean populations.

Table 1. List of MSAP primers and adapters used

	EcoRI	MspI/HpaII
Adapters	EA$_1$:5'-CTC GTA GAC TGC GTA CC-3' EA$_2$:5'-AAT TGG TAC GCA GTC TAC-3'	HMA$_1$:5'-GAT CAT GAG TCC TGC T-3' HMA$_2$:5'-CGA GCA GGA CTC AGA A-3'
Primers for Preamplification	E$_0$:5'-GAC TGC GTA CCA ATT C-3'	HM$_0$:5'-ATC ATG AGT CCT GCT CGG G-3'
Primers for Selective Amplification	E$_1$: 5'-GAC TGC GTA CCA ATT C ACA-3' E$_2$: 5'-GAC TGC GTA CCA ATT C AGT-3' E$_3$: 5'-GAC TGC GTA CCA ATT C AAC-3' E$_4$: 5'-GAC TGC GTA CCA ATT C GTC-3' E$_5$: 5'-GAC TGC GTA CCA ATT C GCT-3'	HM$_1$: 5'-ATC ATG AGT CCT GCT CGG GC TGA-3' HM$_2$: 5'-ATC ATG AGT CCT GCT CGG GC TGT-3' HM$_3$: 5'-ATC ATG AGT CCT GCT CGG GC TAT-3' HM$_4$: 5'-ATC ATG AGT CCT GCT CGG GC TAC-3' HM$_5$: 5'-ATC ATG AGT CCT GCT CGG GC TCA-3' HM$_6$: 5'-ATC ATG AGT CCT GCT CGG GC TCT-3' HM$_7$: 5'-ATC ATG AGT CCT GCT CGG GC TTC-3' HM$_8$: 5'-ATC ATG AGT CCT GCT CGG GC TTA-3'

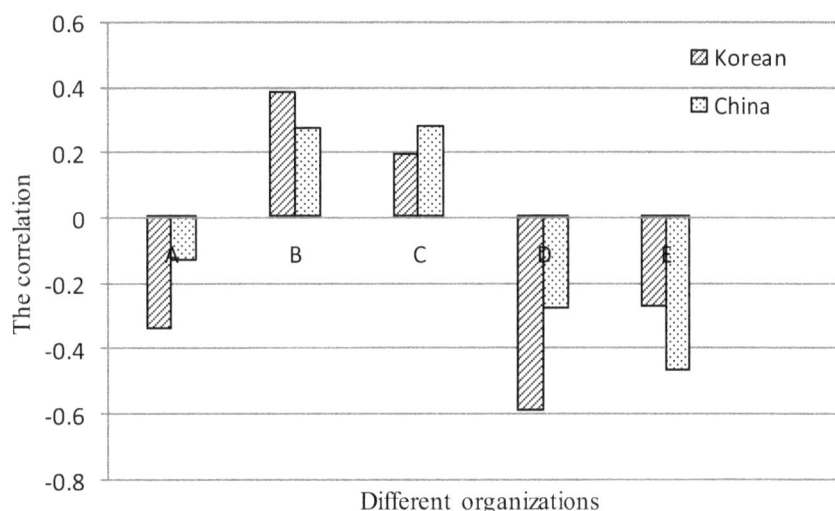

Figure 1. The correlation between methylation rates and phenotypic traits

A: The shell length; B: the shell broadness; C: the shell height; D: the gross weight; E: the weight of soft body

3.2 The Methylation Rates of Parental Lines and Hybrid

The two parental lines, their F1 hybrid were compared using tissue from adductor muscle with the same primers (Table 1). A total of 732 fragments were amplified, each of the fragments represented a recognition site cleaved by one or both of the isoschizomers. Firstly, the two parents showed significantly different degree of methylation (Figure 2): 197 differentially amplified fragments were detected in *Scapharca broughtonii* from Korean and 183 were observed in *Scapharca broughtonii* from China. Thus, approximately 31.6% of 5'-CCGG sites in *Scapharca broughtonii* (from Korean) genome were cytosine-methylated, in *Scapharca broughtonii* from China were 33%. In *Scapharca broughtonii* from Korean, the full methylation rate was 26.58%, the hemimethylation rate was 4.82%, in *Scapharca broughtonii* from China, the full methylation rate was 28.68%, and the hemimethylation rate was 5.39%.The methylation rates of F1was less than the parents (Figure 2), a total of 231 fragments were amplified, and the rates of the methylation were 29.98%. In F1, full methylation of internal cytosines accounted for 79.94% of the methylated sites, and the remaining 20.06% were due to hemimethylation. The F1 inclined to the female parent on the traits, so the methylation rate was close to the *Scapharca broughtonii* from Korean, it was 31.6%.

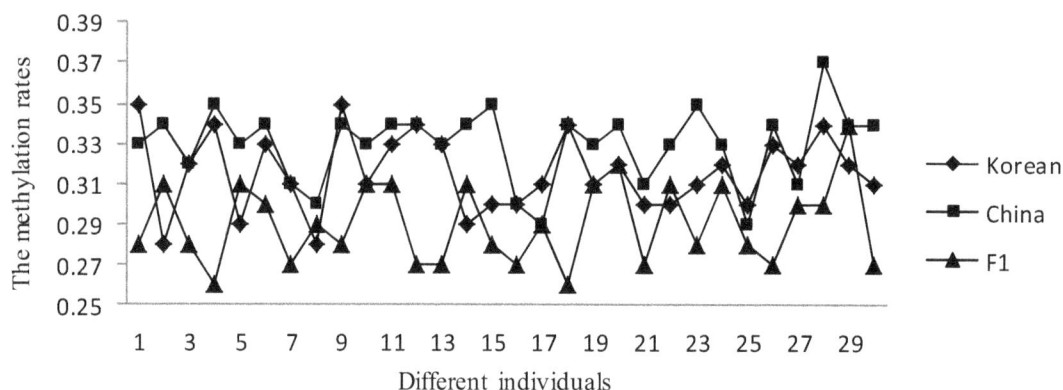

Figure 2. The methylation rates of parental lines and hybrid

Each group chose 30 individuals; each site stands for the average methylation rate of 9 pairs of primes.

3.3 Differential Methylation Patterns among Parental Lines and Hybrid

Four major classes of banding patterns were identified among the differentially amplified fragments (Table 3). In the first class (class A), the same methylation sites were detected in both parents and in the hybrid; these are referred to as monomorphic with respect to cytosine methylation, within the resolving power of this technique. In the four groups, 19 sites detected by 9 primer pairs reflected full methylation of the internal cytosine, and 6 sites were the result of hemimethylation. The second class (class B) showed simple Mendelian inheritance of the methylated bands, irrespective of the enzyme digest: A band that was detected in either parent was also detected in the hybrids. Class B could be divided in to four subclasses, B1, B2, B3 and B4 (Table 2). B1 averagely accounted for 3.1% methylated sites, B2 did 16.7%, B3 did 2.1% and B4 did 26.0% methylated sites. This class accounted for totally 46.9%methylated sites. Class C represents an increase in the level of methylation in the hybrid compared to the parental lines: a site detected in one or both parental lines was not observed in the hybrid (Table 2).Class C could be divided into 3 subclasses, C1 (a band was only revealed in *P. yessoensis*), C2 (a band was only revealed in *C. farreri*), C3 (a band was revealed in both of *C. farreri* and *P. yessoensis*). On the contrary, class D indicated demethylation in the hybrid genome. That is, a band was observed in the hybrids, but not in the parental lines (Table 2).The two classes occupied 26% methylated sites; with 18.7%belong to C and 7.3% to D.

Table 3. Patterns of cytosine methylation in parental lines and their F1 hybrid

Patterns	*Scapharca broughtonii* from Korean		*Scapharca broughtonii* from China		F1 hybrid		Number of sites
	HpaII	MspI	HpaII	MspI	HpaII	MspI	
A1	-	+	-	+	-	+	19
A2	+	-	+	-	+	-	6
B1	-	+	-	-	-	+	3
B2	-	-	-	+	-	+	16
B3	+	+	-	+	+	+	2
B4	-	+	+	+	+	+	25
C1	-	+	-	-	-	-	10
C2	-	-	-	+	-	-	6
C3	-	+	-	+	-	-	2
D1	-	+	-	-	+	+	0
D2	-	-	-	+	+	+	7

"-"stands for no band was found at the site; "+" stands for a band was found at the site.

4. Discussion

In recent years, there has been an increased interest in understanding the role of DNA methylation in regulating gene expression during the growth and development of the organisms (Chen, Ma, Chen, Song, & Zhang, 2009). The methylation sensitive amplified polymorphism (MSAP) technique has been used in various studies on cytosine methylation, and has proven to be a powerful tool for investigating DNA methylation. In this study, we have adapted the MSAP technique for the detection of cytosine methylation in the *Scapharca broughtonii* genome, the results showed that this technique is highly efficient for investigating the cytosine methylation of *Scapharca broughtonii*; all bands detected displayed good stability, reproducibility, and consistency.

The DNA methylation levels are varied in different species: 16.3% of the methylation has been reported to occur in the rice genome(Xiong et al., 1999), 35–43% in *Arabidopsis* of different ecological type(Cervera, Ruiz-Garcia, & Martinez-Zapater, 2002), 33% in wheat(Horváth et al., 2002), 15.7% in the developing seeds from Brassica napus(Lu et al., 2006). In animals, the pigs' methylation ratio is about 10%; the grass carp's methylation is very high, about 75.9%. Our research found that 31.6% of 5'-CCGG sites in *Scapharca broughtonii* from Korean genome were cytosine methylated, in *Scapharca broughtonii* from China were 33%, the DNA methylation level is different significantly to other organism, the differences may come from the detection method(such as the number of primes, the reaction conditions and time),the experiment material (different tissues: adductor muscle, mantle, gill filaments, gut, and gonad; or different collection time), and the genetic factors(the genetic factors play a more important part). Riddle (Riddle & Richards, 2002) found that natural variation in NOR methylation results from a combination of genetic and epigenetic mechanisms.

The purpose of this study was to develop an approach to investigating the possible role of methylation in the expression of heterosis. The genetic basis of heterosis has been debated for decades, but the mechanism underlying heterosis remains mysterious. Before the 1990s, two major hypotheses have been promulgated to explain this phenomenon: the dominance hypothesis and the overdominance hypothesis, but they are ideal and very limited, because they all thought that one trait was controlled by one allele. The fact is that more and more studies show that the appearance of quantitative character was the result of many alleles acting together. Many researchers used QTL method to study the heterosis, and put forth the hypothesis of epistasis, but the results were also not satisfactory (Z. Li et al., 2001; Luo et al., 2001; Xiao, Li, Yuan, & Tanksley, 1995; Yu et al., 1997).

Now, many researchers believe that the molecular basis of heterosis may be attributed to the increased gene expression level in the hybrid(Leonardi et al., 1991; Romagnoli et al., 1990; A. Tsaftaris et al., 1997; S. Tsaftaris, 2006) or to the altered regulation of gene expression in the hybrid either at the global level or for specific classes of genes.

The phenotypic traits that the organism expressed are the results of gene expression, an individual take on an advantage in a trait, it should be the result of over-expression of some genes or under-expression of some genes, that is to say, the over-expression and under-expression of some genes lead to the expression of heterosis together. Romagnoli et al. (Romagnoli et al., 1990) found that heterosis may derive from simple dominant or codominant gene effects in addition to the increased expression of certain loci. Tsaftaris et al.(A. Tsaftaris et al., 1997) found that the transcriptional level of 35 gene were higher than the parents lines. Sun et al.(Sun et al., 2004) found that 30% genes were differentially expressed between hybrids and their parents, which play an important role for hybrids to demonstrate heterosis. Li et al.(X. Li, Wei, Nettleton, & Brummer, 2009) found that nonadditive expression of transcript levels may contribute to heterosis for biomass yield in alfalfa. Zhao (Zhao, Chai, & Liu, 2007), Meng (Meng, Ni, Wu, & Sun, 2005), Użarowska et al.(Użarowska et al., 2007) received the similar results.

Considering effects of DNA methylation on gene expression(Finnegan et al., 2000; Jacobsen et al., 2000; Rangwala & Richards, 2004), Hyper-methylation can lead to the gene silence, and demethylation can lead to the over-expression (Neves, Heslop-Harrison, & Viegas, 1995; Sardana, O'Dell, & Flavell, 1993), there may be a correlation between the DNA methylation and heterosis.

A negative correlation was found between DNA methylation and the economic characters of shell length, shell height, soft body weight and adductor muscle weight, Wan(Wan, 2008) found that, in the Chongqing mountainous cattle and their hybrids, methylation affected heart girth and body weight highly significantly, methylation and body slanting length significantly. Methylation content and heterosis rates of heart girth, body weight showed significant correlation, it just indicated indirectly that DNA methylation have a correlation with the gene expression. Zhang, Shiu, Cal, and Borevitz (2008) found that cytosine methylation alterations immolL-1ediately upstream or downstream of the gene were inversely correlated with the degree of expression variation for that gene. Jin found a direct relationship between cytosine methylation alteration and gene expression variation. So we can say DNA cytosine methylation alteration lead to the expression of heterosis in some extent.

The methylation of the *Scapharca broughtonii* from Korean and China are different, the cytosine methylation ratio of the filial generation will between the parents in theory, but the result was that the filial generation cytosine methylation ratio was lower than any parents. And four classes of patterns of cytosine methylation characterized by differences in degree of methylation between the hybrids and parental lines: (1) the same level of methylation in both parental lines and the hybrids, (2) the same level of methylation in either parent or hybrid,(3) an increased level of methylation in the hybrids compared to the parents, (4) a decreased level of methylation in the hybrids, and the number of demethylation sites were more than the number of the hypermethylation sites. Hepburn found the same result, Tsaftaris(A. Tsaftaris et al., 1997) found that hybrids were less methylated than inbreeds. It was further proposed that differential DNA methylation patterns in hybrids may play an important role in materializing heterosis

It has been widely recognized that, in animals, the inheritance of the epigenetic state through mitotic rounds of cell division is relatively faithful, in development (embryogenesis and gametogenesis) the epigenetic state is reset, that is, erased and reestablished; and parental epigenetic state in plants is often stably inherited to sexual progenies (Cubas, Vincent, & Coen, 1999; Monk, Boubelik, & Lehnert, 1987; Riddle & Richards, 2002). Penterman thought that demethylation processes may be the result of a deficiency in enzymatic maintenance after DNA replication or an active enzymatic process involving plant glycosylases with specific functions in genomic imprinting and to protect genes from potentially deleterious methylation(Grewal & Elgin, 2002; Grewal & Klar, 1996; Penterman et al., 2007). Grewal and Klar(Grewal & Klar, 1996) showed that the epigenetic modification of a reporter gene placed in the mating-type region of Schizosaccharomyces pombe could be inherited through mitosis and meiosis. Furthermore, they showed that loci influencing this process were, either directly or indirectly, involved with the organization of heterochromatin(Grewal & Klar, 1996). More recent work has shown that these modifiers include histone deacetylases, histone methyltransferases and other structural proteins associated with telomeres and centromeres (Grewal & Elgin, 2002).

Anyway, the change of the DNA cytosine methylation took place, though the mechanism is still unknown. Two different alleles brought together in the hybrid may create a combined allelic expression pattern in the hybrids. Alternatively, at some loci, allelic interaction or a change in the spectrum of trans-acting factors causes gene expression in the hybrid to deviate from simple additive allelic expression patterns of the parents(Birchler et al., 2003; Gibson & Weir, 2005).The patterns of the F1 DNA methylation experienced change and adjustments in order to coordinate the expression of the gene from the parents, most of the cytosine sites' methylation patterns could be inherited to next generation stably, some sites' methylation patterns experience hypermethylated and demethylated, the demethylation cause some gene expressing largely, the hypermethylation restrain the expression of some gene, with the participation of the environment, a new methylation patterns was formed which added together the parents' characters and can adapted to the circumstances very well, so the heterosis is the result of the differential expression of some gene, the DNA methylation also have to do with the expression of heterosis.

Generally speaking, after investigating the difference of methylation between the parents and the offspring, we can find out the special methylation sites which may play an important part in the expression of heterosis, and then using the methods of sequences analysis and genetic analysis to find out the inherent factor of heterosis. The genomes of the animals and plants contain a great quantity of CG dinucleotides, and the methylation sequences which control the expression of some functional gene also contain a great quantity of CG dinucleotides, so it is easy to find out the change regularity of the methylation.The advantages of MSAP include simplicity, rapidness, low cost, higher sensitivity, and high polymorphism, however, this technique also has some limitations associated with resolving power.

In some organisms, methylation also occurred in CAGs and CTGs. Furthermore, the types of non-methylation and inner-methylation of a single strand can not be distinguished as both HpaII and MspI are capable of recognizing the sites of non-methylation and inner-methylation of a single strand, thus revealing similar patterns of methylation following PCR amplification. Moreover, outer methylation of double strands can not be detected through MSAP analysis. For these reasons, actual levels of the cytosines methylation are likely to be significantly higher than those detected in this study. Notwithstanding some limits, the results here reported show this technique to be highly efficient for large-scale detection of cytosine methylation in *Scapharca broughtonii*. The ability to isolate and amplify these MSAP fragments may thus make possible direct identification of sequences which play a role during the expression of heterosis.

References

Birchler, J., Auger, D., & Riddle, N. (2003). In search of the molecular basis of heterosis. *The Plant Cell Online, 15*(10), 2236.

Cervera, M., Ruiz-Garcia, L., & Martinez-Zapater, J. (2002). Analysis of DNA methylation in Arabidopsis thaliana based on methylation-sensitive AFLP markers. *Molecular Genetics and Genomics, 268*(4), 543-552.

Chen, X., Ma, Y., Chen, F., Song, W., & Zhang, L. (2009). Analysis of DNA methylation patterns of PLBs derived from Cymbidium hybridium based on MSAP. *Plant Cell, Tissue and Organ Culture, 98*(1), 67-77.

Crow, J. (2000). The rise and fall of overdominance. *Plant breeding reviews, 17*, 225-258.

Cubas, P., Vincent, C., & Coen, E. (1999). An epigenetic mutation responsible for natural variation in floral symmetry. *Nature, 401*(6749), 157-161.

Finnegan, E., Peacock, W., & Dennis, E. (2000). DNA methylation, a key regulator of plant development and other processes. *Current opinion in genetics & development, 10*(2), 217-223.

Gibson, G., & Weir, B. (2005). The quantitative genetics of transcription. *TRENDS in Genetics, 21*(11), 616-623.

Grewal, S., & Elgin, S. (2002). Heterochromatin: new possibilities for the inheritance of structure. *Current opinion in genetics & development, 12*(2), 178-187.

Grewal, S., & Klar, A. (1996). Chromosomal inheritance of epigenetic states in fission yeast during mitosis and meiosis. *Cell, 86*(1), 95-102.

Horváth, E., Szalai, G., Janda, T., Páldi, E., Rácz, I., & Lásztity, D. (2002). Effect of vernalisation and azacytidine on the DNA methylation level in wheat (Triticum aestivum L. cv. Mv 15). *Acta Biologica Szegediensis, 46*(3-4), 35-36.

Jacobsen, S., Sakai, H., Finnegan, E., Cao, X., & Meyerowitz, E. (2000). Ectopic hypermethylation of flower-specific genes in Arabidopsis. *Current Biology, 10*(4), 179-186.

Lamkey, K., & Edwards, J. (1997). *The quantitative genetics of heterosis.*

Leonardi, A., Damerval, C., Hebert, Y., Gallais, A., & Vienne, D. (1991). Association of protein amount polymorphism (PAP) among maize lines with performances of their hybrids. *TAG Theoretical and Applied Genetics, 82*(5), 552-560.

Li, X., Wei, Y., Nettleton, D., & Brummer, E. (2009). Comparative gene expression profiles between heterotic and non-heterotic hybrids of tetraploid Medicago sativa. *BMC Plant Biology, 9*(1), 107.

Li, Z., Luo, L., Mei, H., Wang, D., Shu, Q., Tabien, R., ... Khush, G. (2001). Overdominant epistatic loci are the primary genetic basis of inbreeding depression and heterosis in rice. I. Biomass and grain yield. *Genetics, 158*(4), 1737.

Lippman, Z., & Zamir, D. (2007). Heterosis: revisiting the magic. *TRENDS in Genetics, 23*(2), 60-66.

Lu, G., Wu, X., Chen, B., Gao, G., Xu, K., & Li, X. (2006). Detection of DNA methylation changes during seed germination in rapeseed (Brassica napus). *Chinese Science Bulletin, 51*(2), 182-190.

Luff, B., Pawlowski, L., & Bender, J. (1999). An inverted repeat triggers cytosine methylation of identical sequences in Arabidopsis. *Molecular Cell, 3*(4), 505-511.

Luo, L., Li, Z., Mei, H., Shu, Q., Tabien, R., Zhong, D., ... Paterson, A. (2001). Overdominant epistatic loci are the primary genetic basis of inbreeding depression and heterosis in rice. II. Grain yield components. *Genetics, 158*(4), 1755.

Meng, F., Ni, Z., Wu, L., & Sun, Q. (2005). Differential gene expression between cross-fertilized and self-fertilized kernels during the early stages of seed development in maize. *Plant science, 168*(1), 23-28.

Monk, M., Boubelik, M., & Lehnert, S. (1987). Temporal and regional changes in DNA methylation in the embryonic, extraembryonic and germ cell lineages during mouse embryo development. *Development, 99*(3), 371.

Neves, N., Heslop-Harrison, J., & Viegas, W. (1995). rRNA gene activity and control of expression mediated by methylation and imprinting during embryo development in wheat x rye hybrids. *TAG Theoretical and Applied Genetics, 91*(3), 529-533.

Paszkowski, J., & Whitham, S. (2001). Gene silencing and DNA methylation processes. *Current opinion in plant biology, 4*(2), 123-129.

Penterman, J., Zilberman, D., Huh, J., Ballinger, T., Henikoff, S., & Fischer, R. (2007). DNA demethylation in the Arabidopsis genome. *Proceedings of the National Academy of Sciences, 104*(16), 6752.

Rangwala, S., & Richards, E. (2004). The value-added genome: building and maintaining genomic cytosine methylation landscapes. *Current opinion in genetics & development, 14*(6), 686-691.

Reif, J., Warburton, M., Xia, X., Hoisington, D., Crossa, J., Taba, S., ... Melchinger, A. (2006). Grouping of accessions of Mexican races of maize revisited with SSR markers. *TAG Theoretical and Applied Genetics, 113*(2), 177-185.

Reyna-Lopez, G., Simpson, J., & Ruiz-Herrera, J. (1997). Differences in DNA methylation patterns are detectable during the dimorphic transition of fungi by amplification of restriction polymorphisms. *Molecular and General Genetics MGG, 253*(6), 703-710.

Riddle, N., & Richards, E. (2002). The control of natural variation in cytosine methylation in Arabidopsis. *Genetics, 162*(1), 355.

Romagnoli, S., Maddaloni, M., Livini, C., & Motto, M. (1990). Relationship between gene expression and hybrid vigor in primary root tips of young maize (Zea mays L.) plantlets. *TAG Theoretical and Applied Genetics, 80*(6), 769-775.

Sardana, R., O'Dell, M., & Flavell, R. (1993). Correlation between the size of the intergenic regulatory region, the status of cytosine methylation of rRNA genes and nucleolar expression in wheat. *Molecular and General Genetics MGG, 236*(2), 155-162.

Soppe, W., Jacobsen, S., Alonso-Blanco, C., Jackson, J., Kakutani, T., Koornneef, M., & Peeters, A. (2000). The late flowering phenotype of fwa mutants is caused by gain-of-function epigenetic alleles of a homeodomain gene. *Molecular Cell, 6*(4), 791-802.

Sun, Q., Wu, L., Ni, Z., Meng, F., Wang, Z., & Lin, Z. (2004). Differential gene expression patterns in leaves between hybrids and their parental inbreds are correlated with heterosis in a wheat diallel cross. *Plant science, 166*(3), 651-657.

Tariq, M., & Paszkowski, J. (2004). DNA and histone methylation in plants. *TRENDS in Genetics, 20*(6), 244-251.

Tsaftaris, A., Kafka, M., Polidoros, A., & Tani, E. (1997). *Epigenetic changes in maize DNA and heterosis.*

Tsaftaris, S. (2006). Molecular aspects of heterosis in plants. *Physiologia plantarum, 94*(2), 362-370.

Użarowska, A., Keller, B., Piepho, H., Schwarz, G., Ingvardsen, C., Wenzel, G., & Lübberstedt, T. (2007). Comparative expression profiling in meristems of inbred-hybrid triplets of maize based on morphological investigations of heterosis for plant height. *Plant Molecular Biology, 63*(1), 21-34.

Vos, P., Hogers, R., Bleeker, M., Reijans, M., Lee, T., Hornes, M., ... Kuiper, M. (1995). AFLP: a new technique for DNA fingerprinting. *Nucleic Acids Research, 23*(21), 4407.

Wan, Y. (2008). Study on the Relationship beween DNA Methylation and Heterosis in Beef Cattle. *Master paper, Southwest University,Chongqing,China(in Chinese)*, 25-33.

Xiao, J., Li, J., Yuan, L., & Tanksley, S. (1995). Dominance is the major genetic basis of heterosis in rice as revealed by QTL analysis using molecular markers. *Genetics, 140*(2), 745.

Xiong, L., Xu, C., Saghai Maroof, M., & Zhang, Q. (1999). Patterns of cytosine methylation in an elite rice hybrid and its parental lines, detected by a methylation-sensitive amplification polymorphism technique. *Molecular and General Genetics MGG, 261*(3), 439-446.

Xu, M., Li, X., & Korban, S. (2000). AFLP-based detection of DNA methylation. *Plant Molecular Biology Reporter, 18*(4), 361-368.

Yu, S., Li, J., Xu, C., Tan, Y., Gao, Y., Li, X., ... Maroof, M. (1997). Importance of epistasis as the genetic basis of heterosis in an elite rice hybrid. *Proceedings of the National Academy of Sciences of the United States of America, 94*(17), 9226.

Zhang, X., Shiu, S., Cal, A., & Borevitz, J. (2008). Global analysis of genetic, epigenetic and transcriptional polymorphisms in Arabidopsis thaliana using whole genome tiling arrays. *PLoS Genetics, 4*(3).

Zhao, X., Chai, Y., & Liu, B. (2007). Epigenetic inheritance and variation of DNA methylation level and pattern in maize intra-specific hybrids. *Plant science, 172*(5), 930-938.

Zilberman, D., & Henikoff, S. (2007). Genome-wide analysis of DNA methylation patterns. *Development, 134*(22), 3959.

18

Architecture of Bacterial Promoters;
The case of the *Escherichia coli ogt* Promoter

Aida Ansarikaleibari[1]

[1] School of Biosciences, University of Birmingham, United Kingdom

Correspondence: Aida Ansarikaleibari, School of Biosciences, University of Birmingham, United Kingdom.
E-mail: aida.ansari85@gmail.com

Abstract

All bacteria utilize RNA polymerase enzyme and transcription activator proteins to regulate gene expression in response to internal or external stress. Some bacterial promoters are regulated with only one transcription factor whilst two or more transcription activators regulate some other promoters. NarL is a transcription activator protein that activates the *E. coli yeaR* and *ogt* promoters in response to nitrate and nitrite induction in absence of oxygen. In the present study we have studied *ogt1052* promoter, which is a derivative of *ogt* promoter containing only one NarL binding site very close to -35 element. Therefore, it is considered as class II activator dependent promoter just as *yeaR* promoter. A molecular structure of *ogt1052* promoter was proposed which suggests that NarL binding site is located in opposite face of DNA that contains α-CTD and sigma domain 4 of RNA polymerase enzyme required for promoter recognition. The aim of the present study was to study and test the suggested molecular model by creating point mutations at -35 element and deletion of one base pair in spacer region, to test whether sigma domain 4 is necessary to bind -35 hexamer in order to start transcription initiation, and to test whether NarL activates the promoter by interaction with α-CTD in the opposite face of the DNA. Based on the result achieved, *ogt1052* promoter is a class I promoter "dressed" like a class II promoter.

Keywords: bacterial gene expression regulation, RNA polymerase enzyme, *ogt* promoter, two- component regulatory systems

1. Introduction

1.1 RNA Polymerase Recognition of Bacterial Promoters

All bacteria have DNA dependent multi subunit RNA polymerase enzyme that is essential for bacteria to be able to do transcription when necessary (Naryshkin et al., 2000). RNA polymerase (RNAP) is present within the bacteria in two different compositions, core enzyme and holoenzyme. The core enzyme that is not able to start transcription has a conserved subunit composition of (α_2 β β' ω) with a molecular mass of approximately 400 KDa. However, when bacterial specific initiation factor, σ, binds to the core enzyme, it converts the core enzyme to holoenzyme which is capable to recognize specific sequences of DNA called promoters, and therefore it starts transcription in response to specific stress (Borukhov & Lee, 2005). The α subunit of RNAP is composed of two independently folded domains, α-CTD and α-NTD, that are linked by a flexible linker that is 20 amino acids long. The β and β' subunits are large subunits that are assembled by amino terminal domain of α subunit (α-NTD) and are known as the active site of RNAP enzyme that has locations for DNA templates. The ω subunit is a small subunit that seems to assist the folding of the β' subunit and it has no direct role in transcription (Browning et al., 2004).

The σ subunit or factor consists of four different domains (σ1, σ2, σ3, and σ4) joint by linkers that are responsible for promoter recognition (Murakami et al., 2003; Browning et al., 2004). RNAP recognizes promoters from four different sequences. Two principal sequences are the -35 hexamer (consensus sequence of TTGACA) and -10 hexamer (with consensus sequence of TATAAT), which are located at position 35 and 10 base pairs upstream (before) from the transcription start site respectively. The other two sequences are the UP element (20 base pare long) and extended -10 element (3-4 base per) that are located upstream of -35 hexamer and -10 hexamer, respectively. The hexamers are separated by spacer region, which is a nonspecific DNA sequence and is usually 16-18 base pairs long (Browning, et al., 2004; Borukhov and Lee, 2005). The α subunit of RNAP is responsible to recognize the UP element while Domain 3 of σ is responsible to recognize and bind to the extended -10 element; -10 element is recognized by σ domain 2 and -35 element is recognized by σ domain 4.

In prokaryotes, there are more than one σ factor. For instance, *E. coli* has one main σ factor known as σ^{70} and six other σ factors responsible to bind to RNAP to response to specific stress. Two of these specific response σ factors are σ^H and σ^E that bind to RNAP in response to heat shock stress in cytoplasm and periplasm respectively. The σ subunit ensures the recognition of promoter sequences and locates the RNAP holoenzyme to the recognized sequences and enhances the unwinding of the DNA duplex (Ishihama, 2000; Yura et al., 1999; Browning and Busby, 2004).

1.2 Regulation of Bacterial Gene Expression

Most of the bacteria are versatile organisms that are able to prosper in different environmental conditions. These organisms contain large genetic information that encodes mechanisms necessary for bacterium to cope with variety of challenges. Gene expression in bacteria is expensive, in terms of ATP consumption; therefore, the process is controlled in a way that prevents wasteful synthesis of unnecessary materials. This can be an explanation of why different bacteria produce different numbers of proteins in order to survive in variety of conditions (Dale & Park, 2004). Bacteria produce proteins by making messenger RNA (mRNA) copies using RNAP. mRNA carries the message that is translated into amino acids for producing specific proteins. Transcription is the first step in gene expression, which has three main stages, initiation, elongation and termination. In the initiation stage, RNAP holoenzyme is bound to -35 and -10 hexamers and forms the closed RNAP- promoter complex. Next, the DNA becomes unwound, forming a bubble, by isomerization in which the non-template DNA binds to σ2 of RNAP. This results in formation of a stable open complex. Short RNA products are synthesized and released in presence of NTPs and after, elongation stage starts. In elongation stage, the conformational change of RNAP-promoter open complex leads to loss of RNAP-promoter contact (σ factor dissociates).

Transcription factors are involved in regulation of gene expression at initiation stage. In some promoters, one transcription factor is required in order to activate the promoter (simple promoters), whereas, some promoters are complex and need two or more activators to start transcription initiation. There are three types of promoters being activated by three different mechanisms; Class I, class II and class III activator dependent promoters. In class I activator dependent promoters, there is only one binding site for the activator, upstream of -35 element and therefore, only one transcription factor interacts directly with α-CTD, recruiting RNAP to promoter. A good example of this type of promoter is the *E. coli lac* promoter that is activated by CRP (CAMP receptor protein) in response to lactose availability. In class II activator dependent promoters, there is only one binding site for the activator but located at roughly position -41.5 very close to -35 element and in some promoters, binding site overlaps with -35 element. In such promoters the activator binds to the sequence that overlaps with -35 and can contact to σ4, resulting the recruitment of RNAP to the promoter. For instance, bacteriophage λ PRM promoter is activated by λ CI protein that binds to σ4 of RNAP. In some of the class II activator dependent promoters, activator contacts to α-NTD, while binding to the sequence that overlaps the -35 element. In class III activator dependent promoters, there are two binding sites for the transcription factors, which activate the promoter in response to multiple signals. One site is located upstream of -35 element and the other binding site is positioned very close to -35 element and sometimes overlaps with the element. There are two mechanisms by which promoter is activated by class III mechanism: the first one is when one activator functions as class II and the other one functions as class I. the best example can be the *proP* P2 promoter in which CRP and fis activate the promoter as class I and class II respectively. The second mechanism is when both activators function as class I promoter in both activate promoter-binding sites. For instance, *acs* P2 in which CRP activates the promoter by binding to α-CTD and functions as class I (Busby et al., 1996; Scott et al., 1995; McLeod et al., 2002; Beatty et al., 2003; Browning & Busby, 2004).

1.3 Responses to Nitrate and Accompanying Stress (NarL; NarP; NsrR; NorR)

In pathway of nitrate reduction, reactive nitrogen species (RNS) are produced from which the bacteria need to be protected. Transcription regulators such as NarL, NarP, NsrR and NorR are involved in gene regulation in *E. coli* in response to RNS (Lin & Stewart, 2010; Hartig et al., 1999). A two-component system named the Nar system in *E. coli* is composed of four regulatory proteins, NarX and NarQ which are receptors located in the cell membrane and NarL and NarP which are the response regulators located in the cytoplasm of the *E. coli*. Together, these proteins activate the transcription of genes that are needed in anaerobic respiration condition and presence of nitrate and nitrite that are electron acceptors in the absence of oxygen. NarX and NarQ are activated by extracellular nitrate or nitrite and create a signal by phosphorylation. In turn, phosphorylation causes conformational change in NarL and NarP respond regulators that causes the interaction of NarL and NarP with Nar-regulated promoters to regulate transcription of genes positioned downstream of promoter (Zhang et al., 2003; Lin & Stewart, 2010; Egan & Stewart, 1991; Noriega et al., 2010). NsrR and NorR are activated in response to cytoplasmic nitric oxide (NO) in order to repress transcription. NsrR is a global regulator while NorR is a specific regulator that interacts only by σ^{54} and controls the transcription of *norVW* gene. The product of this gene is an enzyme, which is required in cell for NO detoxification. Another enzyme which is produced by *hmp* gene that is activated by NsrR has a key role in cytoplasmic NO detoxification as well (Patridge et al., 2009). NarL and NarP sense the extracellular RNS and anticipate the DNA or cellular damage. Therefore, they are activated to protect the cell from any upcoming damage. In contrast, NsrR and NorR sense the cytoplasmic NO and act when NO is inside the cell and cell is partially damaged and needs detoxification (Lin & Stewart, 2010; Patridge et al., 2009).

1.4 The ogt Promoter and the ogt1052 Model

The *E. coli ogt* promoter encodes O^6 alkylguanine DNA alkyltransferase (a DNA repair protein) in response to DNA damage caused by reactive nitrate species. This promoter is activated by NarL and NarP transcription factors in absence of oxygen and in presence of nitrate and nitrite (Squire et al., 2009). Most of the promoters regulated by NarL and NarP are regulated by FNR "the master anaerobic regulator" as well. However, studies have shown that NarL and NarP can activate *ogt* promoter independent of FNR (Constantinidou et al., 2006). *ogt* promoter contains two 7-2-7 NarL binding sites centered at positions -78.5 and -45.5. Both locations of NarL in *ogt* promoter must be occupied with NarL proteins in order to activate the promoter. According to this fact, *ogt* promoter is known as a class III activator dependent promoter (Squire et al., 2009).

It was reported that *yeaR* promoter which is a class II dependent promoter and contains only one NarL binding site centered at position -43.5, was expressed 5 folds greater than *ogt* promoter. This indicates that NarL binding site in *yeaR* promoter corresponds better to the consensus sequence (Squire et al., 2009). Based on this finding, it was suggested that *ogt* promoter with single NarL binding site, which corresponds better to the consensus sequence, can have a higher activity compare to the *ogt* promoter with two weak NarL binding sites. Therefore, two *ogt* promoter mutants were constructed by disrupting one NarL binding site and optimizing the other one thus it corresponds better to the consensus sequence. The derivatives were introduced each containing single NarL binding site, one centered at position -78.5 named *ogt1041* and the other one centered at position - 45.5 named *ogt1052*. *ogt1041* functions as class I activator dependent promoter whilst *ogt1052* is known to function as class II activator dependent promoter. It was reported that the activity of *ogt1052* is much higher than the activity of *ogt1041*, which confirms the fact that an activator-binding site located close to the core promoter functions more effectively (Chismon, 2010).

In the present study we have tested this model in order to investigate whether σ4 of RNAP is essential for promoter recognition in *ogt1052* promoter and if NarL dimer interacts with α-CTD in order to activate the promoter in the opposite face of DNA. We have done point mutations at -35 element to test the interaction of σ4 with -35. We have created a mutant of *ogt1052* containing 17bp spacer and compared it with the *yeaR100* promoter that also contains 17bp spacer to test if the mutant has the same promoter activity as the *yeaR100* promoter has. *yeaR* promoter is a class II activator dependent promoter. The NarL binding site in *yeaR* promoter overlaps with -35 element. By deleting 1bp of *ogt1052* spacer, in case *ogt1052* is a class II activator, binding site should overlap -35 element and the activity of both promoters will be nearly the same.

2. Methods

2.1 Bacterial Strains, Plasmids and Promoter Fragments

Escherichia coli K-12 strains and promoter fragments used in this study. Each strain was streaked on an agar plate containing lactose and relevant antibiotic, and was incubated in 37°C with 200rpm shaking in a conical flask containing 2XLB and relevant antibiotics. At all stages of the bacterial growth, aseptic technique was applied. Plasmid used in this study is pRW50, a low-copy number lac expression vector that clones the EcoRI-HindIII fragments as transcriptional fusion to lacZ. The plasmid encodes tetracycline resistance and was used in all β-galactosidase assays.

2.2 Bacterial Media, Solutions, Buffers, Reagents and Antibiotics

Specified amount of bacterial media as described below were dissolved in distilled water and autoclaved for 20 minutes at 120　C, 1 atmosphere.

LB (Luria Broth): It was purchased from Sigma. It contained 10g/L tryptone, 5 g/L yeast extract and 5 g/L NaCl. 1X LB was prepared by dissolving 20 g powder in 1 L of distilled water.

LB agar: It was purchased from Oxoid. LB agar was prepared by dissolving 28 g powder in 1 L of distilled water.

MacConkey lactose agar: MacConkey lactose agar was prepared by dissolving 50 g powder in 1 L of distilled water. It was purchased from Difco.

Table 1 Bacterial strains and promoter fragments used in the study

Strain	Genotype
JCB387	*nir, lac* (Page et al., 1990)
JCB3875	*narP* derivative of JCB387 (Page et al., 1990)
JCB3883	*narL* derivative of JCB387 (Page et al., 1990)
JCB3884	*narLP* derivative of JCB387 (Page et al., 1990)
Promoter fragment	Description
Ogt1052	Derivative of the *ogt100* promoter fragment carrying mutations at -85,-63,-51,-48,-44 and -42.
Ogt1052A1	Derivative of the *ogt1052* promoter fragment carrying mutations at position-36.
Ogt1052A2	Derivative of the *ogt1052* promoter fragment carrying mutations at position -35.
Ogt1052A3	Derivative of the *ogt1052* promoter fragment carrying mutations at position -28.
yeaR100	*E. coli yeaR* promoter fragment carrying nucleotide sequences from -294 to +96.

MS (Minimal Salts) medium: 1 l MS medium contained 4.5 g KH_2PO_4, 10.5 g K_2HPO_4, 1 g $(NH_4)_2SO_4$, 0.5 g Tri-Sodium Citrate, 0.05 g $MgSO_4.7H_2O$, 1 ml *E.Coli* Sulphur-free salt (82 g $MgCl_2.7H_2O$, 10 g $MnCl.4H_2O$, 4 g $FeCl_2.6H_2O$, 1 g $CaCl_2.6H_2O$, 2o ml HCL per 1 l of distilled water), 1 ml sodium selenate (1mM), 1 ml ammonium molybdate (1 mM), 0.1 g caseine digest, 0.05 g yeast digest, 0.32 g sodium fumarate, 8 ml 50% glycerol.

SOC medium: It was purchased from Sigma. It contained 2% (w/v) tryptone, 0.5% (w/v) yeast extract, 8.6 mM NaCl, 3.5 mM KCl, 20mM $MgSO_4$, 20 mM glucose.

All the solid and liquid autoclaved bacterial media were supplemented by tetracycline during the study. Antibiotic was added to the media after it had cooled down enough. Tetracycline was prepared by dissolving 20 mg of tetracycline stock in 1 ml of 100% methanol.

2.3 Measurements of Promoter Activity

β-galactosidase activity assays: β-galactosidase was measured using the Miller protocol. β-galactosidase assay was used to observe the activity of promoter in *vivo* by measuring the amount of β-galactosidase within cell extract. The promoter of interest was cloned in to pRW50 containing *lacZ* gene and was prepared for the assay. Assay was done with two sets of capped tubes containing minimal media only and minimal media supplemented with 20 mM sodium nitrate as final concentration in anaerobic condition. Bacteria containing the cloned promoter were left in 37°C and optical density (OD) was checked until it reached to 0.5 in 650 nm. Tubes were kept in ice in case of having rapid growth compare to other tubes and when all the tubes reached to the OD between 0.480-0.500, 2 ml of each cultures were vortexed with 30 l 1% sodium deoxycholate and 30 l toluene in order to lysate. While prepared tubes were shaking for 20 minutes, assay tubes containing z buffer were warmed at 30 C. After that, 100 l of the lysate was added and mixed in the assay tubes. The reaction was stopped by adding 1M sodium carbonate after yellow color developed. Then, OD 420 was measured and used in the formula to calculate promoter activity.

2.4 molecular Biology Methods

Agarose gel electrophoresis: TBE buffer (0.5X) containing agarose powder was boiled in microwave oven until totally dissolved. The agarose concentration was 0.8% for separating large fragments like plasmids and 1%-1.5% concentration for separating small fragments like Polymerase Chain Reaction (PCR) products. Sybrsafe™ 10,000X (Invitrogen) was added to make the final concentration of 1X before the gel was poured. DNA samples were mixed with 6X Green DNA loading dye (Fermentas) to make a final concentration of 1X. The DNA samples were run together with DNA ladder such as 100 bp ladder or 2-log ladder (NEB). Electrophoresis was done at 100V in 0.5X TBE buffer for about 30 minutes or when fragments separated adequately for observation.

Plasmid isolation: pRW50 is the plasmid used in this study and was isolated through QIA prep Mini-prep Spin Kit following provided protocol from the manufacturer. The day before isolation, a 5 ml 2x LB containing tetracycline culture was inoculated by a single colony containing pRW50 and was shake overnight in 37°C and before isolation cell pellet was collected by centrifugation at 4,000 rpm for 1minute.

DNA purification: In order to gain concentrated plasmid DNAs, the equal volume of phenol/chloroform was added to the sample. The mixture was vortexed until the emulsion was formed then it was centrifuged at 13,000 rpm for 1 minute to separate organic components from the aqueous phase. The aqueous phase was taken to a new micro centrifuge tube. In order to obtain more DNA, TE buffer was added to the organic phase followed by vortex and re-centrifugation. The upper aqueous phase was taken and combined with the previous one (final volume ~400 µl). The next step was alcohol precipitation. The process removes chloroform traces and precipitates concentrated DNA. 44 µl of 3M sodium acetate (pH 5.2), 4 µl of 1M magnesium chloride and 888 µl of iced cold (-20°C) 100% ethanol was added. The DNA sample was incubated for 15 minutes at -80°C. The sample was centrifuged at 14,000 rpm (4°C) for 15 minutes. The supernatant was removed and the pellet was suspended again by 70% iced cold ethanol before re-centrifugation. Supernatant was discarded and the DNA pellet was vacuum-dried for 10 minutes to get rid of ethanol. The last step was to re-suspend the pellet with 50 µλ TE Buffer. The more convenient way to clean up DNA was using QIAquick PCR purification kit. In appropriate pH and high salt condition, the DNA bound to the silica membrane of the spin-column while proteins and free nucleotides could not be absorbed thus passed through the column. The same protocol as described by the manufacturer was applied when DNA gel extraction was done by QIAquick gel extraction kit.

Ligation for DNA recombination: DNA ligation was done to insert the desired DNA fragment into the pRW50 vector. T4 DNA ligase (NEB) was used to connect 5′ phosphate and 3′ hydroxyl groups by forming phosphodiester bonds. The protocol involves 2 µl of dephosphorylated plasmid vector, 1   l of T4 DNA ligase, 2 µl of 10X T4 ligase buffer and 10 to 15 µl of inserted DNA. The total volume was made to 20 µl using sterile water. The mixture was left incubate at room temperature for at least 2 hours until transformation. Also, the control was done using the same protocol but without inserted DNA.

Transformation of DNA: In order to do transformation, competent cells were prepared according to the protocol below:

Five ml of 2X LB media containing tetracycline was inoculated with a single colony picked from the agar plate. The culture was incubated shaking overnight at 37°C. The next day, 100 ml of 2X LB media within 500 ml conical flask was inoculated with 1 ml of the overnight culture. The flask was incubated shaking until the optical density at 650 nm (OD 650) reached around 0.4-0.5. The flask was incubated on ice for another 10 minutes then the content was transferred to two 50 ml centrifuge tubes in order to centrifuge at 4,000 rpm at 4°C for 9 minutes to collect the cell pellet. The supernatants in both tubes were discarded. Thirty ml of TFB1 buffer was added to one tube to re-suspend the pellet then the whole content was transferred to the second tube. After re-suspension, the tube was placed in ice for at least 90 minutes. Centrifugation was done as before and the supernatant was discarded. The pellet was re-suspended with 4 ml of TFB2 buffer. Aliquots of 300 µl were prepared and immediately frozen at 80°C.

Transformation method: Plasmids from mini-prep or ligation reaction were transformed to competent cells. Started by taking competent cells from -80°C and placing them in ice to allow cells to defrost, 40 µl of thawed cells were transferred to the prepared micro-centrifuge tubes containing 5- 10 µl of ligation mix or 1 µl in case of plasmids gained from mini-prep. The tubes were placed in ice for 30 minutes. Heat-shock was done by incubating the tubes in 42°C water bath for 42 seconds before immediately placing the tubes back in ice. Heat-shock made cell walls permeable for plasmids to enter to the cells easily. After 5 minutes, 250 µl of room temperature SOC broth (Sigma) was added to the transformed cells. The tubes were incubated shaking at 37°C for 30 minutes. Finally, 100 µl of the transformation reaction was plated on appropriate media agar. Plates were incubated overnight at 37°C.

DNA restriction: Digestion of DNA was done using <5% (v/v) restriction enzyme with 10X restriction buffer provided with the enzyme. Plasmid DNA was digested at 37°C for at least 3 hours to get plasmid totally cut while PCR product digestion was done for less than 1 hour. After complete digestion, DNA purification was done using PCR purification kit (QIAquick). Then, the required DNA fragments were collected by gel extraction after analyzing by gel electrophoresis. Multiple enzymatic cuts at the same time needed a control digestion in order to check for complete digestion. Prior to digestion, 2µl of DNA-buffer was taken out. The DNA-buffer mixing was separated to two tubes and enzyme was added to each tube. 2 µl of each was taken out. Then, the two tubes were mixed together. After digestion, another 2 µl of DNA-buffer was taken. All control samples were run by gel electrophoresis to compare and ensure that the DNA was cut completely.

DNA vector dephosphorylation: During preparation of the plasmid vector, calf intestinal phosphatase (CIP, from NEB) was additionally added after restriction digestion to prevent re-ligation by removing 5′ phosphates. DNA vector without 5′ phosphates cannot perform re-circularization, thus increasing cloning efficiency and decreasing background problems. CIP is active in NEBuffer 3, which gives the condition of 100mM NaCl, 50 mM Tris-HCl, 10 mM $MgCl_2$ (pH 7.9 at 25°C). It also works well in NEBuffer 2, 4, NEBuffer for EcoRI and BamHI. Preparation of plasmid vector was done by growing 6 cell cultures of JCB387 containing the pRW50 plasmid overnight in 5 ml rich media. After centrifugation at 4,000 rpm for harvesting, the plasmid was extracted separately by mini-prep

(QIA-prep mini-prep Spin Kit). EB buffer was used to elute the plasmid to get final 50   l volume of each, thus total 300 μl was obtained when added altogether. Digestion with appropriate restriction enzymes was done followed by adding 5 μl of CIP to the mixture. The plasmid vector was purified by QIA-quick PCR purification kit after incubating at 37°C for at least 30 minute.

Polymerase Chain Reaction: PCR is a method using DNA polymerase and two oligonucleotides as primers to amplify DNA template. All primers used in this study were synthesized by Alta Bioscience, University of Birmingham; UK. The enzyme used was polymerase (NEB), which contains 5′-3′ DNA polymerase activity and 3′-5′ exonuclease activity. If proof-reading was not required for PCR, Taq polymerase (Bioline) was used instead. The protocol of 50 μl of PCR reaction using Phusion as DNA polymerase contained 10 μl of 5X reaction buffer, 0.1 μM of each primers, 1μl of the template (e.g. plasmid DNA), 4 μl of 2.5 mM dNTP mix, 0.5 μl of Phusion and 15.5 μl SDW. The program was set at 98 °C, 1.30 minutes for initial denaturation. The 30 cycles were set at 98°C, 10 seconds for denaturation; 40-60°C, 20 seconds for annealing depending on annealing temperature of primers used; 72°C, 15 seconds per kb for extension. The final stage was set at 72°C for 5 minutes then immediately cooled down to 4 °C.

2.5 Site-Directed Mutagenesis

In this study, *ogt1052A1* and *ogt1052A2* fragments were constructed by point mutation using mega-primer PCR method. Fragments were cloned to pRW50 plasmid and were synthesized in the first-round PCR. In the first-round PCR, D10527 flanking primer was used with the mutagenic primer. For the second-round PCR, 10-15 μl of mega-primer products of the first-round PCR purified from gel were used as a primer of the second-round PCR together with primer D10520. This step created a full length fragment with mutation at the desired point. After PCR purification, the PCR products were digested with EcoRI and HindIII, and then cloned into pRW50 vector by ligation method that is explained above.

Deletion of DNA using overlapping PCR product: In this study, *ogt1052A3* fragment was constructed with overlapping PCR. Two PCR products were generated separately in the first-round PCR. One PCR used the upstream D10520 flanking primer while the downstream primer (f1052A6) was designed to carry sequence-carrying deletion. Another PCR used the downstream D10527 primer with the upstream primer (e1052A6) carrying deletion.

DNA sequencing: Functional Genomics and Proteomics Laboratory, University of Birmingham sequenced Plasmids. Plasmids were mini-prep and stored as stock. Three μl of plasmids and 3.2 μl of D10527 plasmid and 3.8 μl of distilled water were mixed for sequencing.

3. Results

3.1 Activity of the ogt1052 Promoter

The result clearly indicates that in presence of NarL and NarP activators, the *ogt1052* promoter is highly expressed while in absence of these activators the expression of promoter is highly reduced. It can be concluded from the result that presence of NarL and NarP activators is necessary for activating the *ogt1052* promoter. It can be clearly seen from the Figure 1 that *ogt1052* promoter in JCB387 cells is more activated in growth medial supplemented with nitrate in compare to the promoter activity in growth media without nitrate induction. The expression level of the *ogt1052* promoter in JCB3884 cells with or without nitrate induction is the same as the control vector pRW50. Therefore, comparing the activity of *ogt1052* promoter in JCB387 with promoter activity in JCB3884, it can be concluded that nitrate induction results in high expression level of the *ogt1052* promoter only in the case that NarL and NarP are present in the cell. Based on this result, *ogt1052* is an activator dependent promoter that only is activated in response to extracellular nitrate to protect DNA from any damage caused by RNS.

3.2 Mutations in the ogt1052 Promoter -35 Hexamer

An experiment was planned to see the effect of different mutations in RNA polymerase recognition of sigma domain 4 and to check whether sigma domain 4 is necessary for the promoter activation or to see if C- terminal domain of RNA polymerase alpha subunit (α-CTD) can bind to NarL and activate the promoter *ogt1052* instead. Two point mutations were introduced in -35 element of *ogt1052* promoter fragment. The sequence of the -35 element in this promoter is TGGCTG and the consensus sequence for -35 hexamer is TTGACA (Browning & Busby, 2004). The first mutation was done in order to replace one base in -36 position to make the sequence of the hexamer to be more like the consensus sequence (G◊T) and was named *ogt1052A1*. The second mutation was done to replace one base of the -35 hexamer at position -35 to have mutant, which has the sequence that is less like the consensus sequence and is named *ogt1052A2* (G◊T). The mutants were constructed by mega primer PCR method. Based on the results achieved, -35 element plays an important role in promoter activation with or without presence of NarL activator protein. In summary, recognition of -35 hexamer by sigma domain 4 of RNA polymerase is necessary for transcription activation. NarL is not the only factor for promoter activation and *ogt* promoter can be activated even in the absence of NarL with a sequence that corresponds to the consensus sequence of -35 element.

3.3 Spacer Mutation

NarL can activate both ogt and *yeaR* promoters in response to RNS. Therefore, an experiment was planned to create a mutation in spacer region of ogt1052 promoter to measure the activity of both promoters in the presence and absence of NarL. By deletion of one base pair in spacer region of ogt1052 the NarL, binding site was expected to overlap the -35 element and function like *yeaR* promoter. In order to do this experiment, spacer mutation was done by overlapping PCR and the mutant was named ogt1052A3. The *yeaR* 100 promoter fragment was used along in this experiment to compare the activity of ogt1052A3 promoter. The activity of *yeaR* promoter in the presence of NarL is much higher compare to its activity when NarL is not present in the cell. Activity of promoter when NarL is present is higher with nitrate induction compare to the growth condition that is not supplemented by nitrate. This confirms that *yeaR* promoter is activated with NarL in response to RNS. There is a slight difference between the expression level of *yeaR* with nitrate induction and without nitrate induction in JCB3884 (NarL is not present). This can be due to regulation of NsrR that was discussed previously in section 1.3. Following the experiment, the suggested model of *ogt1052* was tested by *ogt1052A3* mutant. The promoter fragment was cloned into pRW50 vector as promoter::*lacZ* fusion and was transformed into JCB387 and JCB 3884 cells. Cells were grown anaerobically in minimal media only and minimal media supplemented with nitrate to the final concentration of 20 Mm. Based on the results achieved, the spacer mutation does not play an important role in the presence of NarL. Also, the results show that the activation of *ogt1052A3* promoter is more like the *yeaR* promoter. This indicated that the spacer mutation only is important and causes activation of *ogt1052A3* promoter more like the *yeaR* promoter only in absence of NarL activator.

4. Discussion

This experiment shows that *ogt* promoter does not contain FNR binding site. Therefore, it is not activated by FNR. Two weak NarL binding sites identified by Squire et al, makes the promoter to function as class III and each derivative of this promoter with only a single NarL binding site, *ogt1042* and *ogt1052*, to function as class I and class II promoters, respectively. These characteristics of ogt promoter make it a good model for studying promoter regulation in three different classes. Generally, we call a promoter class I when the activator binding site is located upstream of -35 element and the activator directly interacts with α-CTD of RNAP subunit and recruits RNAP to start transcription. In a class II promoter, the activator binding site overlaps with -35 element and it contacts sigma domain 4 of RNAP and enhances RNAP promoter recognition. A good example of class II promoter is the *yeaR* promoter in which there is only one NarL binding site overlapping with -35 element. In *yeaR* promoter, there is another sequence that overlaps with NarL binding that belongs to NsrR repressor. NsrR regulates the promoter by sensing the intracellular nitrate or RNS and activating the transcription of genes, which encodes the enzyme necessary for detoxification. This mechanism protects cell from further DNA damage. Whereas in case of *ogt* promoter, NarL/P involves in two-component regulatory system called Nar as response regulator. The system contains two receptors in cell membrane, NarX and NarQ that sense the RNS and conformational change results in NarL binding to its target sequence at promoter. So we can conclude that *ogt* promoter anticipates the DNA damage by RNS while NsrR in *yeaR* promoter is activated when RNS are in the cell and causing cellular or DNA damages to cell.

In conclusion, if we consider *ogt*1052 promoter as a class II promoter, the NarL binding site should overlap with -35 element. In the present study we obtained results that confirms that binding site is not overlapping with -35 element as it showed different activity compare to *yeaR* promoter. Therefore, the *ogt* 1052 is a class I promoter but seems to be class II at first sight. The NarL binding site is located at position -45.5 but in the other face of the DNA. This explains why even though it is close to -35 element but it's not contacting with sigma domain 4. So if it is a class I promoter NarL should directly bind to α-CTD and activate the promoter. The residues involved in this interaction were identified recently. Two important residues in α-CTD are 288/273 and 293 and one, interacting with these residues in NarL dime is 178/179.

It should be mentioned that the result about the sigma domain 4 experiment clearly showed that RNAP is not able to recognize promoter without sigma domain four, Whereas, the promoter can be activated if containing the -35 hexamer that corresponds better to consensus sequence.

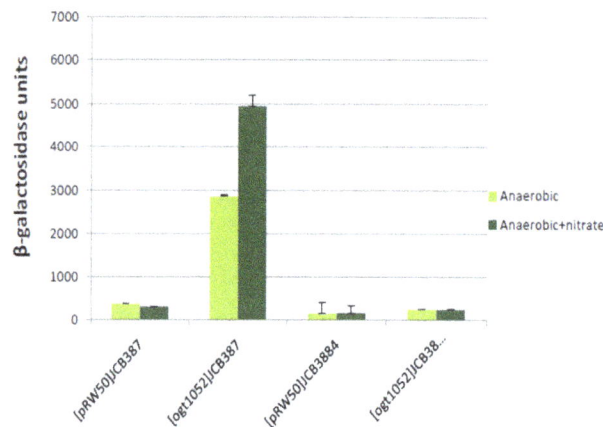

Figure 1. Measured β-galactosidase activity in different E.coli strains

Note. The figure illustrates measured β-galactosidase activity in JCB387 (both NarL and NarP activators are present in the cell) and JCB3884 (neither NarL nor NarP activators are present in the cell) cells carrying pRW50 as control vector and pRW50 containing ogt1052 promoter fragment. Cells were grown in minimal salt media in anaerobic condition and where indicated, sodium nitrate was added to the final concentration of 20 mM. Data shown in figure are averages of three biological repeat for each plasmid. Error bars were put to show the standard deviation from the mean. The activity of ogt1052 promoter in JCB387 is much higher than the activity of the promoter in JCB3884. It shows the necessity of NarL and NarP presence for the activation of ogt1052 promoter. In JCB387 cells, the promoter is highly expressed with nitrate induction while in JCB3884, even though growth media is supplemented by nitrate, the activity is as low as the control vector pRW50. Therefore, the result shows the promoter is highly expressed through nitrate induction only in the presence of NarL/NarP.

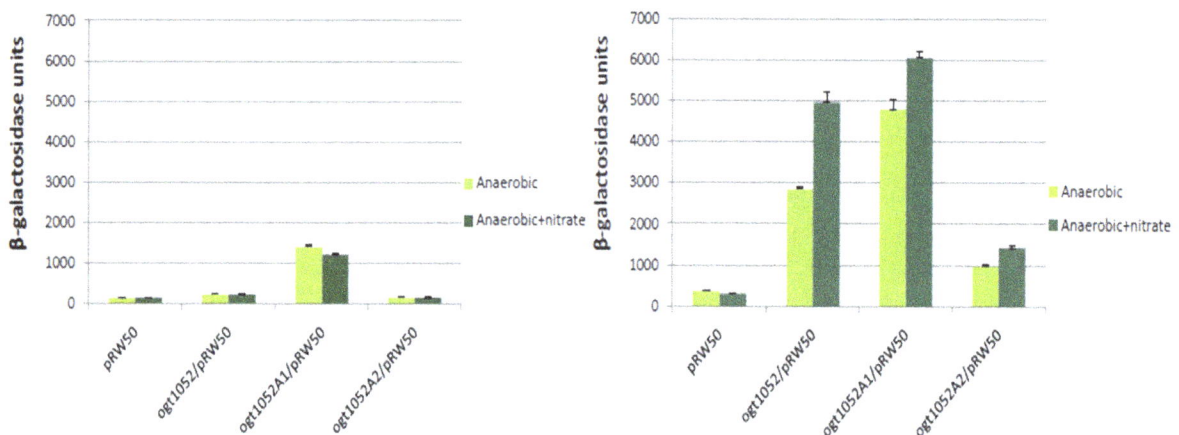

Figure 2. Measured β-galactosidase activity in different E.coli strains containing mutation in -35 element.

Note. The left chart reveals measured β-galactosidase activity in JCB387 (both NarL and NarP activators are present in the cell) cells carrying pRW50 as control vector and pRW50 containing ogt1052, ogt1052A1, *ogt1052A2* promoter fragment respectively. Cells were grown in minimal salt media in anaerobic condition and where indicated, sodium nitrate was added to the final concentration of 20 mM. Data shown in the figure are averages of three biological repeat for each plasmid. Error bars were put to show the standard deviation from the mean. Ogt1052A1 is highly expressed in compare to ogt1052 and ogt1052A2 which shows that -35 hexamer with sequence similar to the consensus sequence can be recognized by RNA polymerase and can activate the promoter. The activation of the promoter in mutant Ogt1052A2 is highly reduced compare to *ogt1052A1* that means NarL is not the only factor, which can activate the promoter.

The right chart shows measured β-galactosidase activity in JCB3884 (neither NarL nor NarP activators are present in the cell) cells carrying pRW50 as control vector and pRW50 containing *ogt1052, ogt1052A1, ogt1052A2* promoter fragment respectively. Cells were grown in minimal salt media in anaerobic condition and where indicated, sodium nitrate was added to the final concentration of 20 mM. Data shown in the figure are averages of three biological repeat for each plasmid. Error bars were put to show the standard deviation from the mean. *ogt1052A1* still is expressed in absence of NarL but the expression is not as high as when NarL is present. The result shows that promoter can be activated by -35 hexamer with near consensus sequence independent from NarL.

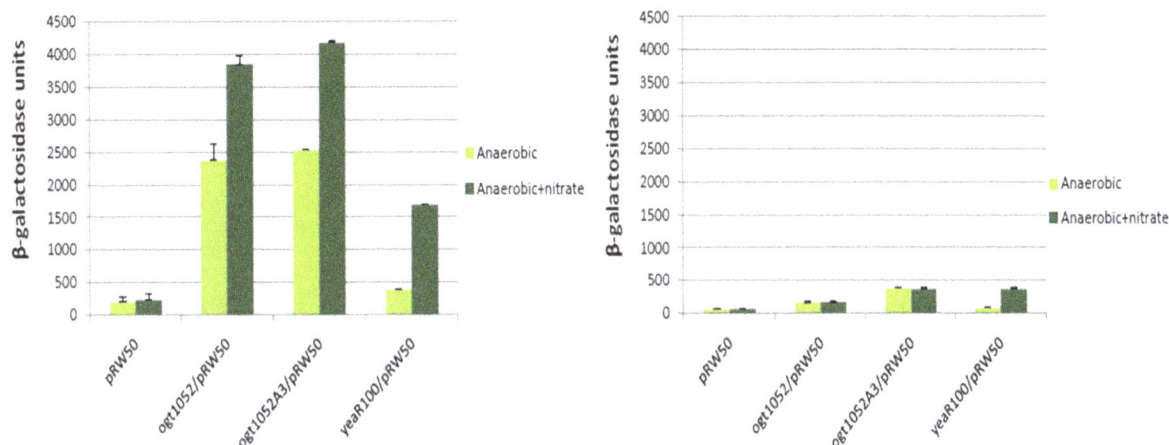

Figure 3. Measured β-galactosidase activity in different E.coli strains containing spacer mutation

Note. The left chart reveals measured β-galactosidase activity in JCB387 (both NarL, and NarP activators are present in the cell) cells carrying pRW50 as control vector and pRW50 containing *ogt1052, ogt1052A3*, yeaR100 promoter fragments, respectively. Cells were grown in minimal salt media in anaerobic condition and where indicated, sodium nitrate was added to the final concentration of 20 mM. Data shown in the figure are averages of three biological repeats for each plasmid. Error bars were put to show the standard deviation from the mean. The activation of *ogt1052A3* promoter is similar to the activation of *ogt1052* promoter, which means that with spacer mutation it still acts like the *ogt1052* promoter than like *yeaR* promoter.

The right chart shows measured β-galactosidase activity in JCB3884 (neither NarL, nor N NarP activators are present in the cell) cells carrying pRW50 as control vector and pRW50 containing *ogt1052, ogt1052A3, yeaR100* promoter fragments, respectively. Cells were grown in minimal salt media in anaerobic condition and where indicated, sodium nitrate was added to the final concentration of 20 mM. Data shown in the figure are averages of three biological repeats for each plasmid. Error bars were put to show the standard deviation from the mean. The activation of *ogt1052A3* promoter is more like the *yeaR* promoter. This indicated that the spacer mutation only is important and causes activation of *ogt1052A3* promoter more like the *yeaR* promoter only in absence of NarL activator.

Acknowledgement

I'm thankful to Prof. Steve Busby for his support in this research. I'm also grateful to Patcharawarin Ruanto and David Lee for their assistance.

References

Beatty, C., Browning, D., Busby, S., & Wolfe, A. (2003). CRP-dependent activation of the Escherichia coil acsP2 promoter by a synergistic class III mechanism. *J. Bacteriol., 185*, 5148-5157.

Bell, A., Gaston, K., Williams, R., Chapman, K., Kolb, A., Buc, H., … Busby, S. (1990). Mutations that affect the ability of the Escherichia coli cyclic AMP receptor protein to activate transcription. *Nucl. Acids Res., 17*, 3865-3874.

Borukhov, S., & Lee, J. (2005). RNA Polymerase structure and function at lac operon. *C.R. Biologies, 328*, 576-587.

Browning, D. F., Beatty, C. M., Sanstad, E. A., Gunn, K. A., Busby, S. J . W., & Wolfe, A. J. (2004). Modulation of CRP-dependent transcription at the Escherichia coli acsP2 promoter by nucleoprotein complexes: anti-activation by the nucleoid proteins FIS and IHF. *Mol. Microbiol., 51*, 241-254.

Busby, S., & Ebright, R. (1997). Transcription activation at class II CAP-dependent promoters. *Mol. Microbiol. 23*, 853-859.

Busby. S., & Ebright, R. (1994). Promoter structure, promoter recognition, and transcription activation in prokaryotes. *Cell, 79*, 743-746.

Busby. S., & Ebright, R. (1999). Transcription activation by catabolite activator protein (CAP). *J. Mol. Biol., 293*, 199-213.

Constantinidou, C., Hobman, J. L., Griffiths, L., Patel, M. D., Penn, C. W., Cole, J. A., & Overton, T. W. (2006). A reassessment of the FNR regulon and transcriptomic analysis of the effects of nitrate, intrite, NarXL, and NarQP as Escherichia coli K12 adapts from aerobic to anaerobic growth. *J. Biol. Chem., 281*, 4802-4815.

Egan, S. M., & Stewart, V. A. L. L. E. Y. (1991). Mutational analysis of nitrate regulatory gene narL in Escherichia coli K-12. *Journal of bacteriology, 173*(14), 4424-4432.

Gaston, K, Bell, A., Buc, H., & Busby, S. (1990). Stringent spacing requirements for transcription activation by CPR. *Cell, 62*, 733-740.

Hartig, E., & Zumft, W. G. (1999). Kinetics of nirS expression (cytochrome cd1 nitrite reductase) in Pseudomonas stutzeri during the transition from aerobic respiration to denitrification: evidence for a denitrification-specific nitrate and nitrate-responsive regulatory system. *J. Bacteriol., 181*, 161-166.

Ishihama, A. (2000). Functional modulation of Escherichia coli RNA polymerase. *Annu. Rev. Microbiol., 54*, 499-518.

Joung, J., Le, L., & Hochschild, A. (1993). Synergistic activation of transcription by Escherichia coli cAMP receptor protein. *Proc. Natl Acad. Sci. USA, 90*, 3083-3087.

McLeod, S. M., Aiyar, S. E., Gourse, R. L., & Johansen, R. C. (2002). The C-terminal domains of the RNA polymerase alpha subunits: Contact site with Fis and localization during co-activation with CRP at the Escherichia coli prop P2 promoter. *J. Mol. Biol., 316*, 517-529.

Naryshkin, N., Revyakin, A., Kim, Y., Mekler, V., & Ebright, R. H. (2000). Structural organization of the RNA polymerase-promoter open complex. *Cell, 101*(6), 601-611.

Niu, W., Kim, Y., Tau, G., Heyduk, T., & Ebright, R. (1996). Transcription activation at class II CAP-dependent promoters: Two interactions between CAP and RNA polymerase. *Cell, 87*, 1123-1134.

Noriega, C. E., Lin, H, Chen, L, Williams, S., & Stewart, V. (2010). Asymmetric cross-regulation between the nitrate-responsive NarX-NarL and NarQ-NarP two-component regulatory systems from Escherichia coli K – 12. *Mol Microbiol, 75*, 394-412.

Savery, N. J., Belyaeva, T., & Busby, S, J. (1996). Introduction to protein: DNA interaction, DNase I footprinting, hydroxyl radical footprinting, permanganate footprinting and supplementary protocols. In K. Docherty (Ed.), *Essential techniques: Gene Transcription* (pp. 1-5; 21-33). BIOS Scientific Publishers, Oxford

Scott, S., Busby, S., & Beacham, I. (1995). Transcriptional coactivation at the ansB promoter: involvement of the activating regions of CRP and FNR when bound in tandem. *Mol, Microbiol., 18*, 521-532.

Squire, D., Xu, M., Cole, J., Busby, S., & Browning, D. (2009). Competition between Narl-dependent activation and Fis-dependent repression controls expression from the Escherichia coli *year* and *ogt* promoters. *Biochem. J., 420*, 249-257.

Yura, T., & Nakahigashi, K. (1999). Regulation of the heat-shock response. *Curr. Opin. Microbiol., 2*, 153-158.

Zhou, Y., Pendergrast, P., Bell, A., Williams, R., Busby, S. & Ebright, R. (1994b). The functional subunit of a dimeric transcription activator protein depends on promoter architecture. *EMBO J., 13*, 4549-4557.

19

Glutathione S- Transferase M1 in the Serum of Patients with Endometriosis

Zahra Sorkhoo[1]

[1] Gilan University, Iran

Correspondence: Zahra Sorkhoo, Gilan University, Iran. E-mail: zsorkhou@gmail.com

Abstract

Endometriosis is a female disease with high incidence and unknown causes because of the presence of endometrial tissue outside of the uterine wall. Glutathione S- transferase (GSTs) is a key enzyme in phase I and II that catalyzes the conjugation of glutathione to genotoxic compounds. Glutathione S- transferase M1 (GSTM1) is a versatile enzyme that plays an important role in detoxification of carcinogenic metabolites. The aim of this study was to evaluate serum levels GSTM1 in patients with endometriosis. In this study, blood samples from 20 women with endometriosis and 20 healthy women were collected. The final concentration of protein in the blood of patients (TPC) and GSTM1 in serum by using "biuret" and enzyme-linked immunosorbent assay "(ELISA) were measured. It was shown that the level of total protein concentration all samples are within the normal range and no significant changes between the two groups were seen. Although significant reduction in serum levels GSTM1 was observed in patients with Endometriosis compared to the control group (P = 0.002). The results of this study showed that GSTM1 is involved pathophysiology of endometriosis.

Keywords: endometriosis, Glutathione S- Transferase, infertility

1. Introduction

Endometriosis is the presence of endometrial tissue outside the uterus is believed to play an important role in infertility (Thornton, Morley, Lilleyman & Onwude, 1997). In general, women with endometriosis suffer from 15-7% of the total population ((Wu, Lu, Chuang & Tsai, 2010. In addition to free radical agents, immunological factors and environmental factors, genetic factors are the main factors in the development of endometriosis due to the following factors are made.

1.1 Glutathione-S-transferase gene M1 (GSTM1)

Glutathione-S-transferase M1 gene is set on chromosome 1p13 and expresses the enzymes GSTM1 (Zhong, Wolf & Spurr, 1992). Glutathione is involved in detoxification xenobiotic and carcinogens (Mannervik et al., 1992). GSTM1 gene deletion on the impact and cause defects in the detoxification enzyme and the risk of exposure to carcinogens and toxic chemicals will increase (Seidegard, Vorachek, Pero & Person, 1988). GSTM1 gene deletion in diseases such as ovarian cancer (Baxter, Thomas & Campbell, 2001), Cystic fibrosis (Baranova et al., 1997), gastric cancer Harada et al., 1992), lung cancer (Nakachi et al., 1993) and bladder cancer (Brockmoller et al., 1994) have been reported. Studies have shown that GSTM1 gene is removed in a high percentage of women with endometriosis (Nisolle et al., 1990).

1.2 Tumor Suppressor Gene p53

This gene creates p53 protein that this protein affects suppressor genes, such as Bax, P21 and plays a role in cell cycle regulation, cell growth and apoptosis. P53 gene suppresses tumor action that mutations in this gene will lead to tumor formation (Hsieh & Lin, 2006).

1.3 Metalloproteinase Matrix Gene -9 (MMP-9)

Enzyme products of these genes are involved in extracellular matrix degradation. The histologic endometrial changes during the menstrual cycle enzymes are involved. These enzymes play an important role in the analysis of extracellular matrix and enables endometrial cells that are integrated into the peritoneum of aggressive endometrial lesions. A study was accomplished on polymorphisms -1562C> T, R279Q, P574R gene (9- MMP) in women. Han suggested that haplotype AC, GG, CA gene (9-MMP) are associated with the progression of the disease. (Han et al., 2009).

Lin. (Hur, Lee, Moon & Chung, 2005) and colleagues in 2003 showed that GSTM1 genotype is associated with risk of developing endometriosis (Lin et al., 2003).

In 2012, Vichi et al. showed GSTM1 gene polymorphisms with risk of developing endometriosis (Vichi et al., 2012). Ertunc and colleagues in 2005 examined that a member of the family GSTP1 GST is associated with risk of developing endometriosis risk. The results show GSTP1 polymorphism associated with the risk of endometriosis risk (Ertunc et al., 2005). In 2004 Ding and his colleagues examined the correlation between GSTM1 and risk of developing endometriosis in population of Sinn Yang, China and they did not observe the correlation (Ding et al., 2004). Tempfer, Schneeberger and Huber reported in 2004 that GSTM1 polymorphisms as genetic factors susceptibility to diseases are benign and malignant gynecologic diseases (Tempfer et al., 2004).

In general, GST family plays an important role in detoxification. Therefore, according to the role of environmental factors in this disease and GST in detoxification; in this study, the GSTM1 was discussed on patients with endometriosis.

2. Materials and Methods

In order to cater to the amount of 5 ml of blood serum, blood was taken from 20 patients and 20 controls. Healthy female control subjects with no symptoms associated with endometriosis and infertility, cysts and endometriosis had no family history. Blood samples for 20 minutes at room temperature or 37 ° C was placed in a water bath. Then, it was centrifuged for 15 min at 6000. Solution -70 ° C was kept until to be used.

Test tubes after cleaning the samples were numbered. Then, 50 ml of standard solution was shed in a tube and 50 ml of biuret reagent was added to the tubes. The tubes were shaken well and then in the water bath for 10 min. °C37 temperature. Then, using absorption spectrophotometer and standard samples was read at 540 nm and was calculated using the following formula:

$$\frac{OD.T}{OD.ST} \ X \ 6 \ Concentration \ \text{of protein} \tag{1}$$

In the present study, to evaluate the GSTM1 in the sera of ELISA kits CASABIO, GSTM1 CSB-EL009979HV catalog numbers were used.

To prepare the samples, we act as follows:

• biotin - antibody (× 1): Before opening it was centrifuged. Biotin - antibody dilution would require 100 times the dilution contains 10 biotin - antibody diluent Lµ 990 plus biotin antibodies.
• avidin HRP- (× 1): jar before opening was centrifuged. Avidin Hrp - requires dilution before us, which is 100 dilution for Lµ 10 HRP- avidin avidin HRP- diluent is added to Lµ 990.
• Washing buffer (× 1): if crystals formed in concentration to warm to room temperature and mix gently until the crystals are completely gone away. 20ml wash buffer concentration (× 25) was diluted with deionized distilled water to wash buffer ml500 (× 1).
• Standard solution GSTM1: GSTM1 diluent was used to prepare a series of diluted solution.

To prepare sample 100 Lµ standard was added to each well. Then, the coated plates were covered with adhesive tape and were incubated for 2 h at 37 ° C. Liquid of each well was replaced, but it was not washed. lµ 100 biotin - antibody (× 1) was added to each well. Again, it was covered with adhesive tape and was incubated for 1 h at 37 ° C. Each well was washed. This washing process was repeated 2 times for a total of 3 washes. By filling each well was washed with washing buffer 200 lµ remain and 2 minutes at room temperature. lµ100 avidin HRP- (× 1) was added to each well and this was covered with sticky strips were incubated for 1 h at 37 ° C. Washing process was repeated 5 times. Lµ90 TMB substrate was added to each well and were incubated at 37 ° C for 30-15 minutes. Lµ50 stop solution was added to each well. At this stage, the wells were placed in the ELISA reader and optical density was read at 450 nm. According to the optical density of a plot device and according to the optical density, the samples was investigated GSTM1 concentration.

3. Findings

Using biuret method, total protein content sera from healthy individuals and patients with endometriosis were measured and it was found containing concentrations of total serum protein in serum of patients with endometriosis than in healthy subjects showed no significant difference (4/0 P =) (Figure 1). The ELISA technique to study protein level GSTM1 in patients with endometriosis compared with healthy subjects showed that the average concentration in the blood serum of women with endometriosis GSTM1 mu / ml 1.35 ± 4.16, compared with the average concentration of serum GSTM1 healthy female blood mu / ml 1.79 ± 5.82 decreased significantly show that at P=0.002 (Figure 2).

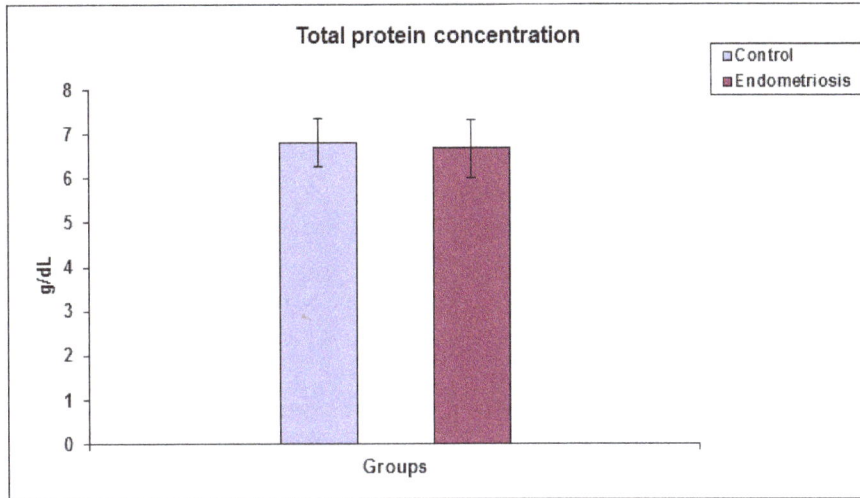

Figure 1. The concentration of serum total protein in the control group and patients with endometriosis that difference is not statistically significant. (P = 0.4)

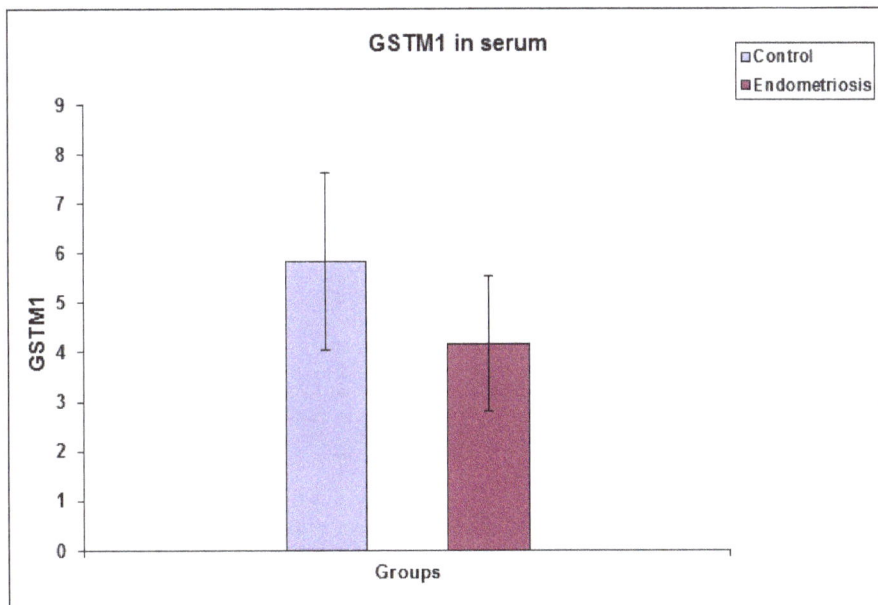

Figure 2. GSTM1 concentration in the blood serum of women with endometriosis and a control group, the difference was statistically significant (P = 0.002).

4. Discussion and Conclusion

The results showed that GSTM1 significantly reduced in the serum of patients with endometriosis than in the control groups, but the concentration of protein in the two groups showed no significant change.

Studies show that removing the GST genes seen in patients with endometriosis. Thus GSTM1 is reduced in the serum of these patients due to removal in GST gene. It was concluded that GSTM1 plays a role in the physiology of endometriosis.

References

Baranova, H., Bothorishvilli, R., Canis, M., Albuisson, E., Perriot, S., Glowaczower, E., … Malet, P. (1997). GlutathioneS-transferase M1 gene polymorphism and susceptibility to endometriosisin a French population. *Molecular Human Reproduction, 3*, 775–780.

Baxter, S. W., Thomas, E. J., & Campbell, I. G. (2001). GSTM1 null polymorphismand susceptibility to endometriosis and ovarian cancer. *Carcinogenesis, 22*, 63-5.

Brockmöller, J. I., Cascorbi, I., Kerb, R., & Roots. I. (1996). Combined analysis of inherited polymorphisms in arylamine N-acetyltransferase 2, glutathione S-transferases M1 and T1, microsomal epoxide hydrolase, and cytochrome P450 enzymes as modulators of bladder cancer risk. *Cancer Research, 56*, 3915-25.

Ding, Y., Chen, Z. F., Lin, R. Y., Wang, X. F., Ding, J. B., Ai, X. Z., & Wen, H. (2004). Relationship between endometriosis and glutathione S-transferase M1, T1 genes of the Uygurs and Hans in Xinjiang. *Zhonghua fu chan ke za zhi, 39*, 101-4.

Ertunc, D., Aban, M., Tok, E. C., Tamer, L., Arslan, M., & Dilek, S. (2005). Glutathione-S-transferase P1 gene polymorphism and susceptibility to endometriosis. *Human reproduction, 20*, 2157-61.

Han, Y. J., Kim, H. N., Yoon, J. K., Yi, S. Y., Moon, H. S., Ahn, J. J., Kim, H. L., & Chung, H. W. (2009). Haplotype analysis of the matrix metalloproteinase-9 gene associated with advanced-stageendometriosis. *Fertility and sterility, 91*, 2324-30.

Hsieh, Y., & lin, C. (2006). P53 codon 11, 72, and 248 gene polymorphisms in endometriosis. *International Journal of Biological Science,* (4), 188-193.

Hur, S. E., Lee, J. Y., Moon, H. S., & Chung, H. W. (2005). Polymorphisms of the genes encoding the GSTM1, GSTT1 and GSTP1 in Korean women: no association with endometriosis. *Molecular Human Reproduction, 11*, 15-9.

Lin, J., Zhang, X., Qian, Y., Ye, Y., Shi, Y., Xu, K., & Xu, J. (2003). Glutathione S-transferase M1 and T1 genotypes and endometriosis risk: a case-controlled study. *Chinese Medical Journal, 116*, 777-80.

Mannervik, B., Awasthi, Y. C., Board, P. G., Hayes, J. D., Dillio, C., Ketterer, B., … Pearson, W. R. (1992). Nomenclature for human glutathione transferases. *Biochemical Journal, 282*, 305–306.

Nisolle, M., Paindaveine, B., Bourdon, A., Berliére, M., Casanas-Roux F., & Donnez, J. (1990). Histologic study of peritoneal endometriosis in infertile women. *Fertility and sterility, 53*, 984-8.

Seidegard, J., Vorachek, W. R., Pero, R. W., & Person W. R., (1988). Hereditary differences in the expression of the human glutathione transferaseavtive on trans-stilbene oxide are due to a gene deletion. *Proceedings of the National Academy of Sciences of the United States of America, 85*, 7293–7297.

Tempfer, C. B., Schneeberger, C., & Huber, J. C. (2004). Applications of polymorphisms and pharmacogenomics in obstetrics and gynecology. *Pharmacogenomics, 5*, 57-65.

Thornton, J. G., Morley, S., Lilleyman, J., & Onwude, J. L. (1997). Currie I and Crompton AC human endometrium: glandular epithelial cell apoptosis is associated with increased expression of bax. *The Journal of clinical endocrinology and metabolism, 82*, 2738-2746.

Vichi, S., Medda, E., Ingelido, A. M., Ferro, A., Resta, S., Porpora, M. G., Abballe, A., Nisticò, L., De Felip, E., Gemma, S., & Testai, E. (2012). Glutathione transferase polymorphisms and risk of endometriosis associated with polychlorinated biphenyls exposure in Italian women: a gene-environment interaction *Fertility and sterility, 97,* 1143-51.e1-3.

Wu, M.Y., Lu, C. W., Chuang, P. C., & Tsai S. J. (2010). Prostoglandin E2: the master of endometriosis? *Experimental biology and medicine (Maywood, N.J.),* 235, 668-677.

Zhong, S., Wolf, C. R., & Spurr, N. K. (1992). Chromosomal assignment linkage analysis of the human glutathione S-transferase mu gene (GSTM1) using specific polymerase chain reaction. *Human genetics, 90*, 435-9.

20

Regulatory T-cell: Regulator of Host Defense in Infection

Mousa Mohammadnia-Afrouzi[1], Mehdi Shahbazi[1], Sedigheh Baleghi Damavandi[2], Ghasem Faghanzadeh Ganji[3], & Soheil Ebrahimpour[2]

[1] Department of Immunology, School of Medicine, Babol University of Medical Sciences, Babol, Iran

[2] Infectious Diseases and Tropical Medicine Research Center, Babol University of Medical Sciences, Babol, Iran

[3] Cardiac surgery Department, Rouhani Hospital, Babol University of Medical Sciences, Babol, Iran

Correspondence: Soheil Ebrahimpour, Infectious Diseases and Tropical Medicine Research Center, Babol University of Medical Sciences, Babol, Iran. Email: drsoheil1503@yahoo.com.

Abstract

Based on diverse activities and production of several cytokines, T lymphocytes and T helper cells are divided into Th1, Th2, Th17 and regulatory T-cell (T regs) subsets based on diverse activities and production of several cytokines. Infectious agents can escape from host by modulation of immune responses as effector T-cells and Tregs. Thus, regulatory T-cells play a critical role in suppression of immune responses to infectious agents such as viruses, bacteria, parasites and fungi and as well as preserving immune homeostasis. However, regulatory T-cell responses can advantageous for the body by minimizing the tissue-damaging effects. The following subsets of regulatory T-cells have been recognized: natural regulatory Tcells, Th3, Tr1, CD8+ Treg, natural killer like Treg (NKTreg) cells. Among various markers of Treg cells, Forkhead family transcription factor (FOXP3) as an intracellular protein is used for discrimination between activated T reg cells and activated T-cells. FOXP3 has a central role in production, thymocyte differentiation and function of regulatory Tcells. Several mechanisms have been indicated in regulation of T reg cells. As, the suppression of T-cells via regulatory T-cells is either mediated by Cell-cell contact and Immunosuppressive cytokines (TGF-β, IL-10) mediated.

Keywords: Infection, Immune system, Lymphocytes, Regulatory T-cell

1. Introduction

The existence of regulatory T-cell was first suggested over two decades ago, but since then a lot of questions about their structure and function remains unanswered. These cells have the ability to regulate innate and adaptive immune responses, especially cellular immunity (Hasanjani et al., 2016; Ranjbar et al., 2012; Yazdani et al., 2015). Although, it has been proven that Th1cells are more susceptible to the suppressive activity of Treg cells than Th2 cells (Cools, Ponsaerts, Van Tendeloo, & Berneman, 2007; Thunberg et al., 2010). These suppressive cells were first recognized by their expression of a phenotype marker called CD25 (IL-2Rα chain) .Howbeit, CD25 is an activation marker, it is also present on the surface of the T helper cells, too. Several the subsets of Tregs include natural regulatory T-cells, Th3 cells, Tr1 regulatory cells, CD8+ Treg and natural killer like Treg (NKTreg) cells.

1.1. Natural Regulatory Tcells (nTregs)

As a subpopulation of immunoregulatory T-cells , Natural Regulatory Tcells constitutes nearly 5%–10% of CD4+ cells circulating in the bloodstream of humans(Sakaguchi, 2005). In spite of , CD25 is an activation marker of T-cells, less than 5% of cells with high expression levels of this marker have been proposed as a regulatory agent(Costantino, Baecher-Allan, & Hafler, 2008). Other markers of nTregs include OX40,CD62 ligand, CTLA4 and Forkhead family transcription factor (FOXP3), among them which only a high level of intracellular FOXP3 is a specific molecular marker for distinction between activated regulatory T-cells and other cells like activated CD4+ T-cells(Hori, Nomura, & Sakaguchi, 2003). It was proven that the transcription factor FOXP3 has a crucial role in generation, thymocyte differentiation and function of nTregs in mice(Ramsdell & Ziegler, 2003). There is also correlations exist between lack of FOXP3 protein expression and scurfy in mice or as well as immune dysregulation, polyendocrinopathy, enteropathy and X-linked syndrome (IPEX) in humans(Mohammadnia-Afrouzi et al., 2015). In other words, the transcription factor FOXP3 is mutated in such like diseases(Gambineri, Torgerson, & Ochs, 2003; van der Vliet & Nieuwenhuis, 2007).Contrary to, some recent studies reject the role of this marker in many case of autoimmune diseases(Costantino et al., 2008).Cytokines are type of soluble glycoproteins produced by a wide variety of cells, mainly leukocytes within the effector periods of immunity and which regulate immune and inflammatory responses. TGF-beta is a necessary cytokine that is capable of inducing the expression of Foxp3. Bone Morphogenetic Proteins (BMPs) are a specific group of proteins within the members of TGF-beta superfamily, which Smight replace with TGF-β during the development of Tregs. As well as, BMPs can also regulate the differentiation, proliferation and apoptosis of various cell types(Ling Lu et al., 2010). Studies on the effect of other cytokines showed that, TNF-α decreases the repressive function of peripheral blood Treg cells in inflammatory bowel diseases (IBD), In so doing, anti-TNF treatment causes to decline in regulatory cells apoptosis or recovery of Tregs percentage(Boschetti et al., 2011; Guidi et al., 2013). IL-15 as a secreted cytokine by mononuclear phagocytes increments expression of regulatory T-cell FOXP3 in the absence of antigenic stimulation. Also it has been shown that FOXP3 expression in the presence of IL-2 and IL-15 compared to IL-7 is significantly increased (Imamichi, Sereti, & Lane, 2008). IL-27Rα is expressed at high levels in the Tregs. IL-27 is capable of binding to IL-27Rα as well as promoting nTreg cell expansion in response to IFN-γ and it is specifically controls of Th1 and as well as expression of CXCR3 on the surface of regulatory cells in intracellular pathogen infections(Hall et al., 2012). IL-2, IL-4 and IL-7 are common gamma chain cytokines. IL-4 and IL-7 are not necessary for the function of nTreg but are important as survival factors(Long & Buckner, 2008). Compared with them, IL-2 is required for both the activity and homeostasis of nTreg (Zorn et al., 2006). Although, a number of researches do not confirm this issue (Davidson, DiPaolo, Andersson, & Shevach, 2007).

1.2. Th3 Regulatory Cells

This subset of Treg was first isolated from Mesenteric lymph node (MLN) of mice tolerated by low dose antigen feeding(X. Wang et al., 2013). Th3 regulatory cells are dependents upon TGF-ß and expresses latency-associated peptide (LAP) on own their surface. In the other words, LAP-expressing Tregs are capable of more further suppressive activity in a TGFβ-dependent manner. LAP as a marker be highly expressed on activated Tregs and platelets.(Howard L Weiner, da Cunha, Quintana, & Wu, 2011) Antigen presenting cells(APCs) induces Th3, however this effect is amplificated by IL-4, IL-10 and TGFβ(Seder et al., 1998). Among them, TGF-β can increase the differentiation of Th3, a process that can be intensified by anti IL-12(Howard L. Weiner, 2001). Albeit, TGF-β can suppress T-cell proliferation and cytotoxicity. TGF-β1 secreted by Th3 is shown to play an incisive role in oral tolerance, as it suppresses of Th1/Th2 balance and cytokine production in Th1(Kim et al., 2004). Th3 cells express another molecule called cytotoxic T-lymphocyte antigen 4 (CTLA-4) on their surface. CTLA-4 cells compete with the binding of B7 ligands to CD28, which inhibit the growth and function of T-cells. Also, CTLA-4 as an effective marker has been shown to induce the production TGF-β, and with increasing proportion of anti–CTLA-4 signaling antibody, TGF-β production is equally reduced(Myint, Leonard, Steinbusch, & Kim, 2005). This regulator cells cause to extensive release of IgA antibodies from plasma cells too much. Overall, Th3 has a regulatory role in homeostasis and gastrointestinal diseases(Akbari, Stock, DeKruyff, & Umetsu, 2003; W. Chen, Jin, & Wahl, 1998).

1.3 Tr1 Regulatory Cells

This subset of regulatory cells is described by a unique cytokine production profile: they produce high levels of TGF-β and IL-10 but low levels of IFN-γ , IL-5 and IL-2(Allan et al., 2008). The cytokine secretion pattern of Tr1 to distinguish it from other T-cells is very helpful. Tr1 cells are capable of controlling immunologic tolerance and auto immune diseases. Some studies showed that IL-10 is a key immunoregulator during viruses, bacteria, parasites infection and decreases immunopathology in many infections. The Th2-associated cytokines IL-4 and IL-13 all perform necessary roles in the pathogenesis of schistosomiasis, with IL-4 and IL-13 directing acute fibrosis. It is has been confirmed that, IL-10 in keeping IL-12 controls Th2 response as described in Schistosoma mansoni involvement(Couper, Blount, & Riley, 2008).The recent studies indicate the effects of IL-10 on the proliferation and differtiation of human peripheral blood T-cells with inhibition of IL-2 and IL-5 secretion or blocking the production of some chemokines in APCs(Moore, de Waal Malefyt, Coffman, & O'Garra, 2001). In addition, IL-10 is effective on T-cells function via suppression of CD28 (O'Garra, Vieira, Vieira, & Goldfeld, 2004). IL-10 inhibits the production of proinflammatory cytokines and CD80/86 molecules in monocytes. IgE synthesis in humoral system could be suppressed by this Tr1 cytokine(Meiler, Klunker, Zimmermann, Akdis, & Akdis, 2008). IL-6 and IL-27 , which are both amplified by TGF-β, boost the production of IL-10 by Tr1 and Th17 cells (Couper et al., 2008). Tr1 cells express some functional markers such as: HLA-DR,CD-40L, IL-15Rα,IL-2Ry and various chemokine receptors CCR5, CCR3,CCR8 and CXCR3(Battaglia, Gregori, Bacchetta, & Roncarolo, 2006). Presence of Tr1 cells in joint fluid and blood specimens from rheumatoid arthritis (RA)patients, confirm this cells effectiveness in immune response(Berthelot & Maugars, 2004).Role of Tr1 in controlling of type 1 diabetes (T1D) in both of human and mouse model was demonstrated, even the deficiency of Tr1 cells in T1D patients has been investigated (Bettini & Vignali, 2009).

Tr1 suppresses the effector cells of allergic inflammation, such as mast cells, eosinophils and basophils. Therefore, some experts emphasize for getting help of Tr1 in developing new strategies for treatment of allergic diseases(K. Wu, Bi, Sun, & Wang, 2007).

1.4. CD8⁺ Regulatory Cells

In recent years, another type of suppressor cells has been introduced. These cells are known as CD8⁺ regulatory cells(H. Jiang & Chess, 2000). They are not part of CD4⁺ regulatory cells and have not cytotoxicity but inhibit the immune response especially TCD4⁺ cells in an Ag-specific manner within MHC class I (Vlad, Cortesini, & Suciu-Foca, 2005). CD8⁺ regulatory cells express FOXP3 (not mouse), high levels of CTLA-4 and low level of CD127, which is helpful for differentiation with CD4⁺ Treg(Fontenot et al., 2005; Scotto et al., 2004). Many studies were showed that CD8⁺ regulatory cells expressed glucocorticoid-induced tumor necrosis factor receptor (GITR) and tumor necrosis factor receptor 2 (TNFR2), which were identified in human thymus(Cosmi et al., 2003). Usefulness of these cells in treatment was first demonstrated in animal models of multiple sclerosis (MS) and experimental autoimmune encephalomyelitis (EAE)(Linrong Lu & Cantor, 2008). Some researches demonstrate that CD8⁺ regulatory cells are presented in the lamina proparia of healthy cases, but not in IBD patients, suggesting that these cells might play a crucial role in immune tolerance in this involvement or in several cancers(Brimnes et al., 2005; Dinesh, Skaggs, Cava, Hahn, & Singh, 2010).

1.5 NKT Regulatory Cells

Several researchers have proposed other names for these cells, including natural T-cells,Vα14i T-cells and iNKT-cells. Although NKT-cells has been accepted and is now used to refer to these suppressive cells, because they simultaneously express some markers of NKcells and T-cells. These markers contain CD161(human), NK1.1(mouse),CD16 and IL-2Rß as NKcells marker and TCRs belonging Tcells (Godfrey, MacDonald, Kronenberg, Smyth, & Van Kaer, 2004). NKT-cells are localized in various tissues like liver, bone marrow, lung and spleen. Thus, these cells observe within account for 50% of T-cells in the liver, 30% in the bone marrow,7% in lung and 3% in spleen(Tarazona et al., 2003). Other types of these cells according to expressing of CD markers are included CD4⁺NKT, CD8⁺NKT and CD4⁻ CD8⁻NKTcells(DN). CD4⁺NKTcells are able to produce cytokines of Th1 and Th2 concurrent, but CD4⁻ CD8⁻NKTcells are only capable of generating cytokines of Th1 cells (Cava, Kaer, & Fu Dong, 2006). NKT regulatory cells can produce high levels of cytokines like IL-17, IL-22، IFN-γ, IL-4 ,IL-13 and TGF-ß thus, effect on Dendritic cells ،NKcells ،helper Tcells and Bcells or regulate innate and acquired immunity(Parekh, Wu, Olivares-Villagómez, Wilson, & Van Kaer, 2013).Proliferation of memory Th1 cells and memory Th2 cells is upregulated by NKT regulatory cells, moreover IL-4 produced by these cells can effect on proliferation and differentiation of Th2. Surprisingly, NKT-cells promote nTregs proliferation by producing of soluble IL-2, but in some cases nTregs regulate NKTcells(S. Jiang, Game, Davies, Lombardi, & Lechler, 2005). Contrary to NKcells, cytotoxicity of NKTcells has not confirmed, though these cells induce high proportions of CD95L and perforin(Smyth et al., 2002). NKT regulatory cells play major role in regulation of immune responses in allergy, infection and cancer. Many studies have showed that the proportion of NKT regulatory cells increased in the lung after Mycobacterial and Streptococcus pneumoniae infections (Iwamura et al., 2012).

2. Mechanism of Suppression

Although, the crucial role of Tregs in regulation of physiologic and pathologic responses including anti tumor and anti bacterial effects has confirmed but, no clearly description has been presented for mechanism of Tregs suppression. Suppression of T-cells is either mediated by cell-cell contact and immunosuppressive cytokines.

2.1 Cell-Cell Contact

Natural (human) Tregs and CD4⁺CD25⁺ (mouse) Tregs express CTLA-4 or CD152. It has been postulated that CD152 is necessary for the function of these cells. CTLA-4 competes with CD28 for binding to B7. The affinity of CTLA4 for B7 was estimated to be about 10 folds higher than the affinity of CD28 for this ligand. T-cell proliferation is inhibited by binding of CTLA-4 to B7, in other words CTLA-4 plays an important role in T-cell peripheral tolerance(Salomon et al., 2000). Tregs stimulate dendritic cells (DCs) to express the enzyme indoleamine 2,3-dioxygenase (IDO) or IDO production is resulted from interactions between CTLA-4 and B7 ligands. Although, these cells may block the expression of B7-1 and B7-2 on some APCs(Sakaguchi, Wing, Onishi, Prieto-Martin, & Yamaguchi, 2009). IDO Serves tryptophan to kynurenine, leading to destruction of T-cells and also to direcT-cell cycle suppression. Many researches were showed that IDO causes to tryptophan degradation and suppress the immunity response to tumors(Soliman, Mediavilla-Varela, & Antonia, 2010). Autoimmune disorders widely occur in CTLA-4 deficient mice widely and the ability of repression in Treg is reduced(Vogel et al., 2015). Moreover, CTLA-4/CD80.86 interactions to the control of colitis were clearly described(Sojka, Huang, & Fowell, 2008).

2.2 Immunosuppressive Cytokine Mediated Suppression

TGF-ß and IL-10 have great role in the suppression of immune system. TGFβ decreases the expression of MHC class II in macrophages and DCs. In astrocyte (a sub-type of glial cells in the central nervous system or CNS), class II MHC transactivator (CIITA), as a suppressor that regulates MHCII expression, is targeted by TGF-β for repression of MHCII expression(Dong, Tang, Letterio, & Benveniste, 2001). Totally, TGF-β suppresses immunity system via in two manners: promoting the function of Treg cells especially in autoimmune diseases and suppression of inflammatory cells. It should be noted that , the role of TGF-β in Tregs suppression is still an issue of argument(Oberle, Eberhardt, Falk, Krammer, & Suri-Payer, 2007). In preliminary mechanism of suppression, IL-10 inhibits the expression of cell surface MHC II by macrophages. In this regard, IL-10 diminishes mRNA as to MHC II and expression of this MHC in peritoneal macrophage (Chadban et al., 1998). Contrary, IFN-γ stimulates the expression of MCHII molecules on mesangial cells and IL-10 and TGF-β regulate this function(Gattoni, Parlato, Vangieri, Bresciani, & Derna, 2006). Other cytokines like IL-35 are produced by Tregs (Bardel, Larousserie, Charlot-Rabiega, Coulomb-L'Herminé, & Devergne, 2008; Collison et al., 2007). In cancer cells, IL-35 increases the expression of gp130 and decreases sensitivity to cytotoxic T-cell (CTL) demolition. More over IL-35 promotes tumor angiogenesis(Z. Wang et al.).

3. Regulatory T-cell and Infection

3.1 Regulatory T-cells in Viral Infections

It is confirmed that infections are associated with various regulatory cells and their cytokines. For example, IL-10 plays a key role in several disorders, including ; hepatitis C, Tuberculosis and Leishmaniasis, as a suppressor of immune systems(Wohlfert & Belkaid, 2008). TCD8$^+$ cell response during acute viral infection is necessary for elimination of virus. But however, Tregs cause to limit in this function and some experiments were showed that clearing the viruses is accelerated without regulatory cells. Virus-specific T-cells can cause to damage to the host tissues, but Treg cells prevent tissues damage and are beneficial. Increase in regulatory T-cells has been shown to suppressive activity in the acute form of HCV patients(Sugimoto et al., 2003). These regulatory cells may also protect the patients from pathologic disorders and viral-induced inflammation in chronic HCV infection(Claassen, de Knegt, Tilanus, Janssen, & Boonstra, 2010). The level of circulating CD4$^+$ CD25$^+$ Treg cells is enhanced in chronic hepatitis B patients since, these lymphocytes are able to inhibit production of cytokines production such as IFN-γ and proliferation of other T-cells. Many researches showed that the frequency of CD4$^+$ CD25$^+$ Treg cells is positively correlated with serum viral load of hepatitis B(Peng et al., 2008). In herpes simplex virus (HSV) infection, the regulatory function of CD4$^+$ CD25$^+$ Treg cells is clearly observed. In the other words, regulatory T-cells impairs immunity to herpes simplex virus which is depended upon CD8$^+$ T-cell response(Rouse, Sarangi, & Suvas, 2006). This was supported by the fact that, depletion of Tregs from cases with HSV infection significantly enhanced the protective immunity(Fernandez et al., 2008). In HIV infection the level of regulatory T-cells in lymph nodes and blood may be significantly high. Treg cells mainly suppress cytokines secretion of CD4$^+$ T-cells and cytolytic activity of CD8$^+$effector T-cells in acute phase of infection more than chronic phase(Thorborn et al., 2010). According to some studies, presence of viraemia in chronic HIV infection is correlated with the reduction of circulating regulatory T-cells which is useful to enhancement of CD4$^+$ T-cells function(Baker et al., 2007). Overall, researches indicate that chronic viral infection cause to induce regulatory T-cells which are implicated in the regulation and impairment of antiviral immune responses.

3.2. Regulatory T-cells in Bacterial Infections

In bacterial infections like active form of tuberculosis (TB), Treg cells meaningfully expand in the lung, blood and lymph nodes relative to healthy or latent cases (Y. E. Wu et al., 2010). CD4$^+$CD25$^+$FoxP3$^+$ T-cells are able to suppressing IL-10 and IFN-γ production in TB patients(Xinchun Chen et al., 2007). Regulatory T-cells prevent from disruption of tubercle bacilli by suppressing efficient CD4$^+$ T-cell response in TB. In these infections like viral disorders, Tregs have some beneficially aspects for host, as could they permit a long-term persistence of M. tuberculosis and thus a relative maintained immunity against re-infection(Majlessi, Lo-Man, & Leclerc, 2008). In Bordetella pertussis infection, the virulence factor of these bactria is toxin (PTx). This toxin causes to enhanced function of regulatory T-cells and promoted in TGF-ß in the serum of cases and thus Bordetella pertussis inhibits Th1 response(Weber et al., 2010). Also other studies showed that PTx cause to down-regulation of the percentage and function of Treg cells and this element is used to induce murine experimental autoimmune encephalomyelitis (EAE)(Xin Chen et al., 2006). During infection with Brucella abortus, similar to other infectious models, there is increase in both effector CD4$^+$ T-cells and regulatory T-cells in the spleen of BALB/c mice(Pasquali et al., 2010). Due to this fact, Treg cells inhibit Th1 activity and used to process of infection (Hasanjani-Roushan, Kazemi, Fallah-Rostami, & Ebrahimpour; Kazemi, Damavandi, & Ebrahimpour, 2015; Roushan, Kazemi, Rostam, & Ebrahimpour,

2014). Furthermore, this is showed that the percentage of regulatory T-cells was shown to be elevated in several forms of human brucellosis, and also, this increment being rather in the chronic patients(Bahador et al., 2014). CD4 $^+$ CD25 $^+$ T regulatory cells, which produce IL-10 mediated alteration in macrophages function, therefore impede CD4$^+$T-cell response. In infection with Salmonella Typhimurium, when the bacterial burden is increasing, Treg cells suppress the activity and proliferation of some immune cells But, when bacterial load is decreased, the suppression ability of Treg cells is declined(Johanns, Ertelt, Rowe, & Way, 2010).

3.3 Regulatory T-cells in Parasitic Infections

In parasitic infection, immune response depends on activity of Th2 and cytokines as IL-4, IL-5 and IL-13. CD4 $^+$ CD25 $^+$ Foxp3$^+$ Treg numbers can expand following parasitic infections. The balance between Treg cells and Th1, Th2 responses is critical in the defense against parasites. Chagas disease is an infection caused by a protozoan parasite called Trypanosoma cruzi and the infection is associated with immunoregulation of cellular and humoral immunity via Treg cells which incapacitate the infected macrophages(Flores-García, Rosales-Encina, Rosales-García, Satoskar, & Talamás-Rohana, 2012). Otherwise, CD4$^+$CD25^{High} Treg cells could be limiting the damage to host and leading to durability of the clinical forms in this infection(Sathler-Avelar, Vitelli-Avelar, Teixeira-Carvalho, & Martins-Filho, 2009). Many studies administrated that CD4$^+$ CD25$^+$ Foxp3 Treg cells are enhanced in peripheral blood of schistosomiasis patients and modulate the activity of effector Th1 and Th2 responses with suppression of the responsiveness of IFNγ, IL-4 and IL-5. Treg cells play significant roles in minimizing pathological effects during schistosomiasis (Nausch, Midzi, Mduluza, Maizels, & Mutapi, 2011).Elevated CD4$^+$CD25$^+$Foxp3$^+$ Tregs producing both IL-10 and TGF-ß contribute to regulation of immune responses during the primary stage of cutaneous leishmaniasis. Moreover, active lesions are presented in patients with polarized Th2 responses and also are correlated with increased IL-10 production. Thus, regulatory T-cells prohibit complete elimination of parasites from the skin of host, during cure of infection (Nylén & Sacks, 2007).

3.4 Regulatory T-cells in Fungal Infections

Aspergillosis is caused by the spores of Aspergillus, and can occur depending on the functions of Th 1, Th2, Th17, and regulatory T-cells. CD4$^+$ T-cells can protect the host against invasive aspergillosis (Bozza et al., 2009). Treg cells prevent innate immune cells and effector Tcells by suppressing the production of inflammatory cytokines. Thus, activation and expansion of Treg cells able to regulate neutrophils within the action of IL-10 on indoleamine 2,3-dioxygenase(IDO) that leads to control of inflammation in this fungal infection(Montagnoli et al., 2006). Some studies showed a higher percentage of CD4$^+$ CD25$^+$ Foxp3$^+$ T-cells and increased GITR expression levels of these cells in cases with active paracoccidioidomycosis (an infection caused by the dimorphic fungus Paracoccidioides brasiliensis) and also indicated that after adequate antifungal therapy, the levels of regulatory T-cells was declined in the circulation (Ferreira, de Oliveira, da Silva, Blotta, & Mamoni, 2010). Furthermore, an important receptor called CCR5 caused to migration of regulatory T-cells to the lesions of paracoccidioidomycosis, leading to suppression of immune response and lasting attendance of Paracoccidioides brasiliensis in the granulomas (Moreira et al., 2008). So, a useful mechanism for avoiding the intensity of infection is regulation of this migration to the lesions. In Candida albicans infection, the immune response contains production of cytokines as IL-1, IL-17A and IFN-γ (Zelante et al., 2007). Th1 cells are associated with protection from this infection. However, the presence of regulatory T-cells inhibits clearance of C. albicans with macrophages. Contrary to , other studies administrated that in C. albicans involvements , elevated percentage of Treg cells correlated to protection from infection, as in some autoimmune diseases with faulty regulatory Tcells, mucocutaneous candidiasis were observed (Whibley et al., 2014).

Table 1. Alteration of Tregs level in different diseases

Disease	Treg	Reference
Viral infection	Increased	(Maizels & Smith, 2011)
Bacterial infection	Increased	(Belkaid & Rouse, 2005; Y. E. Wu et al., 2010)
Parasitic infection	Increased	(Belkaid, 2007)
Fungi infection	Increased	(Ferreira et al., 2010; Romani, 2004)
Multiple sclerosis	Normal or decreased	(Haas et al., 2007; Huan et al., 2005)
Idiopathic thrombocytopenic purpura (ITP)	Decreased	(Lv, Liu, & Wu, 2007)
Type 1 diabetes	Decreased	(Lindley et al., 2005)
Ulcerative colitis	Normal or decreased	(Takahashi et al., 2006)
Cancer	Increased	(Mougiakakos, Choudhury, Lladser, Kiessling, & Johansson, 2010)

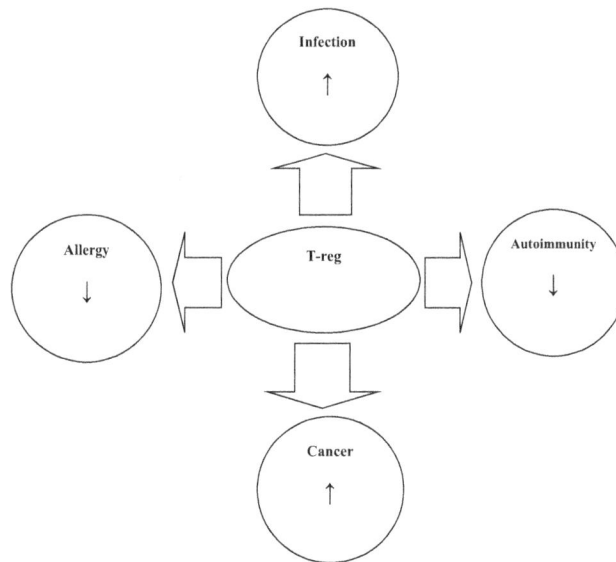

Figure 1. Level of regulatory T-cells in different disorders

In conclusion, in spite of many studies on control and suppression of immune system including cells and tissues involved, many questions regarding to aspects of these issues remain unanswered. According to currently discussed text, Treg cells play a dual role as immune regulatory cells are dual, thus these cells suppress infectious immune response and also moderateing the immunopathologic effects of some disease causing agents. Of course, it seems that inadequacy of the advantage of this simple protection. Such abilities of the Treg cells in association with the infectious immune responses make them remarkable models to achieve new strategies in the treatment of infectious diseases.

Acknowledgments

The authors would like to thank the staffs of departments of immunology and infectious diseases for their sincere cooperation in this project.

References

Akbari, O., Stock, P., DeKruyff, R. H., & Umetsu, D. T. (2003). Role of regulatory T cells in allergy and asthma. *Current Opinion in Immunology, 15*(6), 627-633. doi: http://dx.doi.org/10.1016/j.coi.2003.09.012

Allan, S. E., Broady, R., Gregori, S., Himmel, M. E., Locke, N., Roncarolo, M. G., ... Levings, M. K. (2008). CD4+ T-regulatory cells: toward therapy for human diseases. *Immunological reviews, 223*(1), 391-421.

Bahador, A., Hadjati, J., Hassannejad, N., Ghazanfari, H., Maracy, M., Jafari, S., ... Nejadeh, A. (2014). Frequencies of CD4+ T Regulatory Cells and their CD25high and FoxP3high Subsets Augment in Peripheral Blood of Patients with Acute and Chronic Brucellosis. *Osong Public Health and Research Perspectives, 5*(3), 161-168. doi: http://dx.doi.org/10.1016/j.phrp.2014.04.008

Baker, C., Clark, R., Ventura, F., Jones, G., Guzman, D., Bangsberg, R., & Cao, H. (2007). Peripheral CD4 loss of regulatory T cells is associated with persistent viraemia in chronic HIV infection. *Clinical & Experimental Immunology, 147*(3), 533-539.

Bardel, E., Larousserie, F., Charlot-Rabiega, P., Coulomb-L'Herminé, A., & Devergne, O. (2008). Human CD4+ CD25+ Foxp3+ regulatory T cells do not constitutively express IL-35. *The Journal of immunology, 181*(10), 6898-6905.

Battaglia, M., Gregori, S., Bacchetta, R., & Roncarolo, M.-G. (2006). *Tr1 cells: from discovery to their clinical application.* Paper presented at the Seminars in immunology.

Belkaid, Y. (2007). Regulatory T cells and infection: a dangerous necessity. *Nature Reviews Immunology, 7*(11), 875-888.

Belkaid, Y., & Rouse, B. T. (2005). Natural regulatory T cells in infectious disease. *Nature immunology, 6*(4), 353-360.

Berthelot, J.-M., & Maugars, Y. (2004). Role for suppressor T cells in the pathogenesis of autoimmune diseases (including rheumatoid arthritis). Facts and hypotheses. *Joint Bone Spine, 71*(5), 374-380.

Bettini, M., & Vignali, D. A. (2009). Regulatory T cells and inhibitory cytokines in autoimmunity. *Current opinion in immunology, 21*(6), 612-618.

Boschetti, G., Nancey, S., Sardi, F., Roblin, X., Flourie, B., & Kaiserlian, D. (2011). Therapy with anti-TNFalpha antibody enhances number and function of Foxp3(+) regulatory T cells in inflammatory bowel diseases. *Inflamm Bowel Dis, 17*(1), 160-170. doi: 10.1002/ibd.21308

Bozza, S., Clavaud, C., Giovannini, G., Fontaine, T., Beauvais, A., Sarfati, J., ... Zagarella, S. (2009). Immune sensing of Aspergillus fumigatus proteins, glycolipids, and polysaccharides and the impact on Th immunity and vaccination. *The Journal of immunology, 183*(4), 2407-2414.

Brimnes, J., Allez, M., Dotan, I., Shao, L., Nakazawa, A., & Mayer, L. (2005). Defects in CD8+ regulatory T cells in the lamina propria of patients with inflammatory bowel disease. *The Journal of immunology, 174*(9), 5814-5822.

Cava, A. L., Kaer, L. V., & Fu Dong, S. (2006). CD4+CD25+ Tregs and NKT cells: regulators regulating regulators. *Trends in Immunology, 27*(7), 322-327. doi: http://dx.doi.org/10.1016/j.it.2006.05.003

Chadban, S. J., Tesch, G. H., Foti, R., Lan, H. Y., Atkins, R. C., & Nikolic-Paterson, D. J. (1998). Interleukin-10 differentially modulates MHC class II expression by mesangial cells and macrophages in vitro and in vivo. *Immunology, 94*(1), 72-78.

Chen, W., Jin, W., & Wahl, S. M. (1998). Engagement of cytotoxic T lymphocyte–associated antigen 4 (CTLA-4) induces transforming growth factor β (TGF-β) production by murine CD4+ T cells. *The Journal of experimental medicine, 188*(10), 1849-1857.

Chen, X., Winkler-Pickett, R. T., Carbonetti, N. H., Ortaldo, J. R., Oppenheim, J. J., & Howard, O. (2006). Pertussis toxin as an adjuvant suppresses the number and function of CD4+ CD25+ T regulatory cells. *European journal of immunology, 36*(3), 671-680.

Chen, X., Zhou, B., Li, M., Deng, Q., Wu, X., Le, X., ... Katsanis, E. (2007). CD4+CD25+FoxP3+ regulatory T cells suppress Mycobacterium tuberculosis immunity in patients with active disease. *Clinical Immunology, 123*(1), 50-59. doi: http://dx.doi.org/10.1016/j.clim.2006.11.009

Claassen, M. A., de Knegt, R. J., Tilanus, H. W., Janssen, H. L., & Boonstra, A. (2010). Abundant numbers of regulatory T cells localize to the liver of chronic hepatitis C infected patients and limit the extent of fibrosis. *Journal of hepatology, 52*(3), 315-321.

Collison, L. W., Workman, C. J., Kuo, T. T., Boyd, K., Wang, Y., Vignali, K. M., ... Vignali, D. A. (2007). The inhibitory cytokine IL-35 contributes to regulatory T-cell function. *Nature, 450*(7169), 566-569.

Cools, N., Ponsaerts, P., Van Tendeloo, V. F., & Berneman, Z. N. (2007). Balancing between immunity and tolerance: an interplay between dendritic cells, regulatory T cells, and effector T cells. *Journal of leukocyte biology, 82*(6), 1365-1374.

Cosmi, L., Liotta, F., Lazzeri, E., Francalanci, M., Angeli, R., Mazzinghi, B., ... Romagnani, P. (2003). Human CD8+ CD25+ thymocytes share phenotypic and functional features with CD4+ CD25+ regulatory thymocytes. *Blood, 102*(12), 4107-4114.

Costantino, C. M., Baecher-Allan, C. M., & Hafler, D. A. (2008). Human regulatory T cells and autoimmunity. *European journal of immunology, 38*(4), 921-924.

Couper, K. N., Blount, D. G., & Riley, E. M. (2008). IL-10: the master regulator of immunity to infection. *The Journal of immunology, 180*(9), 5771-5777.

Davidson, T. S., DiPaolo, R. J., Andersson, J., & Shevach, E. M. (2007). Cutting edge: IL-2 is essential for TGF-β-mediated induction of Foxp3+ T regulatory cells. *The Journal of immunology, 178*(7), 4022-4026.

Dinesh, R. K., Skaggs, B. J., Cava, A. L., Hahn, B. H., & Singh, R. P. (2010). CD8(+) Tregs in Lupus, Autoimmunity, and Beyond. *Autoimmunity reviews, 9*(8), 560-568. doi: 10.1016/j.autrev.2010.03.006

Dong, Y., Tang, L., Letterio, J. J., & Benveniste, E. N. (2001). The Smad3 protein is involved in TGF-β inhibition of class II transactivator and class II MHC expression. *The Journal of immunology, 167*(1), 311-319.

Fernandez, M. A., Puttur, F. K., Wang, Y. M., Howden, W., Alexander, S. I., & Jones, C. A. (2008). T regulatory cells contribute to the attenuated primary CD8+ and CD4+ T cell responses to herpes simplex virus type 2 in neonatal mice. *The Journal of immunology, 180*(3), 1556-1564.

Ferreira, M. C., de Oliveira, R. T. D., da Silva, R. M., Blotta, M. H. S. L., & Mamoni, R. L. (2010). Involvement of regulatory T cells in the immunosuppression characteristic of patients with paracoccidioidomycosis. *Infection and immunity, 78*(10), 4392-4401.

Flores-García, Y., Rosales-Encina, J. L., Rosales-García, V. H., Satoskar, A. R., & Talamás-Rohana, P. (2012). Treg Cells Induced by rSSP4 Derived from T. cruzi Amastigotes Increase Parasitemia in an Experimental Chagas Disease Model. *BioMed research international, 2013*.

Fontenot, J. D., Rasmussen, J. P., Williams, L. M., Dooley, J. L., Farr, A. G., & Rudensky, A. Y. (2005). Regulatory T cell lineage specification by the forkhead transcription factor foxp3. *Immunity, 22*(3), 329-341.

Gambineri, E., Torgerson, T. R., & Ochs, H. D. (2003). Immune dysregulation, polyendocrinopathy, enteropathy, and X-linked inheritance (IPEX), a syndrome of systemic autoimmunity caused by mutations of FOXP3, a critical regulator of T-cell homeostasis. *Current opinion in rheumatology, 15*(4), 430-435.

Gattoni, A., Parlato, A., Vangieri, B., Bresciani, M., & Derna, R. (2006). Interferon-gamma: biologic functions and HCV therapy (type I/II) (1 of 2 parts). *Clin Ter, 157*(4), 377-386.

Godfrey, D. I., MacDonald, H. R., Kronenberg, M., Smyth, M. J., & Van Kaer, L. (2004). NKT cells: what's in a name? *Nature Reviews Immunology, 4*(3), 231-237.

Guidi, L., Felice, C., Procoli, A., Bonanno, G., Martinelli, E., Marzo, M., ... Rutella, S. (2013). FOXP3(+) T regulatory cell modifications in inflammatory bowel disease patients treated with anti-TNFalpha agents. *Biomed Res Int, 2013*, 286368. http://dx.doi.org/10.1155/2013/286368

Haas, J., Fritzsching, B., Trübswetter, P., Korporal, M., Milkova, L., Fritz, B., ... Wildemann, B. (2007). Prevalence of newly generated naive regulatory T cells (Treg) is critical for Treg suppressive function and determines Treg dysfunction in multiple sclerosis. *The Journal of immunology, 179*(2), 1322-1330.

Hall, Aisling O. H., Beiting, Daniel P., Tato, C., John, B., Oldenhove, G., Lombana, Claudia G., ... Hunter, Christopher A. (2012). The Cytokines Interleukin 27 and Interferon-γ Promote Distinct Treg Cell Populations Required to Limit Infection-Induced Pathology. *Immunity, 37*(3), 511-523. http://dx.doi.org/10.1016/j.immuni.2012.06.014

Hasanjani-Roushan, M. R., Kazemi, S., Fallah-Rostami, F., & Ebrahimpour, S. Brucellosis Vaccines: An Overview. *Crescent Journal of Medical and Biological Sciences, 1*(4), 118-124.

Hasanjani, R. M., Bayani, M., Soleimani, A. S., Mohammadnia-Afrouzi, M., Nouri, H., & Ebrahimpour, S. (2016). Evaluation of CD4+ CD25+ FoxP3+ regulatory T cells during treatment of patients with brucellosis. *Journal of biological regulators and homeostatic agents, 30*(3), 675.

Hori, S., Nomura, T., & Sakaguchi, S. (2003). Control of regulatory T cell development by the transcription factor Foxp3. *Science, 299*(5609), 1057-1061.

Huan, J., Culbertson, N., Spencer, L., Bartholomew, R., Burrows, G. G., Chou, Y. K., ... Vandenbark, A. A. (2005). Decreased FOXP3 levels in multiple sclerosis patients. *Journal of neuroscience research, 81*(1), 45-52.

Imamichi, H., Sereti, I., & Lane, H. C. (2008). IL-15 acts as a potent inducer of CD4+CD25hi cells expressing FOXP3. *European journal of immunology, 38*(6), 1621-1630. doi: 10.1002/eji.200737607

Iwamura, C., Shinoda, K., Endo, Y., Watanabe, Y., Tumes, D. J., Motohashi, S., ... Nakayama, T. (2012). Regulation of memory CD4 T-cell pool size and function by natural killer T cells in vivo. *Proceedings of the National Academy of Sciences, 109*(42), 16992-16997.

Jiang, H., & Chess, L. (2000). The specific regulation of immune responses by CD8+ T cells restricted by the MHC class Ib molecule, Qa-1. *Annual review of immunology, 18*(1), 185-216.

Jiang, S., Game, D. S., Davies, D., Lombardi, G., & Lechler, R. I. (2005). Activated CD1d-restricted natural killer T cells secrete IL-2: innate help for CD4+ CD25+ regulatory T cells? *European journal of immunology, 35*(4), 1193-1200.

Johanns, T. M., Ertelt, J. M., Rowe, J. H., & Way, S. S. (2010). Regulatory T cell suppressive potency dictates the balance between bacterial proliferation and clearance during persistent Salmonella infection. *PLoS Pathog, 6*(8), e1001043.

Kazemi, S., Damavandi, S. B., & Ebrahimpour, S. (2015). Letter to Editor-macrophages and Brucella Infection. *infection, 1*(3), 69-75.

Kim, Y.-K., Myint, A.-M., Lee, B.-H., Han, C.-S., Lee, H.-J., Kim, D.-J., & Leonard, B. E. (2004). Th1, Th2 and Th3 cytokine alteration in schizophrenia. *Progress in Neuro-Psychopharmacology and Biological Psychiatry, 28*(7), 1129-1134. doi: http://dx.doi.org/10.1016/j.pnpbp.2004.05.047

Lindley, S., Dayan, C. M., Bishop, A., Roep, B. O., Peakman, M., & Tree, T. I. (2005). Defective suppressor function in CD4+ CD25+ T-cells from patients with type 1 diabetes. *Diabetes, 54*(1), 92-99.

Long, S. A., & Buckner, J. H. (2008). Combination of rapamycin and IL-2 increases de novo induction of human CD4+CD25+FOXP3+ T cells. *Journal of Autoimmunity, 30*(4), 293-302. doi: http://dx.doi.org/10.1016/j.jaut.2007.12.012

Lu, L., & Cantor, H. (2008). Generation and regulation of CD8+ regulatory T cells.

Lu, L., Ma, J., Wang, X., Wang, J., Zhang, F., Yu, J., ... Horwitz, D. A. (2010). Synergistic effect of TGF-β superfamily members on the induction of Foxp3+ Treg. *European journal of immunology, 40*(1), 142-152.

Lv, X.-w., Liu, F., & Wu, C.-l. (2007). Eexpression of Foxp3 and CD4^+ CD~ 2~ 5^+ Regulation T Cells in Children with Acute Idiopathic Thrombocytopenic Purpura. *Journal of Applied Clinical Pediatrics, 22*(3), 188.

Maizels, R. M., & Smith, K. A. (2011). Regulatory T cells in infection. *Adv Immunol, 112*, 73-136.

Majlessi, L., Lo-Man, R., & Leclerc, C. (2008). Regulatory B and T cells in infections. *Microbes and Infection, 10*(9), 1030-1035. doi: http://dx.doi.org/10.1016/j.micinf.2008.07.017

Meiler, F., Klunker, S., Zimmermann, M., Akdis, C. A., & Akdis, M. (2008). Distinct regulation of IgE, IgG4 and IgA by T regulatory cells and toll-like receptors. *Allergy, 63*(11), 1455-1463.

Mohammadnia-Afrouzi, M., Zavaran Hosseini, A., Khalili, A., Abediankenari, S., Hosseini, V., & Maleki, I. (2015). Decrease of CD4+ CD25+ CD127low FoxP3+ regulatory T cells with impaired suppressive function in untreated ulcerative colitis patients. *Autoimmunity, 48*(8), 556-561.

Montagnoli, C., Fallarino, F., Gaziano, R., Bozza, S., Bellocchio, S., Zelante, T., ... Romani, L. (2006). Immunity and tolerance to Aspergillus involve functionally distinct regulatory T cells and tryptophan catabolism. *The Journal of immunology, 176*(3), 1712-1723.

Moore, K. W., de Waal Malefyt, R., Coffman, R. L., & O'Garra, A. (2001). Interleukin-10 and the interleukin-10 receptor. *Annual review of immunology, 19*(1), 683-765.

Moreira, A. P., Cavassani, K. A., Massafera Tristão, F. S., Campanelli, A. P., Martinez, R., Rossi, M. A., & Silva, J. S. (2008). CCR5-dependent regulatory T cell migration mediates fungal survival and severe immunosuppression. *J Immunol, 180*(5), 3049-3056.

Mougiakakos, D., Choudhury, A., Lladser, A., Kiessling, R., & Johansson, C. C. (2010). Regulatory T cells in cancer. *Advances in cancer research, 107*, 57-117.

Myint, A.-M., Leonard, B. E., Steinbusch, H. W. M., & Kim, Y.-K. (2005). Th1, Th2, and Th3 cytokine alterations in major depression. *Journal of Affective Disorders, 88*(2), 167-173. doi: http://dx.doi.org/10.1016/j.jad.2005.07.008

Nausch, N., Midzi, N., Mduluza, T., Maizels, R. M., & Mutapi, F. (2011). Regulatory and Activated T Cells in Human <italic>Schistosoma haematobium</italic> Infections. *PLoS One, 6*(2), e16860. http://dx.doi.org/10.1371/journal.pone.0016860

Nylén, S., & Sacks, D. (2007). Interleukin-10 and the pathogenesis of human visceral leishmaniasis. *Trends in Immunology, 28*(9), 378-384. http://dx.doi.org/10.1016/j.it.2007.07.004

O'Garra, A., Vieira, P. L., Vieira, P., & Goldfeld, A. E. (2004). IL-10–producing and naturally occurring CD4+ Tregs: limiting collateral damage. *Journal of Clinical Investigation, 114*(10), 1372.

Oberle, N., Eberhardt, N., Falk, C. S., Krammer, P. H., & Suri-Payer, E. (2007). Rapid suppression of cytokine transcription in human CD4+CD25 T cells by CD4+Foxp3+ regulatory T cells: independence of IL-2 consumption, TGF-beta, and various inhibitors of TCR signaling. *J Immunol, 179*(6), 3578-3587.

Parekh, V. V., Wu, L., Olivares-Villagómez, D., Wilson, K. T., & Van Kaer, L. (2013). Activated invariant NKT cells control central nervous system autoimmunity in a mechanism that involves myeloid-derived suppressor cells. *The Journal of immunology, 190*(5), 1948-1960.

Pasquali, P., Thornton, A. M., Vendetti, S., Pistoia, C., Petrucci, P., Tarantino, M., ... Shevach, E. M. (2010). CD4+ CD25+ T regulatory cells limit effector T cells and favor the progression of brucellosis in BALB/c mice. *Microbes and Infection, 12*(1), 3-10.

Peng, G., Li, S., Wu, W., Sun, Z., Chen, Y., & Chen, Z. (2008). Circulating CD4+ CD25+ regulatory T cells correlate with chronic hepatitis B infection. *Immunology, 123*(1), 57-65.

Ramsdell, F., & Ziegler, S. F. (2003). Transcription factors in autoimmunity. *Current opinion in immunology, 15*(6), 718-724.

Ranjbar, M., Solgi, G., Mohammadnia, M., Nikbin, B., Pourmand, G., Ansaripour, B., & Amirzargar, A. (2012). Regulatory T-cell subset analysis and profile of interleukin (IL)-10, IL-17 and interferon-gamma cytokine-producing cells in kidney allograft recipients with donor cells infusion. *Clinical and experimental nephrology, 16*(4), 636-646.

Romani, L. (2004). Immunity to fungal infections. *Nature Reviews Immunology, 4*(1), 11-24.

Rouse, B. T., Sarangi, P. P., & Suvas, S. (2006). Regulatory T cells in virus infections. *Immunological reviews, 212*(1), 272-286.

Roushan, M. R. H., Kazemi, S., Rostam, F. F., & Ebrahimpour, S. (2014). A Study of Brucella Infection in Humans. *Crescent Journal of Medical and Biological Sciences, 1*(3), 69-75.

Sakaguchi, S. (2005). Naturally arising Foxp3-expressing CD25+ CD4+ regulatory T cells in immunological tolerance to self and non-self. *Nature immunology, 6*(4), 345-352.

Sakaguchi, S., Wing, K., Onishi, Y., Prieto-Martin, P., & Yamaguchi, T. (2009). Regulatory T cells: how do they suppress immune responses? *International immunology*, dxp095.

Salomon, B., Lenschow, D. J., Rhee, L., Ashourian, N., Singh, B., Sharpe, A., & Bluestone, J. A. (2000). B7/CD28 costimulation is essential for the homeostasis of the CD4+ CD25+ immunoregulatory T cells that control autoimmune diabetes. *Immunity, 12*(4), 431-440.

Sathler-Avelar, R., Vitelli-Avelar, D. M., Teixeira-Carvalho, A., & Martins-Filho, O. A. (2009). Innate immunity and regulatory T-cells in human Chagas disease: what must be understood? *Memorias do Instituto Oswaldo Cruz, 104*, 246-251.

Scotto, L., Naiyer, A. J., Galluzzo, S., Rossi, P., Manavalan, J. S., Kim-Schulze, S., ... Suciu-Foca, N. (2004). Overlap between molecular markers expressed by naturally occurring CD4+ CD25+ regulatory T cells and antigen specific CD4+ CD25+ and CD8+ CD28− T suppressor cells. *Human immunology, 65*(11), 1297-1306.

Seder, R. A., Marth, T., Sieve, M. C., Strober, W., Letterio, J. J., Roberts, A. B., & Kelsall, B. (1998). Factors involved in the differentiation of TGF-β-producing cells from naive CD4+ T cells: IL-4 and IFN-γ have opposing effects, while TGF-β positively regulates its own production. *The Journal of immunology, 160*(12), 5719-5728.

Smyth, M. J., Crowe, N. Y., Hayakawa, Y., Takeda, K., Yagita, H., & Godfrey, D. I. (2002). NKT cells—conductors of tumor immunity? *Current opinion in immunology, 14*(2), 165-171.

Sojka, D. K., Huang, Y. H., & Fowell, D. J. (2008). Mechanisms of regulatory T-cell suppression – a diverse arsenal for a moving target. *Immunology, 124*(1), 13-22. http://dx.doi.org/10.1111/j.1365-2567.2008.02813.x

Soliman, H., Mediavilla-Varela, M., & Antonia, S. (2010). Indoleamine 2,3-Dioxygenase: Is It an Immune Suppressor? *Cancer journal (Sudbury, Mass.), 16*(4), 10.1097/PPO.1090b1013e3181eb3343. http://dx.doi.org/10.1097/PPO.0b013e3181eb3343

Sugimoto, K., Ikeda, F., Stadanlick, J., Nunes, F. A., Alter, H. J., & Chang, K. M. (2003). Suppression of HCV-specific T cells without differential hierarchy demonstrated ex vivo in persistent HCV infection. *Hepatology, 38*(6), 1437-1448.

Takahashi, M., Nakamura, K., Honda, K., Kitamura, Y., Mizutani, T., Araki, Y., ... Nawata, H. (2006). An inverse correlation of human peripheral blood regulatory T cell frequency with the disease activity of ulcerative colitis. *Digestive diseases and sciences, 51*(4), 677-686.

Tarazona, R., DelaRosa, O., Peralbo, E., Casado, J., Pena, J., & Solana, R. (2003). Human NKT cells in health and disease. *Immunologia, 22*(4), 359-370.

Thorborn, G., Pomeroy, L., Isohanni, H., Perry, M., Peters, B., & Vyakarnam, A. (2010). Increased sensitivity of CD4+ T-effector cells to CD4+CD25+ Treg suppression compensates for reduced Treg number in asymptomatic HIV-1 infection. *PLoS One, 5*(2), e9254. http://dx.doi.org/10.1371/journal.pone.0009254

Thunberg, S., Gafvelin, G., Nord, M., Grönneberg, R., Grunewald, J., Eklund, A., & Van Hage, M. (2010). Allergen provocation increases TH2-cytokines and FOXP3 expression in the asthmatic lung. *Allergy, 65*(3), 311-318.

van der Vliet, H. J., & Nieuwenhuis, E. E. (2007). IPEX as a result of mutations in FOXP3. *Journal of Immunology Research, 2007*.

Vlad, G., Cortesini, R., & Suciu-Foca, N. (2005). License to heal: bidirectional interaction of antigen-specific regulatory T cells and tolerogenic APC. *The Journal of immunology, 174*(10), 5907-5914.

Vogel, I., Kasran, A., Cremer, J., Kim, Y. J., Boon, L., Van Gool, S. W., & Ceuppens, J. L. (2015). CD28/CTLA-4/B7 costimulatory pathway blockade affects regulatory T-cell function in autoimmunity. *European journal of immunology*.

Wang, X., Sherman, A., Liao, G., Leong, K. W., Daniell, H., Terhorst, C., & Herzog, R. W. (2013). Mechanism of oral tolerance induction to therapeutic proteins. *Advanced Drug Delivery Reviews, 65*(6), 759-773. doi: http://dx.doi.org/10.1016/j.addr.2012.10.013

Wang, Z., Liu, J. Q., Liu, Z., Shen, R., Zhang, G., Xu, J., ... Bai, X. F. *Tumor-derived IL-35 promotes tumor growth by enhancing myeloid cell accumulation and angiogenesis*: J Immunol. 2013 Mar 1;190(5):2415-23. Epub 2013 Jan 23. http://dx.doi.org/10.4049/jimmunol.1202535.

Weber, M. S., Benkhoucha, M., Lehmann-Horn, K., Hertzenberg, D., Sellner, J., Santiago-Raber, M.-L., ... Lalive, P. H. (2010). Repetitive pertussis toxin promotes development of regulatory T cells and prevents central nervous system autoimmune disease. *PLoS One, 5*(12), e16009.

Weiner, H. L. (2001). Oral tolerance: immune mechanisms and the generation of Th3-type TGF-beta-secreting regulatory cells. *Microbes and Infection, 3*(11), 947-954. http://dx.doi.org/10.1016/S1286-4579(01)01456-3

Weiner, H. L., da Cunha, A. P., Quintana, F., & Wu, H. (2011). Oral tolerance. *Immunological reviews, 241*(1), 241-259.

Whibley, N., Maccallum, D. M., Vickers, M. A., Zafreen, S., Waldmann, H., Hori, S., ... Hall, A. M. (2014). Expansion of Foxp3(+) T-cell populations by Candida albicans enhances both Th17-cell responses and fungal dissemination after intravenous challenge. *Eur J Immunol, 44*(4), 1069-1083. http://dx.doi.org/10.1002/eji.201343604

Wohlfert, E., & Belkaid, Y. (2008). Role of Endogenous and Induced Regulatory T Cells During Infections. *J Clin Immunol, 28*(6), 707-715. doi: 10.1007/s10875-008-9248-6

Wu, K., Bi, Y., Sun, K., & Wang, C. (2007). IL-10-producing type 1 regulatory T cells and allergy. *Cell Mol Immunol, 4*(4), 269-275.

Wu, Y. E., Peng, W. G., Cai, Y. M., Zheng, G. Z., Zheng, G. L., Lin, J. H., ... Li, K. (2010). Decrease in CD4+ CD25+ FoxP3+ T reg cells after pulmonary resection in the treatment of cavity multidrug-resistant tuberculosis. *International Journal of Infectious Diseases, 14*(9), e815-e822.

Yazdani, Y., Mohammadnia-Afrouzi, M., Yousefi, M., Anvari, E., Ghalamfarsa, G., Hasannia, H., ... Jadidi-Niaragh, F. (2015). Myeloid-derived suppressor cells in B cell malignancies. *Tumor Biology, 36*(10), 7339-7353.

Zelante, T., De Luca, A., Bonifazi, P., Montagnoli, C., Bozza, S., Moretti, S., ... Mosci, P. (2007). IL-23 and the Th17 pathway promote inflammation and impair antifungal immune resistance. *European journal of immunology, 37*(10), 2695-2706.

Zorn, E., Nelson, E. A., Mohseni, M., Porcheray, F., Kim, H., Litsa, D., ... Soiffer, R. J. (2006). IL-2 regulates FOXP3 expression in human CD4+ CD25+ regulatory T cells through a STAT-dependent mechanism and induces the expansion of these cells in vivo. *Blood, 108*(5), 1571-1579.

21

Impact of Biostimulant and Synthetic Hormone Gibberellic acid on Molecular Structure of *Solanum melongena* L.

Ganash Magdah A.[1]

[1] Biology Department, Faculty of Science, King Abdulaziz University, Jeddah, King Saudi Arabia

Correspondence: Ganash Magdah A., Biology Department, Faculty of Science, King Abdulaziz University, Jeddah, King Saudi Arabia. E-mail: mganash@kau.edu.sa

Abstract

A comparison study between the application of gibberellic acid (GA3) and *Saccharomyces cerevisiae* as a biostimulant on the growth and molecular structures of Eggplant (*Solanum melongena* L.) has been carried out through a pot experiment. Growth of *Solanum melongena* L. increased with exogenous application of GA3 followed by yeast. Chlorophyll contents of plant were enhanced with yeast treatment compared with GA3 application and control. Activitiy of antioxidant enzymes, catalase and peroxidase was increased with increasing concentration of GA3 and *S. cerevisiae* application particularly with using GA3. HPLC analysis showed the highest concentration of salicylic acid in plant treated with GA3 (104.20 mg) followed by *S. cerevisiae* (70.00 mg) application compared with the untreated plant (57.86 mg). Six common polypeptide bands were observed in treated and untreated *S. melongena* plants, their molecular weights were 16, 17, 34, 90, 120 and 150 KDa. While the untreated *S. melongena* plant is characterized by the presence of 8 polypeptide bands, their molecular weights were 19, 24, 32, 33, 36, 50, 109 and 133 KDa. Yeast treatment increased the number of protein bands to 12 instead of 8 in the control plant with molecular weights 18, 125, 74, 69, 62, 31, 30, 27, 25, 23, 20 and 18 KDa. Three polypeptide bands with molecular weights 25, 72 and 125 KDa were detected in *S. cerevisiae* and GA3 treated plants. PCR analysis showed that total of 16 amplified fragments was visualized in the tested samples. Eight fragments with different molecular weights, four of them are monomorphic bands while the others are polymorphic unique bands. Plant sample sprayed with yeast showed 5 fragments range in molecular weight between 426 to 1766 bp. Only one of these fragments was unique polymorphic fragment. Four monomorphic fragments range in molecular weight from 426 to 1213 bp were showed up in plant sample sprayed with gibberellic acid.

Key words: Biostimulant, gibberellic acid , molecular structure, *Solanum melongena* L.

1. Introduction

Eggplant (*Solanum melongena* L.) is one of the Solanaceae plants and considered as one of the cultivated vegetable crops in many regions of the world, including tropical regions. It is considered as a rich crop in carbohydrates, protein and minerals (Tsao and Lo, 2006). Extensive research is going on throughout the world to find out natural sources to be used as plant growth promoter. In this connection, yeasts have been reported to be a rich source of phytohormones, vitamins, enzymes, amino acids and minerals (Mahmoud, 2001). It was reported about its stimulatory effects on cell division, protein, nucleic acid and chlorophyll synthesis (Castelfranco and Beale, 1983). Also, they have been found to produce auxins including indole-3-acetic acid and indole-3-pyruvic acid, gibberellins and polyamines; albeit induced *in vitro*. So called plant growth promoting yeasts include genera such as *Candida*, *Cryptococcus*, *Rhodotorula*, *Sporobolomyces*, *Trichosporon*, *Williopsis* and *Yarrowia* (El-Tarabily and Sivasithamparam, 2006; Roland et al., 2015). Numerous studies indicate that plant growth may be directly or indirectly promoted by microorganisms as well as yeasts (Xin et al., 2009; Botha 2011; El-Tarabily and Sivasithamparam 2006; Abd El-Ghany et al., 2013; Abd El-Ghany et al., 2015a,b,c). Likewise, Reda and Ismail (2008) stated that foliar application with active yeast extract induced significant enhancement the vegetative growth of river red gum plant (*Eucalyptus camaldulensis Dehn.*). Many studies have found the increase in germination, length and biomass of seedlings after seed treated with yeast strains, enhance plant growth, increase of photosynthesis productivity (Amprayn et al., 2012; Agamy et al., 2013; Hu and Qi 2013). Recently, Kahlel (2015) stated that the addition of bread yeast near the plants roots led to a significant increase in the number of stems, leaf area and fresh and dry weight of the plant compared with the control.

Plant growth is regulated and controlled by different action of the small molecules called plant hormones, which may play either close to or remote from their sites of biosynthesis to mediate genetically programmed developmental changes or responses to stimulation of environment (Colebrook et al., 2014). Gibberellins (GAs, also referred to as gibberellic acid) are a phytohormone chemically belongs to the group of tetracyclic diterpenoids, which produced by angiosperms plant and some fungi (Sun 2011; Colebrook et al., 2014). There are different group of bioactive GAs, however the important once which include GA1, GA3, GA4 and GA7, which derived from a basic diterpenoid carboxylic acid skeleton, and generally have a C3- hydroxyl group (Davière and Achard, 2013). The sparse literature on GAs indicates that this group of phytohormones does contribute to the growth and development of the plants. GAs are generally involved in growth and development; they control seed germination, leaf expansion, stem elongation and flowering (Magome et al., 2004). During the last decade, much progress has been made to understand the mechanism of Gibberellic acid signaling. It is well known that GAs promote plant growth by inducing the degradation of the nuclear family of transcription factors known as DELLA proteins. Thus, DELLA proteins restrain growth, while GAs induce their disappearance (Jiang and Fu, 2007). Exogenous application of GAs to plants causes the increase in the activities of many key enzymes (Aftab et al., 2010.) and enhanced the primary root elongation (Bidadi et al., 2010). However, several studies on impact of GAs on animal indicated that they induce the level of DNA breakage in human blood cells (Abou-Eisha, 2001), hepatocellular carcinomas (El-Mofty and Sakr, 1988), reductions of catalase, superoxide dismutase and glutathione peroxidase (Soliman et al., 2010) in animals. Therefore the aim of the study was to explore the effects of yeast, as an environmentally and healthy safe method on molecular level compared with synthetic plant growth hormone gibberellic acid on Eggplant (*Solanum melongena* L.).

2. Material and Methods

2.1 Plant Material, Hormone and Yeast Treatments

In the present study, Eggplant (*Solanum melongena* L.) seeds were used. Prior to sowing, the seeds were surface sterilized for 10 min with 10:1 water/bleach (commercial NaOCl) solution and then washed few times with distilled water. After the sterilization, the seeds were soaked in distilled water about 5 h and then sowed in pots (20-cm diameter) with 10 seeds per pot. During the period of experiment the moist of the soil were kept at fixed percentage of available water. Seedlings were grown in a growth chamber under controlled environmental conditions for 26 days. The used hormone GA3 obtained from the Sigma-Aldrich Co., USA sprayed twice during the plant life on the leaves of 12 and 18 day-old plants at different concentrations (100, 200 and 400 ppm). Also at the same condition brewers' yeast (*S. cerevisiae*) was applied at three doses including 2, 4 and 6 g/L distilled water. Distilled water was used to spray control plants. Plants were harvested at 22 days and used to measure some biochemical and molecular markers.

2.2 Quantitative Determination of Chlorophylls and Antioxidant Enzymes

Chlorophyll content was determined according to Vernon and Seely (1966) using the following equations:

Chlorophyll a (mg)/ tissue (g) = 11.63 (A 665) -2.39 (A 649).

Chlorophyll b (mg)/ tissue (g) = 2.11 (A 649) -5.18 (A 665).

Where A denotes the reading of the optical density

Antioxidant enzymes including catalase and peroxidase of plant were detected according to Kar and Mishra (1976).

2.3 Extraction and Quantification of Salicylic Acid

Salicylic acid (SA) was extracted and quantified as described by Malamy et al. (1992). Treated and untreated plants were washed several times with distilled water to remove the remnants of any dusts. Then 25 g of each dried plant were ground in 25 mL of 90% methanol and centrifuged at 6000 rpm for 15 min. The pellet was re-extracted with 3 ml of 100% methanol and centrifuged. Methanol extracts were combined, centrifuged for 10 min and dried at 40 °C under vacuum. For each sample, the dried methanol extract was re-suspended in 5 ml of water at 80 °C, and an equal volume of 0.2 M acetate buffer (pH 4.5) was added and incubated at 37 °C over night. After digestion, samples were acidified with HCl to pH 1 and SA was extracted and back extracted with 2 volumes of cyclopentane/ethyl acetate/isopropanol (50:50:1, v/v/v). The organic extract was dried under nitrogen, re-suspended in 50 ml of 100% methanol and analyzed by HPLC at Regional Center of Mycology and Biotechnology (RCMB), Al Azhar University, Cairo, Egypt.

2.4 DNA and RNA Detection

Determination of nucleic acids DNA and RNA were extracted following the method was similar to that described by Morse and Carter (1949). RNA was estimated colorimetrically by the orcinol reaction as described by Dishe (1953) while DNA was estimated by diphenylamine (DPA) color reaction as described by Burton (1956).

2.5 Induction of indole-3-acetic acid (IAA) Production by Yeast

Yeast was cultured in Conical flasks containing Yeast Extract-Peptone-Dextrose medium supplemented with 0.1 % (w/v) L-tryptophan at pH 6.0 and incubated in the dark on a shaker at 25 °C and 150 rpm for 4 days. One millilitre of the medium was centrifuged at 3000 rpm for 5 min, and 0.5 mL of the supernatant was mixed with 0.5 mL of Salkowski's reagent which consists of 2 mL of 0.5 M iron (III) chloride and 98 mL of 35 % perchloric acid according to Gordon and Weber (1951). After 30 min, the developed red colour of the sample was quantified using a spectrophotometer at 530 nm.

2.6 Protein Gel-Electrophoresis

Five grams of each dried plant was ground in 0.1 mL Sodium dodecyl sulfate (SDS) sample buffer cracking solution. Extracts were added in 1.5 cm eppendorf centrifuge tube according to Laemmli (1970). Homogenates were heated at 95°C for 5 min then briefly centrifuged at 12000 rpm to pellet cellular debris. The resulting supernatants (total protein extracts) were stored at -70 °C until analysis by PAGE. The extract was separated by electrophoresis on 1mm thick 12.5% acrylamide slab gels. Gels were stained with Coomassie blue at RCMB.

2.7 Genetic Characterization of Treated Plant

Plant samples were ground in a mortar using a sterile pestle, and the samples were placed in eppendorf tubes (1.5 ml) DNA extraction was conducted using DNeasy kit (Qiagen-Germany).

2.7.1 DNA Amplification by RAPD-PCR

For PCR amplification, ONE 10-mer random primer was selected; primer 3:(5'-GTAGACCCGT-3'). The amplification was performed in a thermal cycler program:Template DNA was initially denatured at 92 °C for 3 min, followed by 35 cycles of PCR amplification under the following parameters: denaturation for 1 min at 92 °C, primer annealing at 36 °C for 1 min, and primer extension at 72 °C for 2 min. A final incubation for 10 min at 72 °C was performed to ensure that the primer extension reaction proceeded to completion. PCR products were analyzed by gel electrophoresis on 1.5 % agarose and detected by ethidium bromide staining. A 3K-bp DNA ladder (GeneRuler, Fermentas) was used as the molecular standard in order to confirm the appropriate RAPD markers. After washing of gel, the photograph was taken with UV transilluminator. For atcheiving DNA profile, DNA bands on gels were amalgamated using unweighted pair-group average method analysis (UPGAMA) using Statistica for windows, release 4.5f, state Soft, Inc.1993 software. RAPD bands were treated as binary characters and coded accordingly (presence = 1, absence = 0). Euclidean distances (similarity matrix) were used as the distance metric in both as well as dice coefficient as the calculation method. UV visualization, imaging and cluster analysis of the DNA bands were carried out using Quantity one 4.0.3 software of the gel doc.2000 system (BioRad).

3. Results and Discussion

3.1 Plant Growth and Biochemical Characterization

Recent different strategies have been employed to improve crop growth and development with using natural growth promoter like yeasts but very few studies have been carried out to compare the effect of yeast and GA3 applications on endogenous and molecular structures of plants. It is obvious from fig. (1) that exogenous application of GA3 and yeast increased the growth of seedlings of *Solanum melongena* L. compared with the control (untreatment plant), however GA3 application was more effective on plant length compared with yeast application. On the other hand chlorophyll "a" and chlorophyll "b" contents of plant were enhanced with yeast treatment where their content was 6.31 and 2.26 mg/g fresh weight compared with control, 5.68 and 2.11 mg/g fresh weight respectively. Surprisingly chlorophyll"a" and chlorophyll "b" contents were depressed at low concentrations 100 and 200 ppm of GA3 while increased at high concentration 400ppm (Table 1). Chlorophyll a content was enhanced by increase in GA3 concentration up to 250 mg L -1 in *Ficus benjamina* (Salehi sardoei et al., 2014). The obtained results are in agreement with what has been reported (EL-Ghamriny et al., 1999; Sarhan et al., 2011) that spraying bread yeast on the plants increasing the vegetative parameters of plants. Tartoura (2001) explained the role of yeast in increasing the vegetative growth of plants, it may be due to the content of yeast to many important which is necessary for plant biological processers especially photosynthesis and cell division. In addition to, yeast produce growth regulators such as auxin, gibberellin and cytokinin which stimulate biological processes in plants and led to an increase in the vegetative growth of the plant (EL-Ghamriny et al., 1999; Twfiq 2010; Tarek and Hassan 2014). Sarhan et al. (2011) revealed that the growth and total chlorophyll of *Solanum melongena* L. was significantly increased with dry bread yeast. Recently, Nassar et al. (2016) revealed that leucaena plants grown under salt stress and treated with active yeast extract had better growth behavior than those untreated. Activities of antioxidant enzymes, catalase (CAT) and peroxidase (POX) were increased with increasing dose of GA3 and yeast application particularly with using GA3 (Table 1). In this regard, it can be suggested that exogenous applications of these treatments might enhance plant resistance to stress conditions by inducing antioxidant enzymes activity. The increase in CAT and POX activity in *Solanum melongena* L might lead to accumulation of toxic amount of H_2O_2. Erdal and Dumlupinar (2010) demonstrated that progesterone stimulated antioxidant enzyme activities in germinating seeds of chickpea, maize and bean. Yeast used in the current study was tested to produce IAA with or without addition of tryptophan as precursor of IAA, their productivity increased with addition of tryptophan (Fig. 2). According to Lyudmila et al. (2015) IAA production increased when medium was supplemented with the tryptophan. Recently, Liu et al. (2016) found that almost all *Saccharomyces* yeasts produced IAA when cultured in medium supplemented with the primary precursor L-tryptophan.

Figure 1. Seedlings of *Solanum melongena* L. treated with 6g/L yeast and 400ppm giberellic acid

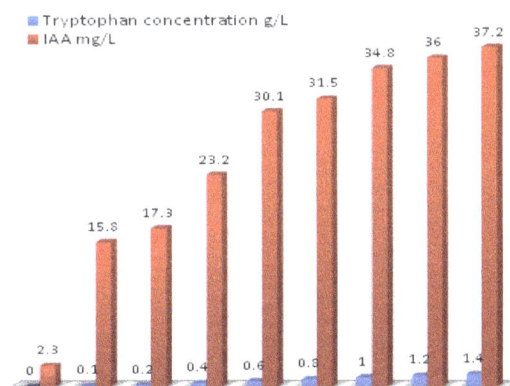

Figure 2. Effect of L-tryptophan concentration on IAA production by yeast.

Table 1. Chlorophyll content (mg/g fresh weight) and antioxidant enzyme of treated plant with different concentrations of gibberellic acid and yeast

Plant Treatment	Concentration	Chlorophyll contents (mg/g fresh weight)		Antioxidant enzymes (U/ml)	
		Chlorophyll "a"	Chlorophyll "b"	Catalase	Peroxidase
Control	0.0	5.68±0.09	2.11±0.32	19.02	0.79
Gibberilic acid	100ppm	4.78±0.10	1.96±0.12	20.87	0.90
Gibberilic acid	200ppm	4.59±0.11	1.82±0.05	21.86	0. 96
Gibberilic acid	400ppm	5.91±0.04	2.89±0.20	22.76	0.97
Yeast	2g/L	5.68±0.02	2.21±0.30	20.54	0.88
Yeast	4g/L	6.31±0.10	2.26±0.12	21.50	0.85
Yeast	6g/L	6.05±0.21	2.12±0.20	21.08	0.83

Salicylic acid (SA) is a phenolic phytohormone known for its primary function as an endogenous signal mediating plant defense responses against pathogens, as well as influencing responses to a biotic stresses and other important aspects of plant growth and development (Vlot et al., 2009). HPLC analysis chromatogram (Fig. 3) approves the presence of highest concentration of SA in plant treated with GA3 (104.209 mg) followed by yeast (70.00 mg) application compared with the untreated plant (57.868 mg). Muhammad and Muhammad (2013) who reported that the priming with GA3 was very effective in enhancing SA concentration in wheat cultivars. During recent years there has been increasing evidence on the role of SA in elicitation of plant defense mechanism in several a biotic stress conditions (Horvath et al., 2007), although information about the onset of defense mechanisms mediated by SA at the level of seed germination is very scarce (Rajjou et al., 2006). The accumulation of SA has been proposed as an endogenous marker for plant resistance (Klessig and Malamy, 1994).

It is clear that in yeast treatment of *Solanum melongena* L. plants, the resultant increase DNA, as well as RNA synthesis must be an important part of the growth response. While DNA and RNA were reduced with GA3 application and there is no difference in the detected amount of DNA and RNA with any concentration of GA3 (Table 2). Mady (2009) stated that yeast as a natural source of cytokinins-stimulates cell division and enlargement, synthesis of protein and nucleic acids. Also, Shalaby and El-Nady (2008) reported that the increase in photosynthetic pigments could be attributed to the role of yeast cytokinins in delaying the aging of leaves by reducing the degradation of chlorophyll and enhancing the protein and RNA synthesis. The present findings in case application of GA3 are generally in agreement with those reported by (Abdel-Hamid and Mohamed 2014) how found that DNA and RNA decreased from 133 to 132 and from 6.20 to 2.65 respectively in *Hordeum vulgare* as a result of GA3 application. It may therefore be suggested that yeast exerts its control over plant growth by controlling RNA and hence protein metabolism.

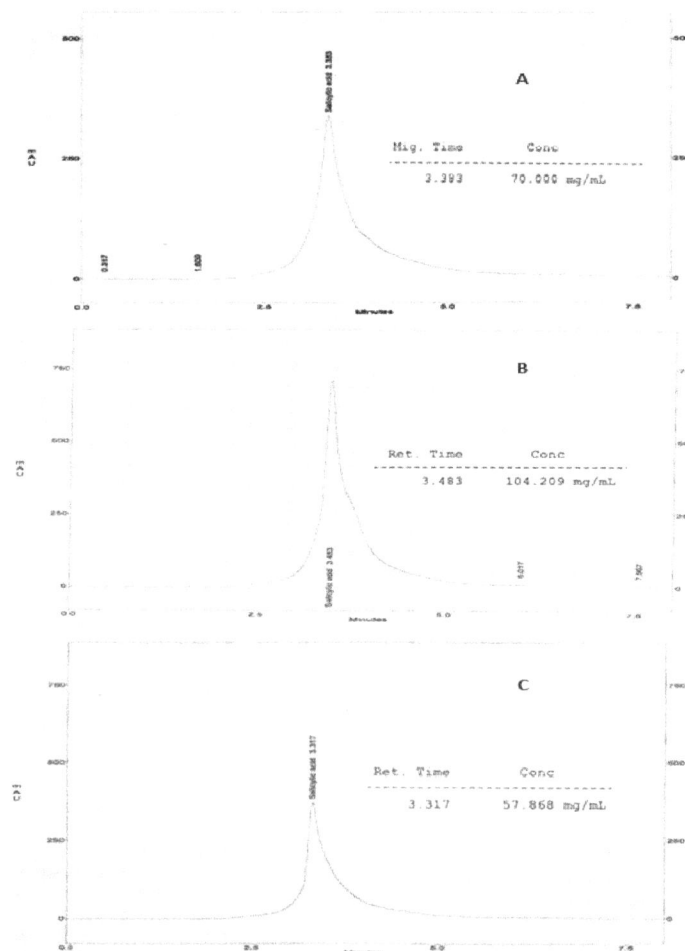

Figure 3. HPLC typical chromatogram of detected Salcylic acid in plant treated with 6g/L yeast (A), 400ppm gibberellic acid (B) and controlwithout treatment (C)

Table 2. DNA and RNA contents (µg g-1 dry weight) of *Solanum melongena* L. seedlings in response to Yeast and Gibberellic acid

Treatment		Nucleic acid content µg g-1 dry weight	
	Concentration	DNA	RNA
Control	0.0	50.56	2.70
Gibberilic acid	100 ppm	48.00	1.89
Gibberilic acid	200 ppm	48.60	1.50
Gibberilic acid	400 ppm	49.68	1.08
Yeast	2g/L	50.56	1.87
Yeast	4g/L	52.08	1.89
Yeast	6g/L	57.00	2.72

3.2 Changes in Protein Profiles

The electrophoretic pattern of *Solanum melongena* plants (Fig.4 and Table 3) was used in the present investigation to differentiate between the treatment with yeast and GA3. Six common polypeptide bands were observed in treated and untreated *S. melongena* plants, their molecular weights were 16, 17, 34, 90, 120 and 150 KDa. While the untreated control *S. melongena* plant is characterized by the presence of 8 polypeptide bands their molecular weights were 19, 24, 32, 33, 36, 50, 109 and 133 KDa. Yeast treatment increased the number of protein bands to 12 instead of 8 in the control plant with molecular weights 18, 125, 74, 69, 62, 31, 30, 27, 25, 23, 20 and 18 KDa. Three polypeptide bands with molecular weights 25, 72 and 125 KDa were detected in yeast and GA3 treated plants. Protein content in plants is an important indicator of reversible and irreversible changes in metabolism, therefore according to Dogra and Kaur (1994), these parameters are considered to be markers of growth and development. Stefanov et al. (1998) reported that sodium dodecyl sulfate polyacrylamide gel electrophoresis of soluble proteins revealed quantitative differences between treated plants with gibberellic acid. In the current study, the appearance of high molecular weight protein (125 KDa) in plant treated with yeast and gibberellic acid may be due to similar effect of this treatment. The SDS-electrophoretic pattern showed that one new band having molecular weight 77.5 KDa that appeared by the effect of biostimulants *Rhizobium* with Phosphobacteria treated seeds of *Vigna mungo* L (Selvakumar et al., 2012). Shehata and EL-Khawas (2003) stated that treated sunflower plant with biostimulant stimulate new protein with low molecular weight 2.1 Kda

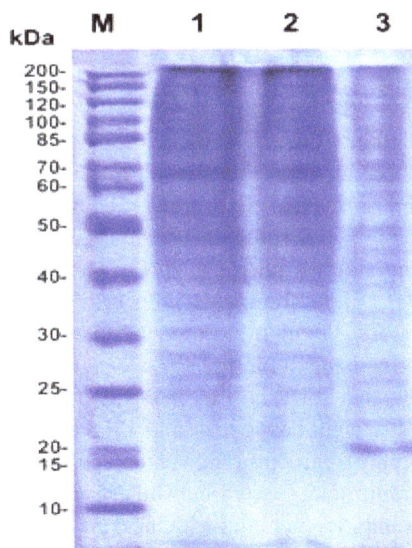

Figure 4. Protein electrophoresis of *Solanum melongena* plant after 26 days from sowing. M, Lane Marker represents Genei molecular marker (200–10 kDa); 1 untreated plant; 2 treated with 6g/L yeast; 3 treated with 400 ppm gibberellic acid

Table 3. Molecular weight (KDa) of separated protein in of *Solanum melongena* plant treated with 400 ppm gibberellic acid and 6g/L yeast after 26 days from sowing

BANDS (KDa)	Control	Yeast	GA3	BANDS (KDa)	Control	Yeast	GA3
155KD	-	-	-	40KD	-	-	-
150KD	+	+	+	39KD	+	+	-
142KD	-	+	-	38KD	-	-	-
133KD	+	-	-	37KD	-	-	+
125KD	-	+	+	36KD	+	-	-
120KD	+	+	+	35KD	-	-	-
109KD	+	-	-	34KD	+	+	+
104KD	+	+	-	33KD	+	-	-
100KD	-	-	+	32KD	+	-	-
93KD	-	-	-	31KD	-	+	-
90KD	+	+	+	30KD	-	+	-
85KD	-	-	+	29KD	+	+	+
80KD	+	+	-	28KD	-	-	+
78KD	-	-	+	27KD	-	+	-
74KD	-	+	-	26KD	-	-	-
72KD	-	+	+	25KD	-	+	+
69KD	-	+	-	24KD	+	-	-
66KD	-	-	+	23KD	-	+	-
62KD	-	+	+	22KD	-	-	+
60KD	-	-	+	21KD	-	-	-
57KD	+	+	-	20KD	-	+	-
56KD	-	-	-	19KD	+	-	-
54KD	-	-	+	18KD	-	+	-
52KD	+	+	-	17KD	+	+	+
51KD	-	-	+	16KD	+	+	+
50KD	+	-	-	15KD	+	-	+
49KD	-	+	-	14KD	+	-	-
48KD	-	+	+	13KD	+	-	-
46KD	-	-	-	12KD	+	+	-
45KD	+	+	+	11KD	+	+	+
43KD	-	+	-	10KD	-	+	-
42KD	-	+	+	9KD	-	+	+
41KD	-	-	-	7KD	+	-	-

3.3 RAPD- PCR Analysis of Plant Treated with Yeast and Gibberellic Acid

Plants sprayed with yeast extract and gibberellic acid showed morphological and/or physiological changes when compared to control samples. These changes might be a result of physiological effect or genetic modifications. Hence, Random amplified polymorphic DNA analysis were carried out for control (untreated) and treated plants with 6 g/L yeast and gibberellic acid (Fig.5 and Table 4). DNA was extracted and purified from the samples and their amplification was carried out using Single primer 3: (5'-GTAGACCCGT-3'). Total of 16 amplified fragments were visualized in the tested samples. Eight fragments with different molecular weights, four of them are monomorphic bands while the others are polymorphic unique bands. Monomorphic fragments could be plant markers. Control sample produced 7 fragments with molecular weights range from 426 bp to 3063 bp. Three of these are polymorphic unique fragments that don't show up in the treated samples. Plant sample sprayed with yeast showed 5 fragments range in molecular weight between 426 to 1766 bp. Only one of these fragments was unique polymorphic fragment. Four monomorphic fragments range in molecular weight from 426 to 1213 bp were showed up in plant sample sprayed with gibberellic acid. Generally in all cases, polymorphisms were due to the loss and/or gain of amplified bands in the treated samples compared with the control. DNA polymorphisms can become useful markers in general fingerprinting (Kawakami et al., 1999; Shehata & EL-Khawas, 2003). Also, Yang and Quiros (1993) postulated that quantitative changes could be explained on the basis of alterations of some DNA sequences. Increases in SA levels have been correlated to changes in gene expression. Precisely how SAR leads to resistance is not completely understood, but several of the pathogenesis-related genes (PR genes) expressed during the development of resistance are antagonistic to pathogens (Navarrea & Mayoa, 2004). In the present study, the appearance and disappearance of some bands may be attributed to the structural genes.

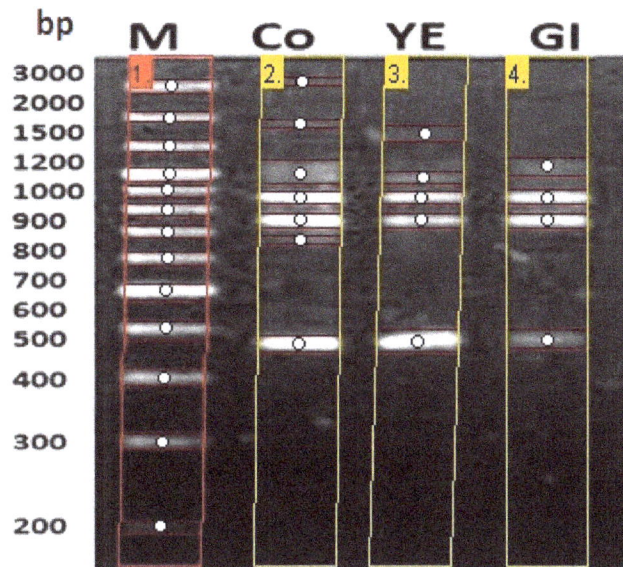

Figure 5. PCR-amplification Lane 1: Marker; Lane 2: control plant ; Lane 3: plant treated with 6g/L yeast; Lane 4: plant treated with 400 ppm gibberellic acid

Table 4. Amplification products polymorphism of *Solanum melongena* L. treated with yeast and giberellic acid

MW	Control	Yeast	Giberellic acid	Polymorphism
3063	+	-	-	poly-unique
1953	+	-	-	poly-unique
1766	-	+	-	poly-unique
1213	+	+	+	mono
972	+	+	+	mono
813	+	+	+	mono
709	+	-	-	poly-unique
426	+	+	+	mono

4. Conclusion

From the above mentioned results it can be concluded that, the application of the biostimulants has been effected on several metabolic processes, enhances plant growth and development via the increasing of photosynthesis, endogenous hormones salicylic acid. Similar proteins bands with the same molecular weights were detected with yeast and gibberellic acid treatments; this indicates the yeast plays the same role of gibberellic acid in plant development. PCR showed16 amplified fragments visualized in the tested samples. Eight fragments with different molecular weights, four of them are monomorphic bands while the others are polymorphic unique bands. Monomorphic fragments could be plant markers. Control sample produced 7 fragments with molecular weights range from 426 bp to 3063 bp. Three of these are polymorphic unique fragments that don't show up in the treated samples.

Acknowledgements

To King Abdulaziz university, Jeddah, Saudi Arabia and to Regional Center of Mycology and Biotechnology, Al Azhar University, Cairo, Egypt.

References

Abd El-Ghany, T. M., Masrahi, Y. S., Mohamed, A., Abboud, A., Alawlaqi, M. M., & Elhussieny, N. I. (2015a). Maize (*Zea Mays* L.) Growth and Metabolic Dynamics with Plant Growth-Promoting Rhizobacteria under Salt Stress. *Journal of Plant Pathology & Microbiology*, 2015. http://dx.doi.org/10.4172/2157-7471.1000305

Abd El-Ghany, T. M., Masrahi, Y. S., Mohamed, A., Abboud, A., & Alawlaqi, M. M. (2015b). Rhizosphere and rhizoplane bacteria isolated from subtropical region of Jazan in Saudi Arabia. *J. Biol. Chem. Research., 32*(2), 934-944. 41.

Abd El-Ghany, T. M., Mohamed, A. A. A., Abdel-Rahman, M. S., & Moustafa, E. N. (2015c). Rhizosphere as inducers for phytoremediation A review. *International J. Bioinformatics and Biomedical Engineering, 1*(1), 7-15.

Abd El-Ghany, T. M., Mohamed A. A. A., & Alawlaqi, M. M. (2013). Role of biofertilizers in agriculture: a brief review. *Mycopath, 11*(2), 95-101.

Abdel-Hamid, A. M., & Mohamed, H. I. (2014). The effect of the exogenous gibberellic acid on two salt stressed barley cultivars. *European Scientific Journal, 10*(6), 1857- 7881.

Abou-Eisha, A. (2001). Evaluation of cytogenetic and DNA damage induced by Gibberlic acid. *Toxicol In Vitro, 20*(5), 601-607.

Aftab, T., Khan, M. M. A., Idrees, M., Naeem, M., Singh, M., & Ram, M. (2010). Stimulation of crop productivity, photosynthesis and artemisinin production in *Artemisia annua* L. by triacontanol and gibberellic acid application. *Journal of Plant Interactions, 5*, 273–281. http://dx.doi.org/10.1080/17429141003647137

Agamy R., Hashem M., & Alamri S. (2013). Effect of soil amendment with yeasts as bio-fertilizers on the growth and productivity of sugar beet. *Afr. J. Agric. Res., 8*(1), 46–56. http://dx.doi.org/10.5897/AJAR12.1989

Amprayn K., Rose M. T., & Kecskés M. (2012). Plant growth promoting characteristics of soil yeast (Candida tropicalis HY) and its effectiveness for promoting rice growth. *Appl. Soil Ecol., 61*, 295–299. http://dx.doi.org/10.1016/j.apsoil.2011.11.009

Bidadi H., Yamaguchi S., Asahina M., & Satoh S. (2010). Effects of shoot-applied gibberellin/gibberellin-biosynthesis inhibitors on root growth and expression of gibberellin biosynthesis genes in Arabidopsis thaliana. *Plant Root, 4*, 4-11. http://dx.doi.org/10.0.3117/plantroot.4.4

Botha A. (2011). The importance and ecology of yeasts in soil. *Soil Biol Biochem, 43*, 1-8. http://dx.doi.org/10.1016/j.soilbio.2010.10.001

Burton K. (1956). A study of the conditions and mechanism of the diphenylamine reaction for the colorimetric estimation of deoxyribonucleic acid. *Biochem J. Feb., 62*(2), 315–323.

Castelfranco, P. A., & Beale, S. I. (1983). Chlorophyll biosynthesis: recent advances and area of current interest. *Ann. Rev. Plant Physiol., 34*, 241–278. http://dx.doi.org/10.1146/annurev

Colebrook E. H., Thomas S. G., Phillips A. L., & Hedden, P. (2014). The role of gibberellin signalling in plant responses to a biotic stress. *J. Experimental Biol., 217*(1), 67-75. http://dx.doi.org/10.1242/jeb.089938

Davière J. M. 1., & Achard, P. (2013). Gibberellin signaling in plants. *Development, 140*(6), 1147-51. http://dx.doi.org/10.1242/dev.087650

Dishe, L. L. (1953). Physiological studies on the herbicide cotoran. *Journal of the American Chemical Society, 22*, 3014-3022.

Dogra, R., & Kaur, A. (1994). Effect of steroids on some growth and biochemical parameters of *Triticum aestivum* L. during germination. *Crop Res., 8*, 611–620.

EL-Ghamriny, E. A., Arisha, H. M. H., & Nour, K. A. (1999). Studies in tomato flowering, fruit set, yield and quality in summer season, I. Spraying with thiamine, ascorbic acid and yeast, *Zagazig. J. Agric. Res., 26*, 1345-1364.

El-Mofty, M. M., & Sakr, S. A. (1988). Induction of neoplasms in the Egyption toad Bufo regularis By gibberellin-A3. *Oncology, 45*, 61-4.

El-Tarabily, K. A., & Sivasithamparam, K. (2006). Potential of yeasts as biocontrol agents of soil-borne fungal plant pathogens and as plant growth promoters. *Mycoscience, 47*(1), 25–35. http://dx.doi.org/10.1007/s10267-005-0268-2

Erdal, S., & Dumlupinar, R. (2010). Progesterone and β-estradiol stimulate the seed germination in chickpea by causing important changes in biochemical parameters. *Z Naturforsch C, 65*, 239–244.

Gordon S. A., & Weber, R. P. (1951). Colorimetric estimation of indoleacetic acid. *Plant Physiology, 26*, 192-195.

Horvath, E., Szalai, G., & Janda, T. (2007). Induction of abiotic stress tolerance by salicylic acid signaling. *J. Plant Growth Regul., 26*, 290–300. http://dx.doi.org/ 10.1007/s00344-007-9017-4

Hu, C., & Qi, Y. (2013). Long-term effective microorganisms application promote growth and increase yields and nutrition of wheat in China. *Eur. J. Agron., 46*, 63–67. http://dx.doi.org/10.1016/j.eja.2012.12.003

Jiang, C., & Fu, X. (2007). GA action: turning on de-DELLA repressing signaling. *Curr. Opin. Plant Biol.*, 10461–465. http://dx.doi.org/ 10.1016/j.pbi.2007.08.011

Kahlel, A. S. (2015). Effect of Organic Fertilizer and Dry Bread Yeast on Growth and Yield of Potato (*Solanum tuberosum* L.). *J. Agric. Food. Tech.*, *5*(1), 5-11.

Kar, M., & Mishra, D. (1976). Catalase, peroxidase and polyphenoloxidase activity during rice and leaf senescence. *Plant physiol.*, *57*, 315-319.

Kawakami, K., Yasuda, J., Kayama, T., Doi, K., & Sokiyaa, T. (1999). Structures of primer-template hybrides in arbitrarily primed polymerase chain reaction. *Genetic Analysis: Biomolecular Engineering, 15*, 5–8. http://dx.doi.org/10.1016/S1050-3862(98)00013-8

Klessig, D. F., & Malamy, J. (1994). The salicylic acid signal in plants. *Plant Mol. Biol., 26*, 1439-1458. http://dx.doi.org/ 10.1007/978-94-011-0239-1_12

Laemmli, U. K. (1970). Cleavage of structural protein during the assembly of the head of bacteriophage T4. *Nature (London), 227*, 680-685. http://dx.doi.org/ 10.1038/227680a0

Liu Y. Y., Chen, H. W., & Chou, J. Y. (2016). Variation in Indole-3-Acetic Acid Production by Wild *Saccharomyces cerevisiae* and *S. paradoxus* strains from Diverse Ecological Sources and Its Effect on Growth. *PLoS ONE, 11*(8)e0160524. http://dx.doi.org/ 10.1371/journal.pone.0160524

Lyudmila, V. I., Yelena, V. B. , Ramza, Z. B., & Togzhan, D. M. (2015). Plant growth-promoting and antifungal activity of yeasts from dark chestnut soil. *Mirobiological Research, 175*, 78–83. http://dx.doi.org/10.1016/j.micres.2015.03.008

Mady, M. A. (2009). Effect of foliar application with yeast extract and zink on fruit setting Faba bean (*Vicia faba* L). *J. Biol. Chem. Environ. Sci., 4*(2), 109-127.

Magome, H., Yamaguchi, S., Hanada, A., Kamiya, Y., & Odadoi, K. (2004). Dwarf and delayed-flowering 1, a novel Arabidopsis mutant deficient in gibberellin biosynthesis because of over expression of a putative AP2 transcription factor. *Plant J., 37*, 720–729.

Mahmoud, T. R. (2001). Botanical studies on growth and germination of Magnolia (*Magnolia grandiflora L.*). *Plants. M.Sc. Thesis Fac. Agric.*, Moshtohor Zagazig Univ., 103.

Malamy, J., Hennig, J., & Klessig, D. F. (1992). Temperature-dependent induction of salicylic acid and its conjugates during the resistance response to tobacco mosaic virus infection. *Plant Cell, 4*, 359-366. http://dx.doi.org/10.1105/tpc.4.3.359

Morse, M. L., & Carter, C. F. (1949). The synthesis of nucleic acid in cultures of *Escherchia coli* strains B and B/R. *Journal of Bacteriology, 58*, 317-323.

Muhammad, I., & Muhammad, A. (2013). Gibberellic acid mediated induction of salt tolerance in wheat plants: Growth, ionic partitioning, photosynthesis, yield and hormonal homeostasis. *Environmental and Experimental Botany, 86*, 76- 85. http://dx.doi.org/10.1016/j.envexpbot.2010.06.002

Nassar, R. M. A., Nermeen, T., & Shanan, F. M. R. (2016). Active yeast extract counteracts the harmful effects of salinity stress on the growth of leucaena plant. *Scientia Horticulturae, 201*, 61–67. http://dx.doi.org/10.1016/j.scienta.2016.01.037

Navarrea D. A., & Mayoa, D. (2004). Differential characteristics of salicylic acid-mediated signaling in potato. *Physiol. Mol. Plant Pathol., 64*, 179-188. http://dx.doi.org/10.1016/j.pmpp.2004.09.001

Rajjou, L., Belghazi, M., Huguet, R., Robin, C., Moreau, A., Job, C., & Job, D. (2006). Proteomic investigation of the effect of salicylic acid on Arabidopsis seed germination and establishment of early defense mechanisms. *Plant Physiol., 141*, 910–923. http://dx.doi.org/10.1104/pp.106.082057

Reda, F. M., & Ismail, F. M. (2008). The influence of spraying with active yeast extract on vegetative growth and volatile oil of river red gum plant (*Eucalyptus camaldulensis* Dehn.). *J. Agric. Sci. Mansoura Univ., 33*(1), 425–434.

Roland, M. H., Teresita, U. D, Edna, Y. A., Maria, L. O., & Virginia, C. C. (2015). Endophytic Yeasts Possibly Alleviate Heavy Metal Stress in their host Phragmitesaustralis Cav. (Trin.) ex Steud. through the Production of Plant growth Promoting Hormones. *Bull. Env. Pharmacol. Life Sci., 4*(3), 82-86.

Salehi, S. A., Rahbarian, P., & Fallah, I. A. (2014). Stimulatory Effect of gibberellic acid and benzyladenine on Growth and Photosynthetic pigments of *Ficus benjamina* L. Plants. *International Journal of Advanced Biological and Biomedical Research, 2*(1), 34-42.

Sarhan, T. Z., Smira, T. A., & Rasheed, S. M. S. (2011). Effect of bread yeast application and seaweed extract on cucumber (Cucumis sativus L.) plant growth, yield and fruit quality. *Mesopotamia J. of Agric., 39*(2), 26-32.

Selvakumar, G., Reetha, S., & Thamizhiniyan, P. (2012). Response of Biofertilizers on Growth, Yield Attributes and Associated Protein Profiling Changes of Blackgram (*Vigna mungo* L. Hepper). *World Applied Sciences Journal, 16*(10), 1368-1374.

Shalaby, M. E., & El-Nady, M. F. (2008). Application of *Saccharomyces cerevisiae* as a biocontrol agent against Fusarium infection of sugar beet plants. *Acta Biologica Szegediensis, 52*(2), 271-275.

Shehata, M. M., & EL-Khawas, S. A. (2003). Effect of two bio-fertilizers on growth parameters, yield characters, nitrogenous components, nucleic acids content, minerals, oil content, protein profiles and DNA banding pattern of sunflower yield. *Pakistan J. Biol. Sci., 6*(14), 1257-1268.

Soliman H. A. E., Mona M., M., & Hany M. H. (2010). Biochemical and Molecular Profiles of Gibberellic Acid Exposed Albino Rats. *Journal of American Science, 6*(11), 18-23.

Stefanov, B. J., Iliev, K., & Popova, N. I. (1998). Influence of GA3 and 4-PU-30 on Leaf Protein Composition, Photosynthetic Activity, and Growth of Maize Seedlings. *BiologiaPlantarum, 41*(1), 57–63. http://dx.doi.org/10.1023/A:1001756315472

Sun, T. P. (2011). The Molecular Mechanism and Evolution of the   Review   GA–GID1–DELLA   Signaling Module in Plants. *Current Biolog., 21*(9), 340. http://dx.doi.org/10.1016/j.cub.2011.02.036

Tarek, A. S., & Hassan, E. (2014). Effect of foliar application of bio-stimulants on growth, yield, components, and storability of garlic (*Allium sativum* L.). *AJCS, 8*(2), 271-275.

Tartoura, E. A. A. (2001). Response of pea plant to yeast extract and two sources of N-fertilizers. *J. Agric. Sci. Mansora Univ., 26*(12), 7887-7901.

Tsao and Lo, Vegetables (2006). Types and Biology. In H. Y. Hui (Eds.), *Handbook of Food Science Technology and Engineering.* CRC Press: USA.

Twfiq, A. A. (2010). Estimation levels of Indole acetic acid (IAA) and Gibberellic acid (GA3) in dry bakery yeast *Saccharomycescereviciae. J. biotechnology research center, 4*(2), 94-100.

Vernon, L. P., & Seely, G. R. (1966). The chlorophylls. Academic Press, New York, London.

Vlot, A. C., Dempsey, D. A., & Klessig, D. F. (2009). Salicylic Acid, a multifaceted hormone to combat disease. *Annu Rev Phytopathol., 47*, 177-206. http://dx.doi.org/ 10.1146/annurev.phyto.050908.135202

Xin, G., Glawe, D., & Doty, S. L. (2009). Characterization of three endophytic: indole-3-acetic acidproducing yeasts   occurring   in   Populus   trees.   *Mycol   Res.,   113*(9),   973-80. http://dx.doi.org/10.1016/j.mycres.2009.06.001

Yang, X., & Quiros, C. (1993). Identification and classification of celery cultivars with RAPD markers. *Theoretical and Applied Genetics, 86*(2), 205-212. http://dx.doi.org/ 10.1007/BF00222080

Permissions

All chapters in this book were first published in JMBR, by Canadian Center of Science and Education; hereby published with permission under the Creative Commons Attribution License or equivalent. Every chapter published in this book has been scrutinized by our experts. Their significance has been extensively debated. The topics covered herein carry significant findings which will fuel the growth of the discipline. They may even be implemented as practical applications or may be referred to as a beginning point for another development.

The contributors of this book come from diverse backgrounds, making this book a truly international effort. This book will bring forth new frontiers with its revolutionizing research information and detailed analysis of the nascent developments around the world.

We would like to thank all the contributing authors for lending their expertise to make the book truly unique. They have played a crucial role in the development of this book. Without their invaluable contributions this book wouldn't have been possible. They have made vital efforts to compile up to date information on the varied aspects of this subject to make this book a valuable addition to the collection of many professionals and students.

This book was conceptualized with the vision of imparting up-to-date information and advanced data in this field. To ensure the same, a matchless editorial board was set up. Every individual on the board went through rigorous rounds of assessment to prove their worth. After which they invested a large part of their time researching and compiling the most relevant data for our readers.

The editorial board has been involved in producing this book since its inception. They have spent rigorous hours researching and exploring the diverse topics which have resulted in the successful publishing of this book. They have passed on their knowledge of decades through this book. To expedite this challenging task, the publisher supported the team at every step. A small team of assistant editors was also appointed to further simplify the editing procedure and attain best results for the readers.

Apart from the editorial board, the designing team has also invested a significant amount of their time in understanding the subject and creating the most relevant covers. They scrutinized every image to scout for the most suitable representation of the subject and create an appropriate cover for the book.

The publishing team has been an ardent support to the editorial, designing and production team. Their endless efforts to recruit the best for this project, has resulted in the accomplishment of this book. They are a veteran in the field of academics and their pool of knowledge is as vast as their experience in printing. Their expertise and guidance has proved useful at every step. Their uncompromising quality standards have made this book an exceptional effort. Their encouragement from time to time has been an inspiration for everyone.

The publisher and the editorial board hope that this book will prove to be a valuable piece of knowledge for researchers, students, practitioners and scholars across the globe.

List of Contributors

Aide Wang
Department of Horticulture, Cornell University, New York State Agricultural Experiment Station, Geneva, New York, USA
Present address: Department of Horticulture, Shenyang Agricultural University, Shenyang, Liaoning, China

Kenong Xu
Department of Horticulture, Cornell University, New York State Agricultural Experiment Station, Geneva, New York, USA

M. M. Corley and J. Caviness
Agriculture Research Station, Virginia State University, P.O. Box 9061, Virginia 23806, USA

Luca Nardo, Nicola Camera and Edoardo Totè
Department of Science and High Technology, University of Insubria, Italy

Maria Bondani
Institute for Photonics and Nanotechnology, National Research Council, Italy

Roberto S. Accolla and Giovanna Tosi
Department of Surgical and Morphologic Sciences, University of Insubria, Italy

Ricardo Martí-Arbona, Tuhin S. Maity, John M. Dunbar, Clifford J. Unkefer and Pat J. Unkefer
Los Alamos National Laboratory, Los Alamos, NM 87545, United States

Sachiyo Aburatani
Computational Biology Research Center (CBRC), National Institute of Advanced Industrial Science and Technology, Tokyo, Japan

Wataru Fujibuchi
Center for iPS Research and Application, Kyoto University, Kyoto, Japan

Tommy Rodriguez
Pangaea Biosciences, Department of Research & Development, Miami, FL, USA

Ibraimov A. I.
Institute of Balneology and Physiotherapy, Bishkek, Kyrgyz Republic
Kazakh National Medical University after name S. D. Asfendyarov, Almaty, Kazakhstan

José L. Fernández-García and María P. Vivas Cedillo
Genetics and Animal Breeding, Veterinary Faculty, Universidad de Extremadura, 10071 Cáceres, Spain

Richard S Hall, Ricardo Martí-Arbona, Scott P Hennelly, Tuhin S Maity, Fangping Mu, John M Dunbar, Clifford J Unkefer and Pat J Unkefer
Los Alamos National Laboratory, Los Alamos, NM 87545, United States

Obinaju, Blessing Ebele
Centre for Biophotonics, Lancaster University, Lancaster, United Kingdom

Biao Wu, Aiguo Yang, Xiujun Sun, Zhihong Liu and Liqing Zhou
Key Laboratory of Sustainable Development of Marine Fisheries, Ministry of Agriculture, Yellow Sea Fisheries Research Institute, Chinese Academy of Fishery Sciences, Qingdao 266071, PR China

Ningning Cheng
Key Laboratory of Sustainable Development of Marine Fisheries, Ministry of Agriculture, Yellow Sea Fisheries Research Institute, Chinese Academy of Fishery Sciences, Qingdao 266071, PR China
College of Fisheries and Life Science, Shanghai Ocean University, Shanghai, 201306, PR China

J. Spencer Johnston
Department Entomology, Texas A & M University, College Station, TX, USA

Molly Schoener
Department Biochemistry & Biophysics, Texas A & M University, College Station, TX, USA

Dino P. McMahon
Department of Zoology, University of Oxford, Oxford, UK

Fauzia Siddiq and Fulvio Lonardo
John D. Dingell VA Medical Center, Departments of Otolaryngology, Surgery, and Karmanos Cancer Institute, Wayne State University, Detroit, MI, USA

Harvey I. Pass
Division of Cardiothoracic Surgery, New York University Cancer Center, New York, USA

Arun K. Rishi
Department of Oncology, Wayne State University, John D. Dingell VA Medical Center, Research & Development Section, Detroit, MI, USA

Anil Wali
John D. Dingell VA Medical Center, Departments of Otolaryngology, Surgery, and Karmanos Cancer Institute, Wayne State University, Detroit, MI, USA
Center to Reduce Cancer Health Disparities, National Cancer Institute, National Institutes of Health, Rockville, MD, USA

Oluduro Anthonia Olufunke, Aregbesola Oladipupo Abiodun and Fashina Christina Dunah
Department of Microbiology, Faculty of Science, Obafemi Awolowo University, Ile-Ife 220005, Nigeria

Fernando Rosas-Sanchez, Lenin Ochoa-de la Paz, Ricardo Miledi and Ataúlfo Martínez-Torres
Instituto de Neurobiología, Campus Juriquilla - Universidad Nacional Autónoma de México, Querétaro, México

Angélica López-Rodríguez
Instituto de Neurobiología, Campus Juriquilla - Universidad Nacional Autónoma de México, Querétaro, México
National Institute of Neurological Disorders and Stroke, National Institutes of Health, Bethesda, USA

Carlos Saldaña
Facultad de Ciencias Naturales, Universidad Autónoma de Querétaro, Querétaro, México

Hailin Sun
Changdao Fisheries Research Institute, Changdao, China

Yanxin Zheng, Chunnuan Zhao, Tao Yu and Jianguo Lin
Changdao Enhancement and Experiment Station, Chinese Academy of Fishery Sciences, Changdao, China

Aida Ansarikaleibari
School of Biosciences, University of Birmingham, United Kingdom

Zahra Sorkhoo
Gilan University, Iran

Mousa Mohammadnia-Afrouzi and Mehdi Shahbazi
Department of Immunology, School of Medicine, Babol University of Medical Sciences, Babol, Iran

Sedigheh Baleghi Damavandi and Soheil Ebrahimpour
Infectious Diseases and Tropical Medicine Research Center, Babol University of Medical Sciences, Babol, Iran

Ghasem Faghanzadeh Ganji
Cardiac surgery Department, Rouhani Hospital, Babol University of Medical Sciences, Babol, Iran

Ganash Magdah A.
Biology Department, Faculty of Science, King Abdulaziz Univtersity, Jeddah, King Saudi Arabia

Index

www.ingramcontent.com/pod-product-compliance
Lightning Source LLC
Chambersburg PA
CBHW080659200326
41458CB00013B/4920